Plant Cell Monographs

Volume 17

Series Editor: David G. Robinson
Heidelberg, Germany

For further volumes:
http://www.springer.com/series/7089

Plant Cell Monographs
Recently Published Titles

Cell Biology of Metals and Nutrients
Volume Editor: Hell, R., Mendel, R. R.
Vol. 17, 2010

Lipid Signaling in Plants
Volume Editor: Munnik, T.
Vol. 16, 2010

Cell Biology of Plant Nematode Parasitism
Volume Editor: Berg, R. H., Taylor, C. G.
Vol. 15, 2009

Functional Organization of the Plant Nucleus
Volume Editor: Meier, I.
Vol. 14, 2009

The Chloroplast
Interactions with the Environment
Volume Editors: Sandelius, A. S.
Aronsson, H.
Vol. 13, 2009

Root Hairs
Volume Editors: Emons, A. M. C.
Ketelaar, T.
Vol. 12, 2009

Plant Microtubules
Development and Flexibility
2nd Edition
Volume Editor: Nick, P.
Vol. 11, 2008

Plant Growth Signalling
Volume Editors: Bögre, L., Beemster, G.
Vol. 10, 2008

Cell Division Control in Plants
Volume Editors: Verma, D. P. S., Hong, Z.
Vol. 9, 2008

Endosperm
Volume Editor: Olsen, O.-A.
Vol. 8, 2007

Nitric Oxide in Plant Growth Development and Stress Physiology
Volume Editors: Lamattina, L., Polacco, J.
Vol. 6, 2007

The Expanding Cell
Volume Editors: Verbelen, J.-P., Vissenberg, K.
Vol. 5, 2007

The Plant Endoplasmic Reticulum
Volume Editor: Robinson, D. G.
Vol. 4, 2006

The Pollen Tube
A Cellular and Molecular Perspective
Volume Editor: Malhó, R.
Vol. 3, 2006
Vol. 5/L, 2006

Somatic Embryogenesis
Volume Editors: Mujib, A., Samaj, J.
Vol. 2, 2006

Plant Endocytosis
Volume Editors:
Šamaj, J., Baluška, F., Menzel, D.
Vol. 1, 2005

Rüdiger Hell • Ralf-Rainer Mendel
Editors

Cell Biology of Metals and Nutrients

Springer

Editors
Prof. Dr. Rüdiger Hell
Universität Heidelberg
Heidelberger Institut für
Pflanzenwissenschaften
Im Neuenheimer Feld 360
69120 Heidelberg
Germany
rhell@hip.uni-heidelberg.de

Prof. Dr. Ralf-Rainer Mendel
Technische Universität Braunschweig
Institut für Pflanzenbiologie
Humboldtstr. 1
38106 Braunschweig
Germany
r.mendel@tu-bs.de

Series Editor
David G. Robinson
Ruprecht-Karls-University of Heidelberg
Heidelberger Institute for Plant Sciences (HIP)
Department Cell Biology
Im Neuenheimer Feld 230
69120 Heidelberg
Germany

ISSN 1861-1370 e-ISSN 1861-1362
ISBN 978-3-642-26275-3 e-ISBN 978-3-642-10613-2
DOI 10.1007/978-3-642-10613-2
Springer Heidelberg Dordrecht London New York

Cover photo: The upper part of the cover was created by Steffen Rump. It shows a mitochondrion which is essential for the biosynthesis of two metal-containing prosthetic groups: the molybdenum cofactor (left) and iron sulfur clusters (middle).

Cover design: WMXDesign GmbH, Heidelberg, Germany

Printed on acid-free paper

Springer is part of Springer Science+Business Media (www.springer.com)

About the Editors

Rüdiger Hell studied Biology at the Technical University of Darmstadt, Germany, and completed his PhD at the University of Cologne, Germany, in 1989. From 1990 to 1992, he worked at the University of California in Berkeley as a postdoctoral researcher. After returning to Germany, he completed his postdoctoral thesis at the University of Bochum, in 1998, and held a position at the Leibniz Institute for Plant Genetics and Crop Plant Research in Gatersleben. During that time, he developed his ongoing interest in molecular mechanisms of plant nutrition, especially in sulfur metabolism and cellular redox control. In 2003, he was appointed chair at the Heidelberg Institute for Plant Sciences. He served as Dean of the Faculty of Biosciences at Heidelberg University from 2005 to 2007, and is currently the managing director of the university's Plant Sciences Institute.

Ralf-Rainer Mendel studied biochemistry at the Humboldt University in Berlin, completed his PhD at the Martin-Luther-University Halle in 1979 and his postdoctoral thesis, in 1985. During that time he held a position at the Institute for Plant Genetics and Crop Plant Research in Gatersleben.

In 1992, he was appointed as Full Professor of Botany at the (now) Institute of Plant Biology of the Braunschweig University of Technology, Germany. He has been the director of the Institute since 1993 and also served as Dean of Biosciences at Braunschweig from 1997 to 1999. His research focuses on the cell biology and biochemistry of molybdenum in plants and humans.

Preface

Plants are composed of 17 essential elements and these must be taken up as nutrients to allow for growth and cell division. Macronutrients are defined by their large amounts in plants (>0.1% of dry mass), while micronutrients are much less abundant (<0.01% of dry mass). C, H and O provide the bulk of biomass but only N, P, K, S, Ca and Mg (in decreasing order of abundance) are considered macronutrients in a strict sense. The micronutrients (Fe, Zn, Cu, Mn, B, Mo, Ni and Cl) are much more heterogeneous with respect to chemistry and biological function: the presence of Ni appears to be required in only one or two processes while the content of Mo is the lowest of the essential nutrients found in a plant. Metals are found in both groups and reflect their multiple catalytic and electron transfer functions in interaction with high and low molecular organic compounds in cells. Hence, the terms metals and nutrients are partially overlapping without precisely characterizing macro- and micronutrients. In addition, so-called beneficial elements are known to promote growth in only some taxa. This term refers to Al, C, Na, Se, and Si, the latter particularly being required by monocotyledonous plants. Other elements are taken up by plants but have no proven function or are even deleterious such as heavy metals.

Much effort has been devoted to study the physiology and biochemistry of metals and nutrients in plants. The aspect of cell biology, however, is an emerging new field, and even a new Gordon Research Conference (Cell Biology of Metals) has been launched recently. While N and S are structural and functional components of proteins, P is crucial for nucleic acids and energy metabolism, Ca, K, Cl, B and the metals are of key importance for a large number of biological reactions. These include photosynthesis, cell pressure, metabolic catalysis, oxygen transport, anabolism and catabolism, redox reactions, cell signalling, etc. Both the regulatory networks and the nutrient aspect of these elements have been successfully investigated using nutrient-deficiency experiments and transcriptome, proteome and metabolome tools as well as targeted approaches. However, much needs to be learnt about nutrient sensing, long-distance communication within the plant and cellular signal transduction chains in response to environmental stress.

With respect to metals, major progress has been made in recent years in our understanding of the biosynthesis of metal-containing prosthetic groups. Complex and multistep biosynthetic pathways were uncovered in greater detail. Progress has also been made in the identification of cellular sites of prosthetic groups' synthesis. However, our knowledge is scarce with respect to the steps subsequent to biosynthesis. It is largely unknown how these prosthetic groups are allocated, transported and directed to various cellular destinations, ultimately finding their way into the correct cognate proteins. There must be a plethora of transporters, chelators, metal cofactors, protein folding chaperones, metallochaperones, carrier proteins, storage proteins, and insertases involved. Another important objective is to understand how protein trafficking pathways and metal trafficking pathways are integrated and intersect. Further areas of focus are mechanisms of metal ion homeostasis, how cells sense diverse metal ions, how homeostatic mechanisms for different metal ions are integrated at the molecular and cellular level, and how the finely tuned interplay of these components ensures safe transport. Clearly, cellular malfunction, and consequently, disease are the result if any of the various key steps in metals allocation and insertion are perturbed and reduced cell division, growth proliferation and overall performance result from deficiencies of these nutrients.

This book emphasizes the cellular biology of metals and nutrients to provide a new concept that reaches beyond plant nutrition and plasmalemma transport but into cellular physiology. Each chapter contains basic information about uptake, physiological function, deficiency and toxicity syndromes, long-distance communication, and intracellular transport. The chapters are devoted to metals and nutrients where recent progress has been made and highlight the aspects of homeostasis and sensing, signalling, and regulation with cross-references to other organisms including humans. Where gaps are identified in our knowledge, these are pointed out and future research directions envisaged.

October 2009

Rüdiger Hell
Ralf-Rainer Mendel

Contents

Contributors

Oliver Batistič Universität Münster, Institut für Botanik und Botanischer Garten, Schlossplatz 4, 48149 Münster, Germany, oliver.batistic@uni-muenster.de

Florian Bittner Department of Plant Biology, Technical University of Braunschweig, Humboldtstrasse 1, 38106 Braunschweig, Germany

Alejandra Chacon-López Departamento de Ingeniería Genética de Plantas, Centro de Investigación y de Estudios Avanzados del Instituto Politécnico Nacional, Irapuato, Guanajuato, México

Stephan Clemens Universität Bayreuth, Lehrstuhl Pflanzenphysiologie, Universitätsstrasse 30, 95440 Bayreuth, Germany

Christopher M. Cohu Biology Department, Colorado State University, Fort Collins, CO 80523-1878, USA

Françoise Daniel-Vedele Institut Jean-Pierre Bourgin (IJPB), Institut National de Recherche Agronomique (INRA), route de St Cyr, F-78026 Versailles, France, vedele@versailles.inra.fr

Toru Fujiwara Biotechnology Research Center, University of Tokyo, Yayoi, Bunkyo-ku, Tokyo 113-8657, Japan; SORST, JST, Japan, atorufu@mail.ecc.u-tokyo.ac.jp

Mary Lou Guerinot Department of Biological Sciences, Dartmouth College, Hanover, NH 03755, Guerinot@Dartmouth.edu

Rüdiger Hell University of Heidelberg, Heidelberg Institute for Plant Sciences, Im Neuenheimer Feld 360, 69120 Heidelberg, Germany, rhell@hip.uni-heidelberg.de

Luis Herrera-Estrella Laboratorio Nacional de Genómica para la Biodiversidad, Centro de Investigación y de Estudios Avanzados del Instituto Politécnico Nacional, Irapuato, Guanajuato, México, lherrera@ira.cinvestav.mx

Werner M. Kaiser Julius-von-Sachs-Institut für Biowissenschaften, Lehrstuhl für Botanik I, Julius-von-Sachs-Platz 2, D-97082 Würzburg, Germany, kaiser@botanik.uni-wuerzburg.de

Alison J. Karley Scottish Crop Research Institute, Invergowrie, Dundee DD2 5DA, UK

M. Sayyar Khan Department of Plant Breeding and Genetics, NWFP Agricultural University Peshawar, Pakistan

Anne Krapp Institut Jean-Pierre Bourgin (IJPB), Institut National de Recherche Agronomique (INRA), route de St Cyr, F-78026 Versailles, France, krapp@versailles.inra.fr

Jörg Kudla Universität Münster, Institut für Botanik und Botanischer Garten, Schlossplatz 4, 48149 Münster, Germany, jkudla@uni-muenster.de

Ralf-Rainer Mendel Department of Plant Biology, Technical University of Braunschweig, Humboldtstrasse 1, 38106 Braunschweig, Germany, r.mendel@tu-bs.de

Kyoko Miwa Hokkaido University, Creative Research Initiative Sousei, Kita-ku, Sapporo, miwakyoko@cris.hokudai.ac.jp

Claudia-Anahí Pérez-Torres Laboratorio Nacional de Genómica para la Biodiversidad, Centro de Investigación y de Estudios Avanzados del Instituto Politécnico Nacional, Irapuato, Guanajuato, México

Marinus Pilon Biology Department, Colorado State University, Fort Collins, CO 80523-1878, USA, pilon@lamar.colostate.edu

Elizabeth A.H. Pilon-Smits Biology Department, Colorado State University, Anatomy/Zoology Building, Fort Collins, CO, 80523, USA, epsmits@lamar.colostate.edu

Jon K. Pittman Faculty of Life Sciences, University of Manchester, Michael Smith Building, Oxford Road, Manchester M13 9PT, UK, jon.pittman@manchester.ac.uk

Colin F. Quinn Biology Department, Colorado State University, Anatomy/Zoology Building, Fort Collins, CO, 80523, USA, cquinn@lamar.colostate.edu

Lenin Sánchez-Calderón Laboratorio Nacional de Genómica para la Biodiversidad, Centro de Investigación y de Estudios Avanzados del Instituto Politécnico Nacional, Irapuato, Guanajuato, México

Philip J. White Scottish Crop Research Institute, Invergowrie, Dundee DD2 5DA, UK, philip.white@scri.ac.uk

Lorraine E. Williams School of Biological Sciences, University of Southampton, Southampton SO16 7PX, UK, lew@soton.ac.uk

Markus Wirtz University of Heidelberg, Heidelberg Institute for Plant Sciences, Im Neuenheimer Feld 360, 69120 Heidelberg, Germany

Role of Boron in Plant Growth and its Transport Mechanisms

Kyoko Miwa and Toru Fujiwara

Abstract In the last decade, molecular understandings of boron transport and boron function in plants have greatly advanced. Crosslinking of pectic polysaccharide rhamonogalacturonan-II (RG-II) via boron was shown to be essential for normal plant growth. Two types of boron (B) transport molecules, BORs and NIPs, localized to plasma membrane were identified from *Arabidopsis thaliana*. BOR1 was identified as the first borate/boric acid transporter in the biological systems. It efficiently exports B into xylem, and BOR4, a BOR1 homolog, excludes high concentrations of B out of the cells for high B tolerance. NIP5;1 and NIP6;1 are channels for boric acid to facilitate B flow from soil into root cells and the preferential distribution of B to growing shoot tissues, respectively. In recent years, compelling evidence has accumulated for the fact that B is also required for some bacteria and animals. *A. thaliana* BOR1 led to the identification of a Na^+-coupled borate transporter NaBC1, a mammalian BOR1 homolog. Further understanding of physiological function of B and transport mechanisms will expand our perspectives of B in biology.

1 Introduction: Specialty of B

Of the 17 elements essential for plant growth and development, B (boron) was considered to be required only for plants but not for animals. Requirement of B in plant cell walls coincided with this understanding; however, increasing evidence suggests physiological roles of B in other organisms including humans.

K. Miwa
Hokkaido University, Creative Research Initiative Sousei Address Kita-ku, Sapporo
e-mail: miwakyoko@cris.hokudai.ac.jp

T. Fujiwara (✉)
Biotechnology Research Center, University of Tokyo, Yayoi, Bunkyo-ku, Tokyo 113-8657, Japan
SORST, JST, Chiyoda-Ku, Tokyo, Japan
e-mail: atorufu@mail.ecc.u-tokyo.ac.jp

R. Hell and R.-R. Mendel (eds.), *Cell Biology of Metals and Nutrients*,
Plant Cell Monographs 17,
DOI 10.1007/978-3-642-10613-2_1, © Springer-Verlag Berlin Heidelberg 2010

In soil solutions at physiological pH, B is present as a charge-neutral molecule, boric acid $B(OH)_3$. Among essential elements of plants that are taken up from the soil, B is the only element that is taken up by plants not as an ion but as a noncharged molecule. As boric acid has a high permeability to lipid bilayer, passive diffusion was considered to be the only transport mechanism until 1990s. Introduction of highly sensitive analytical method, inductively coupled plasma mass spectrometry (ICP-MS) and molecular genetics with a model plant, *A. thaliana*, enabled us to identify the first B transporter in the biological systems. This finding led to shed light on the physiological function and homeostasis in mammals and expanded our views of B in biology.

2 Physiological Significance of B

2.1 B Essentiality in Plants and Animals

Boron (B) was established as an essential trace element for higher plants more than 85 years ago (Warington 1923). Warington (1923) reported that broad bean plants showed reduced root elongation and dwarf and dark green leaves when they were grown in the medium without boric acid. This growth reduction was restored by supply of boric acid to the medium. Similar phenomena have been observed with a broad range of vascular plants species. B is also required in cyanobacteria. In *Anabaena*, depletion of B inhibited growth when they depended on N_2 fixation, indicating B requirement for N_2 fixation (Mateo et al. 1986).

Essentiality of B in animals and non-photosynthesizing microorganisms was not recognized for a long time. To our knowledge, occurrence of B deficiency in nature is not reported except for the cases in plants. This is presumably because B is contaminated in the environment and requirement of B in organisms other than plants are very low. Increasing evidence suggests that B is beneficial or essential in microorganisms and animals. Bennett et al. (1999) demonstrated B significantly stimulated cell growth rate and cell density in the stationary phase in *Saccharomyces cerevisiae*. In zebrafish (*Danio rerio*), fertilization normally occurred, but 92% zygotes failed to survive in B-depleted solution, showing B requirement for embryogenesis (Rowe et al. 1998, Rowe and Eckhert 1999). In *Xenopus laevis*, frogs fed with the low B diet produced a high proportion of necrotic eggs (54%). Fertilized embryos from the low-B diet-fed frogs showed a high frequency of abnormal gastrulation (26.8%), and more than 80% of the embryos died before 96 h of development (Fort et al. 1998, 1999). B supply stimulated growth in trout (Eckhert 1998). B deficiency impaired embryonic development in rodents (Lanoue et al. 1998). B supplementation to semi-purified feeds improved performance of feed efficiency in pigs (Armstrong et al. 2000). In addition, Ahmed et al. (2007) isolated a novel bacteria species, *Bacillus boroniphilus* sp. nov., from naturally high-B containing soils, and which showed tolerance up to 450 mM boric acid. It is noteworthy that growth of the bacteria was inhibited without addition of boric acid to the media and improved by the supply of more than 20 mM boric acid, suggesting the requirement of B for optimum growth. Despite of a number of physiological descriptions on the effects of B deficiency, molecular mechanisms of B functions for cell development and growth in these organisms remain unknown.

2.2 B Deficiency and Toxicity Symptoms in Plants

A number of physiological studies suggested that B deficiency affects the growing portions of plants rather than mature tissues. The symptoms of B deficiency include the cessation of root elongation, reduced leaf expansion, and the loss of fertility (Loomis and Durst 1992; Marschner 1995; Dell and Huang 1997). It has been suggested that B deficiency affects cell elongation rather than cell division (Dell and Huang 1997). Roles of B in cell structures were predicted from swollen cell shapes observed under B deficiency. It is also described that B deficiency affects membrane function and metabolic activities (Bolanos et al. 2004). Bassil et al. (2004) demonstrated that boronic acids, which compete with boric acid for binding to cis-diols, caused the disruption of cytoplasmic strands and cell-to-cell wall detachment in tobacco cultured cells, and proposed that B functions for structure in cytoskeleton.

High concentrations of B are toxic to plants (Nable et al. 1997). Typical visible symptoms of B toxicity include necrosis of marginal regions of leaves, presumably resulting from B accumulation along the transpiration stream. Decreased chlorophyll concentrations, reduced growth, decreased CO_2 fixation were also reported.

3 RG-II-B in Cell Wall and its Requirement for Plant Growth

3.1 Chemical Properties of B: Possible Binding Sites of B in Cell

B in solution is mostly present as a boric acid $B(OH)_3$, a weak Lewis acid accepting OH^- group to form the borate ion $B(OH)_4^-$ (p*Ka* 9.25) in water ($B(OH)_3 + H_2O = B(OH)_4^- + H^+$). Boric acid reacts with alcohols to form borate esters in a pH-dependent equilibrium manner and with a rapid kinetics (Power and Woods 1997). Since the most stable borate esters are formed with *cis*-diols on a furanoid ring (e.g., ribose, apiose), compounds including ribonucleotides such as NADH (NAD^+), RNA, are possible binding targets of boric acid.

There are reports on biomolecules that interact with B. In microorganisms, borophycin and boromycin, antibiotics containing B were isolated from cyanobacterium *Nostoc spongiaeforme* var. tenue and *Streptomyces griseus*, respectively (Dembitsky et al. 2002). Chen et al. (2002) found that autoinducer-2 for quorum-sensing contained B in marine bacteria *Vibrio harveyi*. Amin et al. (2007) isolated vibrioferrin, a B-containing siderophore produced by the particular marine bacteria which required the toxic, bloom forming dinoflagellate *Gymnodinium catenatum* for growth.

Although RG-II in plant cell wall is the only B-compound that is required for the physiological function of B (details are described in Sect. 3.2), demonstration of essentiality of B in other organisms that do not contain RG-II implies presence of other essential B containing or interacting compounds.

3.2 Identification of RG-II-B Complex in Plant Cell Walls

In the 1990s, presence of B-polysaccharide complex in plant cell wall was described. Under conditions of limited B supply, the major portion of B is distributed in the cell wall of tobacco-cultured cells (Matoh et al. 1992), squash plants (Hu and Brown 1994), and *A. thaliana* (Noguchi et al. 2000), suggesting that the cell wall is likely to be the site of B function in cells. Matoh et al. (1993) isolated B-polysaccharide complexes as tetrahedral 1:2 borate-diol diestel from cell walls of radish roots, and then Kobayashi et al. (1996) identified the B-polysaccharides as RG-II. Further, removal of B by acid-hydrolization reduced molecular weight by one-half and it is proposed that boric acid links two RG-II molecules. Ishii and Matsunaga (1996) also isolated B-polysaccharide from cell walls of sugar beet and identified it as RG-II.

The presence of RG-II was demonstrated not only in angiosperms and gymnosperms but also in pteridophyte and bryophyte (Matsunaga et al. 2004; O'Neill et al. 2004), however, essentiality of B in fern and moss was not established.

3.3 Synthesis of RG-II and Physiological Roles of RG-II

RG-II is highly complex polysaccharides containing 12 different glycosyl residues linked together by more than 20 different glycosyl linkages (O'Neill et al. 2004). Although the B-polysaccharide in cell walls was identified as RG-II, a physiological role(s) of dRG-II-B remained unclear at the time. Through the analysis of the mutant required for RG-II synthesis, cellular roles of dRG-II-B have begun to be elucidated.

3.3.1 *A. thaliana* MUR1 for Fucose Synthesis Essential for Efficient Formation of dRG-II-B

O'Neill et al. (2001) provided compelling evidence that efficient formation of dRG-II-B is essential for normal leaf expansion using *A. thaliana* mutant *mur1*. *MUR1* encodes an isoform of GDP-D-mannose-4, 6-dehydratase for synthesis of

GDP-L-fucose (Bonin et al. 1997). Fucosyl residues are present in RG-II and RG-II of the *mur1* mutant plants and the mutant has an abnormal glycosyl composition (O'Neill et al. 2001; Reuhs et al. 2004). The *mur1* mutant plants exhibited reduced rosette leaf expansion under normal B conditions. The extents of crosslinking of RG-II by borates in *mur1* were lower than those of wild-type plants. The growth defects of *mur1* mutant plants were recovered by supplying high levels of B, and in such mutant plants, crosslinking of RG-II were close to the levels in wild type plants. Taken all together, it was suggested that crosslinking of RG-II by borate is essential for normal leaf expansion (O'Neill et al. 2001). Ryden et al. (2003) demonstrated that dRG-II-B was important for tensile strength of cell walls in hypocotyls. It is considered that cross-linking of RG-II by borate establishes pectic network, which provides the physical and chemical properties of cell walls.

3.3.2 NpGUT1, Glucuronyltransferase 1, for Cell Adhesion and Attachment

NpGUT1 was identified as a responsible gene of *nolac-H18* callus mutant of *Nicotiana pumbaginifolia*. The *nolac-H18* is defective in intercellular attachment (Iwai et al. 2002). The RG-II of this mutant callus lacks disaccharides in side chain A. It was shown that 56% of total RG-II in *nolac-H18* was present as mRG-II, whereas more than 95% of RG-II was present as dimer in normal callus. mRG-II of *nolac-H18* is less able to form dRG-II-B. This study suggests involvement of dRG-II-B formation in cellular attachment.

Further, Iwai et al. (2006) described significant roles of NpGUT1 in reproductive stage. *NpGUT1* is strongly expressed in the tapetum of flower buds and in the pollen, pollen tube tips, and transmitting tissue of the pistils of flowers in addition to meristematic tissues. Repression of *NpGUT1* expression in flowers caused sterility with aberrant development of pollen, and transmitting tissue and inhibition of pollen tube elongation. It was also demonstrated an apical gradient of B concentration in a pollen tube of wild type tobacco plants, and abolishment of this gradient in *NpGUT1*-antisense transgenic plants. To support a rapid tip growth of pollen tube, active vesicle-trafficking system transports secretory vesicles including synthetic materials toward the apex and they were delivered by exocytosis. It is considered that RG-II polysaccharides are newly synthesized in Golgi apparatus and the vesicles are secreted out of tube tip in the direction of elongation, leading to accumulation of RG-II-B dimer in apical region of pollen tubes.

3.4 Changes in RG-II Properties in Response to B Nutrition

Ishii et al. (2001) showed that B conditions in media did not affect RG-II sugar composition in plants. This suggests that synthesis of RG-II is not strongly affected by B nutrition.

Baluska et al. (2002) demonstrated using antibody against RG-II, that dRG-II-B, and partially esterified homogalacturonan (HG) were internalized into brefeldin A compartment in meristematic cells of maize roots, suggesting that pectins containing dRG-II-B were internalized via endocytosis. In addition, this internalization of dRG-II-B was inhibited upon B deprivation in maize and wheat root apices (Yu et al. 2002). It was proposed that cell walls are the site of the primary reaction after deprivation of B and that this primary reaction leads to B deficiency responses. This endocytosis-mediated pectin signaling is likely to be involved in signaling through putative cell wall-plasma membrane-cytoskeleton continuum.

It remains unknown in many aspects for borate cross-linking of RG-II from synthesis and delivery. It is not clear whether an enzyme(s) catalyzes this reaction, where this cross-linking occurs in cells.

4 Molecular Mechanism of B Transport in Plants

For B to be utilized by plants, it must be transported from the soil through the roots to the shoots. Three mechanisms of B transport in plants were shown to be present: (1) passive diffusion across lipid bilayers, (2) channel mediated B transport, and (3) active B transport system by transporters.

4.1 Passive Diffusion

Most of the physiological data performed before the 1990s, indicated that amount of B uptake is proportional to the B concentration in the medium (Marschner 1995). Therefore B had been long considered to be transported only by passive diffusion across membranes and distributed within the plant body along transpiration streams. Indeed, theoretical permeability coefficient of boric acid was in the order of 10^{-6} cm s^{-1} (Raven 1980) and the experimental coefficient in artificial liposomes was 4.9×10^{-6} cm s^{-1}(Dordas and Brown 2000). These values were likely to be high enough to account for B uptake and transport in plants.

However, a permeability coefficient of boric acid of plant membranes was found to be smaller than those values: 3.9×10^{-7} cm s^{-1} in plasma membrane of squash roots by Dordas et al. (2000), and 4.4×10^{-7} cm s^{-1} in charophyte alga *Chara coralline* (Stangoulis et al. 2001).

4.2 Channel-mediated B Transport for Facilitation

Dordas and Brown (2000, 2001) experimentally demonstrated the presence of channel-mediated B transport system to promote B flux in plasma membrane.

A non-specific channel blocker partially inhibited boric acid permeation of the plasma membrane vesicles from squash roots. Expression of a major intrinsic protein (MIP), maize PIP1 in *Xenopus laevis* oocytes resulted in a 30% increase in B permeability. It is also showed that non-charged molecule with a similar molecular size to boric acid, such as urea and glycerol, inhibited B uptake.

4.2.1 *A. thaliana* NIP5;1, A Channel for Boric Acid for B Uptake Under B Limitation

NIP5;1, a member of major intrinsic proteins (MIPs) was identified as a boric acid channel for efficient B uptake in *A. thaliana* roots (Takano et al. 2006). MIPs are membrane protein superfamily known as aquaporins. MIPs were reported to transport non-charged small molecule such as glycerol, urea, and ammonia in addition to water (Tyerman et al. 2002). NIP (Nod 26-like Intrinsic Protein) is one of four subfamilies in MIPs. Soybean NOD26 is localized to symbiosome membrane and is capable of transporting water, glycerol and ammonia (Wallace et al. 2006).

NIP5;1 mRNA accumulation in roots increased more than ten folds at 24 h after the initiation of low B treatment (Takano et al. 2006). PromoterNIP5;1-GUS expression was strongly observed in the root under B limitation. Under high B conditions, GUS staining was only slightly detected, indicating that *NIP5;1* accumulation was regulated by the promoter and 5'UTR.

NIP5;1 protein was localized to plasma membrane, and NIP5;1 exhibited transport activity of boric acid in *Xenopus* oocyte. The two independent mutants for *NIP5;1* showed severe growth reduction both in shoots and roots only under limited conditions of B. Increase of B uptake into roots was observed in the wild type but not in the *nip5;1-1* mutant under low B, compared to those under high B. These observations demonstrated that NIP5;1 is essential for B uptake into root cells to support normal plant growth under B limitation (Takano et al. 2006). Kato et al. (2009) reported that enhanced expression of NIP5;1 improved root elongation under low B conditions, further confirming the role of NIP5;1 for efficient utilization of B.

4.2.2 *A. thaliana* NIP6;1, for Preferential Distribution of B into Growing Shoot Tissues

NIP6;1 is the gene most similar to *NIP5;1* among nine NIP genes present in the *A. thaliana* genome. NIP6;1 is also localized to plasma membrane and facilitated boric acid uptake in oocytes but did not show water channel activity, in contrast to NIP5;1. *NIP6;1* mutants exhibited inhibition of the young leaf expansion and lower B concentrations in these tissues under B limitation. *NIP6;1* promoter activity was strongly detected in the vascular of the node of the stem. It was proposed that NIP6;1 functions in xylem-phloem transfer for preferential distribution of B into young growing tissues (Tanaka et al. 2008).

4.3 Active B Transport System Under Limited Supply of B

To transport B from roots to shoots, B is needed to be loaded into xylem vessels. This process, known as xylem loading, is a transport process from symplasm to apoplasm as xylem vessel is in apoplast.

Dannel et al. (2000) reported the presence of low-B inducible active B transport systems both in root uptake and xylem loading in sunflower. Dannel et al. (2001) revealed that this transport system followed the Michaelis-Menten kinetics (Km 15 μM , Vmax 30 nmol g_{rootFW}^{-1} h^{-1}). Similar results were reported in B uptake of charophyte alga *Chara coralline* which followed kinetics with an apparent Km of about 2 μM and Vmax of about 135 pmol m^{-2} s^{-1}. (Stangoulis et al. 2001). These results clearly indicated the presence of high affinity B transport systems.

4.3.1 *A. thaliana BOR1*, the First Borate Transporter Identified in the Biological Systems

The breakthrough of the isolation of boron transporter came from the isolation of an *A. thaliana bor1-1* mutant. *bor1-1* is more sensitive to B deficiency than the wild type. In *bor1-1*, leaf expansion in upper leaves was severely inhibited when plants were grown under low B conditions (Noguchi et al. 1997). B concentrations in root cell saps were almost the same as the B concentrations in the media, both in wild type plants and the *bor1-1* mutant (Takano et al. 2002), but the concentration of B in xylem sap was lower in the *bor1-1* mutant. When B concentrations of root saps and xylem saps were determined in plants exposed to various concentrations of boron, the combination of a linear and a saturatable curve was present in B concentrations in xylem saps in wild type plants (Takano et al. 2002). In the *bor1-1* mutant the relationship was linear, suggesting that the corresponding gene is essential for boron transport against concentration gradients. Furthermore, Takano et al. (2001) demonstrated that *bor1-1* is defective in preferential B translocation into young leaves.

BOR1 was identified by positional cloning. BOR1 showed a high similarity to anion exchanger proteins, including well-characterized Band3 protein in erythrocyte in animal (Takano et al. 2002). *BOR1* is expressed predominately in root pericycles, cells surrounding xylem, and localized to the plasma membrane. In the *S. cerevisiae,* a single gene is present that is similar to *BOR1*. When Arabidopsis BOR1 is expressed in *S. cerevisiae* strain that lacks yeast *BOR1* gene, B concentrations in yeast cells were reduced, suggesting that BOR1 is a B exporter. BOR1 represents the first B transporter identified in the biological systems.

Under B-limited conditions, combination of NIP5;1 for uptake and BOR1 for efflux into xylem is likely to be a key for efficient B transport across root cells. BOR1 is likely to generate concentration gradient between root cells and the media. This concentration gradient is essential for NIP5;1 to facilitate B uptake into root cells because NIP5;1 is likely to be a passive channel for boric acid. Similar combination of a channel and a transporter is reported in the silicon transport system, Lsi1 and Lsi2, in rice (Ma et al. 2006, 2007).

4.3.2 BOR1 Degradation Via Endocytosis in Response to High B Supply

BOR1 mRNA levels are relatively unaffected by boron condition in the media, while BOR1 protein accumulation decreased under high B supply both in roots and shoots. BOR1 accumulation is regulated at the level of intracellular membrane trafficking into vacuoles (Takano et al. 2005). BOR1-GFP recycles between plasma membrane and early endsomes irrespective of B conditions. BOR1-GFP was accumulated to plasma membrane under low B conditions, and internalized via endocytic pathway within 1 h, and degraded in vacuoles when exposed to high B. BOR1 degradation upon high B supply is likely to be important to avoid over-accumulation of B in aerial portions of the plants when B concentrations in the media is high.

Overexpression of BOR1 resulted in increased B translocation to shoots and shoot apex and improved plant growth and fertility under limited B supply, (Miwa et al. 2006). The BOR1-overexpressing transgenic plants showed no detrimental effect on plant growth under normal or toxic level of B supply, possibly due to degradation of BOR1 under high B conditions.

4.4 Active B Transport System Under Toxic Level of B

4.4.1 Cellular B Distribution Under Adequate and Toxic Level of B

Under limited supply of B, B is preferentially distributed to water-insoluble residues which represents cell wall fraction as described in Sect. 3.2. In sunflower, excess B supply (1,000 μM) in a short time increased soluble B concentration but not insoluble B concentration assumingly present in cell walls. B concentration of cell sap is higher than that of apoplastic fluid under normal to toxic levels of B supply (1–1,000 μM) (Dannel et al. 1999). B is preferentially distributed to cell walls, and most of the additional B is likely to be present as a soluble form in cytoplasm.

4.4.2 BOR1 Homologs Involved in High B Tolerance Through B Efflux in Plants

Plants need to manage the toxic level of B supply. Hayes and Reid (2004) found that Sahara, a high-B tolerant barley cultivar, maintains lower B concentrations in roots than those of Schooner, a susceptible cultivar under high boron conditions. This reduction is likely to result in decrease of B concentrations in xylem and leaves, and improvement of high B tolerance in Sahara. Reid (2007) isolated *BOR1* homologs,

HvBOR2 and *TaBOR2*, from barley and wheat. Among cultivars, B tolerance was correlated with the reduction of B concentrations in roots. Positive correlations were found between mRNA levels and high B tolerance both in barley and wheat cultivars.

Sutton et al. (2007) also described the involvement of a barley BOR1 homolog, *HvBOT1* (*HvBOR2*) in high B tolerance. They performed QTL mapping between the two cultivars, Sahara and Clipper, the high B tolerant and susceptible barley cultivars and found that *HvBOR2* resided in the genetically mapped region. *HvBOR2* mRNA level was highly detected in roots and was higher in Sahara than Clipper.

Further, overexpression of AtBOR4 conferred high B tolerance in the transgenic *A. thaliana* lines (Miwa et al. 2007). Unlike the case of AtBOR1, AtBOR4 protein was not degraded under high concentrations of B and BOR4-GFP was localized to the distal side of the plasma membrane of root epidermis. This polar localization is likely to enable directional transport of B out of the cells.

All of these three studies support that BOR homologs contribute to high B tolerance through the reduction of cellular B concentrations. The BOR family genes contain two different types of efflux B transporters; BOR1 type for exporting B into the xylem to transport the required B and BOR4 type for excluding high concentrations of B to reduce B toxicity.

4.5 Retranslocation of B

Typical symptoms of B deficiency occur in growing portions of the plants. In contrast, B toxicity symptoms are mainly observed as necrosis in mature leaves. These observations indicate that B is hardly a mobile element.

Oertli (1993) clearly demonstrated B immobility in tomato. When tomato plants exhibiting B toxicity symptoms, were transferred to the B-depleted media, little new top growth was observed although plants contained high amount of B in the mature portions.

Plant species producing sugar alcohol for sugar translocation are an exception. It is well established that B can be readily re-translocated through phloem as B-sugar alcohol complex (Brown and Hu 1996). Hu et al. (1997) determined B complex forms as mannitol-B-mannnitol complex in phloem sap of celery, and either fructose-B-fructose, sorbitol-B-sorbitol, or fructose-B-sorbitol in the extrafloral nectary exudate of peach.

5 B Transport Mechanisms in Yeast and Mammals

A. thaliana BOR1 led to identification of borate transporters in *S. cerevisiae* and human.

5.1 Function of a BOR1 Homolog in S. cerevisiae

YNL275w, a *BOR1* homolog in *S. cerevisiae* was previously characterized as an anion transporter localized to plasma membrane, and reported to potentially transport HCO_3^-, I^-, Br^-, NO_3^-, and Cl^- (Zhao and Reithmeier 2001). YNL275w (Bor1p) also encodes an efflux-type B transporter (Takano et al. 2002). Knockout of *BOR1* conferred elevated sensitivity to B toxicity, most likely to have resulted from an increase of B concentration in cells (Takano et al. 2007). This revealed that one physiological function of Bor1p was B exclusion.

Additionally when FPS1 (a glycerol channel) and DUR3 (a urea transporter) were disrupted and overexpressed, B concentrations in yeast cells, and B tolerance were altered (Nozawa et al. 2006), suggesting possible involvement of these genes in B transport.

5.2 Function of a BOR1 Homolog in Animals

Little attention was paid to the physiological function of B and B homeostasis in animals until BOR1 was identified in *A. thaliana*. Among members of the SLC4 HCO_3^- transporter family, one protein, NaBC1, shows a high similarity to *A. thaliana* BOR1 (Park et al. 2004). NaBC1 was found to be a conductive transporter permeable to Na^+ and OH^- (H^+) in the absence of borate and an electrogenic, voltage-regulated Na^+-$B(OH)_4^-$ cotransporter in the presence of borate. Immunoblot analysis revealed NaBC1 expression in the rat kidney, parotid gland, submandibular gland, pancreas, liver, and spleen. It was demonstrated that B has mitogenic effects on human cells and NaBC1 is a Na^+-$B(OH)_4^-$ cotransporter for B homeostasis.

6 Conclusions and Foresights

In the last decade, much progress has been made in B biology at cellular and molecular levels. One of the physiological roles of B, formation of RG-II-B complex, in plants was addressed, and the channels and the transporters were identified. Considering that boron is shown to be essential/beneficial to organisms without cell wall, the next challenge will be identification of another novel physiological function of B possibly in the cytoplasm. This function, if any, could be a common function among the biological systems. Another question includes sensing mechanisms of boron in plants. It is true that plants behave in response to B conditions at the transcriptional (*NIP5;1*) and post-transcriptional levels (BOR1). What the plants sense as a signal of the B environments and how they transduce the signals are important questions. For this, three-dimensional protein structure of the boron channels and transporters may enable us to reveal the stereoscopic recognition and binding to boric acid.

References

Ahmed I, Yokota A, Fujiwara T (2007) A novel highly boron tolerant bacterium, *Bacillus boroniphilus* sp. nov., isolated from soil, that requires boron for its growth. Extremophiles 11:217–224

Amin SA, Kupper FC, Green DH, Harris WR, Carrano CJ (2007) Boron binding by a siderophore isolated from marine bacteria associated with the toxic dinoflagellate *Gymnodinium catenatum*. J Am Chem Soc 129:478–479

Armstrong TA, Spears JW, Crenshaw TD, Nielsen FH (2000) Boron supplementation of a semipurified diet for weanling pigs improves feed efficiency and bone strength characteristics and alters plasma lipid metabolites. J Nutr 130:2575–2581

Baluska F, Hlavacka A, Samaj J, Palme K, Robinson DG, Matoh T, McCurdy DW, Menzel D, Volkmann D (2002) F-actin-dependent endocytosis of cell wall pectins in meristematic root cells. Insights from brefeldin A-induced compartments. Plant Physiol 130:422–431

Bassil E, Hu H, Brown PH (2004) Use of phenylboronic acids to investigate boron function in plants. Possible role of boron in transvacuolar cytoplasmic strands and cell-to-wall adhesion. Plant Physiol 136:3383–3395

Bennett A, Rowe RI, Soch N, Eckhert CD (1999) Boron stimulates yeast (*Saccharomyces cerevisiae*) growth. J Nutr 29:2236–2238

Bolanos L, Lukaszewski K, Bonilla I, Blevins D (2004) Why boron? Plant Physiol Biochem 42:907–912

Bonin CP, Potter I, Vanzin GF, Reiter WD (1997) The MUR1 gene of Arabidopsis thaliana encodes an isoform of GDP-D-mannose-4, 6-dehydratase, catalyzing the first step in the de novo synthesis of GDP-L-fucose. Proc Natl Acad Sci USA 94:2085–2090

Brown PH, Hu H (1996) Phloem mobility of boron is species dependent: Evidence for phloem mobility in sorbitol-rich species. Ann Bot 77:497–506

Chen X, Schauder S, Potier N, Van Dorsselaer A, Pelczer I, Bassler BL, Hughson FM (2002) Structural identification of a bacterial quorum-sensing signal containing boron. Nature 415:545–549

Dannel F, Pfeffer H, Römheld V (1999) Distribution within the plant or compartmentation does not contribute substantially to the detoxification of excess boron in sunflower (*Helianthus annuus*). Aust J Plant Physiol 26:95–99

Dannel F, Pfeffer H, Römheld V (2000) Characterization of root boron pools, boron uptake and boron translocation in sunflower using the stable isotope ^{10}B and ^{11}B. Aust J Plant Physiol 156:756–761

Dannel F, Pfeffer H, Walch-Liu P, Romheld V (2001) Plant nutrition-Food security and sustainability of afro-ecosystems. Kluwer, Dordrecht, pp 162–163

Dell B, Huang LB (1997) Physiological response of plants to low boron. Plant Soil 193:103–120

Dembitsky VM, Smoum R, Al-Quntar A, Ali HA, Pergament I, Srebnik M (2002) Natural occurrence of boron-containing compounds in plants, algae and microorganisms. Plant Sci 163:931–942

Dordas C, Brown PH (2000) Permeability of boric acid across lipid bilayers and factors affecting it. J Membr Biol 175:95–105

Dordas C, Brown PH (2001) Expression of the major intrinsic protein (MIP) PIP1 in *Xenopus laevis* oocytes. Evidence for channel mediated transport of boric acid in squash (*Cucurbita pepo*). Plant Soil 235:95–103

Dordas C, Chrispeels MJ, Brown PH (2000) Permeability and channel-mediated transport of boric acid across membrane vesicles isolated from squash roots. Plant Physiol 124:1349–1362

Eckhert CD (1998) Boron stimulates embryonic trout growth. J Nutr 128:2488–2493

Fort DJ, Propst TL, Stover EL, Strong PL, Murray FJ (1998) Adverse reproductive and developmental effects in Xenopus from insufficient boron. Biol Trace Elem Res 66:237–259

Fort DJ, Stover EL, Strong PL, Murray FJ, Keen CL (1999) Chronic feeding of a low boron diet adversely affects reproduction and development in *Xenopus laevis*. J Nutr 129:2055–2060

Hayes JE, Reid RJ (2004) Boron tolerance in barley is mediated by efflux of boron from the roots. Plant Physiol 136:3376–3382

Hu H, Brown PH (1994) Localization of boron in cell walls of squash and tobacco and its association with pectin (evidence for a structural role of boron in the cell wall). Plant Physiol 105:681–689

Hu H, Penn SG, Lebrilla CB, Brown PH (1997) Isolation and characterization of soluble boron complexes in higher plants. The mechanism of phloem mobility of boron. Plant Physiol 113:649–655

Ishii T, Matsunaga T (1996) Isolation and characterization of a boron-rhamnogalacturonan-II complex from cell walls of sugar beet pulp. Carbohydr Res 284:1–9

Ishii T, Matsunaga T, Hayashi N (2001) Formation of rhamnogalacturonan II-borate dimer in pectin determines cell wall thickness of pumpkin tissue. Plant Physiol 126:1698–1705

Iwai H, Masaoka N, Ishii T, Satoh S (2002) A pectin glucuronyltransferase gene is essential for intercellular attachment in the plant meristem. Proc Natl Acad Sci USA 99:16319–16324

Iwai H, Hokura A, Oishi M, Chida H, Ishii T, Sakai S, Satoh S (2006) The gene responsible for borate cross-linking of pectin Rhamnogalacturonan-II is required for plant reproductive tissue development and fertilization. Proc Natl Acad Sci USA 103:16592–16597

Kato Y, Miwa K, Takano J, Wada M, Fujiwara T (2009) Highly boron deficiency-tolerant plants generated by enhanced expression of NIP5;1, a boric acid channel. Plant Cell Physiol 50:58–66

Kobayashi M, Matoh T, Azuma JI (1996) Two chains of rhamnogalacturonan II are cross-linked by borate-diol ester bonds in higher plant cell walls. Plant Physiol 110:1017–1020

Lanoue L, Taubeneck MW, Muniz J, Hanna LA, Strong PL, Murray FJ, Nielsen FH, Hunt CD, Keen CL (1998) Assessing the effects of low boron diets on embryonic and fetal development in rodents using in vitro and in vivo model systems. Biol Trace Elem Res 66:271–298

Loomis WD, Durst RW (1992) Chemistry and biology of boron. Biofactors 3:229–239

Ma JF, Tamai K, Yamaji N, Mitani N, Konishi S, Katsuhara M, Ishiguro M, Murata Y, Yano M (2006) A silicon transporter in rice. Nature 440:688–691

Ma JF, Yamaji N, Mitani N, Tamai K, Konishi S, Fujiwara T, Katsuhara M, Yano M (2007) An efflux transporter of silicon in rice. Nature 448:209–212

Marschner H (1995) Mineral nutritin of higher plants, 2nd edn. Academic, San Diego, CA

Mateo P, Bonilla I, Fernández-Valiente EE, Sanchez-Maeso E (1986) Essentiality of boron for dinitrogen fixation in *Anabaena* sp. PCC 7119. Plant Physiol 81:430–433

Matoh T, Ishigaki K, Mizutani M, Matsunaga W, Takabe K (1992) Boron nutrition of cultured tobacco BY-2 cells. 1. Requirement for and intracellular localization of boron and selection of boron and selection of cells that tolerate low levels of boron. Plant Cell Physiol 33:1135–1141

Matoh T, Ishigaki KI, Ohno K, Azuma JI (1993) Isolation and characterization of a boron-polysaccharide complex from radish roots. Plant Cell Physiol 34:639–642

Matsunaga T, Ishii T, Matsumoto S, Higuchi M, Darvill A, Albersheim P, O'Neill MA (2004) Occurrence of the primary cell wall polysaccharide rhamnogalacturonan II in pteridophytes, lycophytes, and bryophytes. Implications for the evolution of vascular plants. Plant Physiol 134:339–351

Miwa K, Takano J, Fujiwara T (2006) Improvement of seed yields under boron-limiting conditions through overexpression of BOR1, a boron transporter for xylem loading, in *Arabidopsis thaliana*. Plant J 46:1084–1091

Miwa K, Takano J, Omori H, Seki M, Shinozaki K, Fujiwara T (2007) Plants tolerant of high boron levels. Science 318:1417

Nable RO, Bañuelos GS, Paull JG (1997) Boron toxicity. Plant Soil 193:181–198

Noguchi K, Yasumori M, Imai T, Naito S, Matsunaga T, Oda H, Hayashi H, Chino M, Fujiwara T (1997) Bor1-1, an Arabidopsis thaliana mutant that requires a high level of boron. Plant Physiol 115:901–906

Noguchi K, Dannel F, Heidrun P, Römheld V, Hayashi H, Fujiwara T (2000) Defect in root-shoot translocation of boron in Arabidopsis thaliana mutant bor1-1. J Plant Physiol 156:756–761

Nozawa A, Takano J, Kobayashi M, von Wirén N, Fujiwara T (2006) Roles of BOR1, DUR3, and FPS1 in boron transport and tolerance in *Saccharomyces cerevisiae*. FEMS Microbiol Lett 262:216–222

Oertli JJ (1993) The mobility of boron in plants. Plant Soil 156:301–304

O'Neill MA, Eberhard S, Albersheim P, Darvill AG (2001) Requirement of borate cross-linking of cell wall rhamnogalacturonan II for Arabidopsis growth. Science 294:846–849

O'Neill MA, Ishii T, Albersheim P, Darvill AG (2004) Rhamnogalacturonan II: structure and function of a borate cross-linked cell wall pectic polysaccharide. Annu Rev Plant Biol 55:109–139

Park M, Li Q, Shcheynikov N, Zeng WZ, Muallem S (2004) NaBC1 is a ubiquitous electrogenic Na + -coupled borate transporter essential for cellular boron homeostasis and cell growth and proliferation. Mol Cell 16:331–341

Power PP, Woods WG (1997) The chemistry of boron and its speciation in plants. Plant Soil 193:1–13

Raven JA (1980) Short- and long-distance transport of boric acid in plants. New Phytol 84:231–249

Reid R (2007) Identification of boron transporter genes likely to be responsible for tolerance to boron toxicity in wheat and barley. Plant Cell Physiol 48:1673–1678

Reuhs BL, Glenn J, Stephens SB, Kim JS, Christie DB, Glushka JG, Zablackis E, Albersheim P, Darvill AG, O'Neill MA (2004) l-Galactose replaces l-fucose in the pectic polysaccharide rhamnogalacturonan II synthesized by the l-fucose-deficient *mur1* Arabidopsis mutant. Planta 219:147–157

Rowe RI, Eckhert CD (1999) Boron is required for zebrafish embryogenesis. J Exp Biol 202:1649–1654

Rowe RI, Bouzan C, Nabili S, Eckhert CD (1998) The response of trout and zebrafish embryos to low and high boron concentrations is U-shaped. Biol Trace Elem Res 66:261–270

Ryden P, Sugimoto-Shirasu K, Smith AC, Findlay K, Reiter WD, McCann MC (2003) Tensile properties of Arabidopsis cell walls depend on both a xyloglucan cross-linked microfibrillar network and rhamnogalacturonan II-borate complexes. Plant Physiol 132:1033–1040

Stangoulis JC, Reid RJ, Brown PH, Graham RD (2001) Kinetic analysis of boron transport in *Chara*. Planta 213:142–146

Sutton T, Baumann U, Hayes J, Collins NC, Shi BJ, Schnurbusch T, Hay A, Mayo G, Pallotta M, Tester M, Langridge P (2007) Boron-toxicity tolerance in barley arising from efflux transporter amplification. Science 318:1446–1449

Takano J, Yamagami M, Noguchi K, Hayashi H, Fujiwara T (2001) Preferential translocation of boron to young leaves in *Arabidopsis thaliana* regulated by the *BOR1* gene. Soil Sci Plant Nutr 47:345–357

Takano J, Noguchi K, Yasumori M, Kobayashi M, Gajdos Z, Miwa K, Hayashi H, Yoneyama T, Fujiwara T (2002) *Arabidopsis* boron transporter for xylem loading. Nature 420:337–340

Takano J, Miwa K, Yuan L, von Wirén N, Fujiwara T (2005) Endocytosis and degradation of BOR1, a boron transporter of *Arabidopsis thaliana*, regulated by boron availability. Proc Natl Acad Sci USA 102:12276–12281

Takano J, Wada M, Ludewig U, Schaaf G, von Wirén N, Fujiwara T (2006) The Arabidopsis major intrinsic protein NIP5;1 is essential for efficient boron uptake and plant development under boron limitation. Plant Cell 18:1498–1509

Takano J, Kobayashi M, Noda Y, Fujiwara T (2007) *Saccharomyces cerevisiae* Bor1p is a boron exporter and a key determinant of boron tolerance. FEMS Microbiol Lett 267:230–235

Tanaka M, Wallace IS, Takano J, Roberts DM, Fujiwara T (2008) NIP6;1 is a boric acid channel for preferential transport of boron to growing shoot tissues in Arabidopsis. Plant Cell 20: 2860–2875

Tyerman SD, Niemietz CM, Bramley H (2002) Plant aquaporins: multifunctional water and solute channels with expanding roles. Plant Cell Environ 25:173–194

Wallace IS, Choi WG, Roberts DM (2006) The structure, function and regulation of the nodulin 26-like intrinsic protein family of plant aquaglyceroporins. Biochim Biophys Acta 1758: 1165–1175

Warington K (1923) The effect of boric acid and borax on the broad bean and certain other plants. Ann Bot 37:629–672

Yu Q, Hlavacka A, Matoh T, Volkmann D, Menzel D, Goldbach HE, Baluska F (2002) Short-term boron deprivation inhibits endocytosis of cell wall pectins in meristematic cells of maize and wheat root apices. Plant Physiol 130:415–421

Zhao RM, Reithmeier RA (2001) Expression and characterization of the anion transporter homologue YNL275w in *Saccharomyces cerevisiae*. Am J Physiol Cell Physiol 281:C33–C45

Calcium: Not Just Another Ion

Oliver Batistič and Jörg Kudla

Abstract Calcium (Ca^{2+}) represents very likely the most versatile ion in living organisms. It is involved in nearly all aspects of plant development and participates in a plethora of regulatory processes. Calcium is an important signaling compound, regulates cellular metabolism and is important for endocytosis and exocytosis. Calcium can easily form complexes with proteins, membranes and other organic acids rendering this ion a versatile signaling constituent and simultaneously a toxic cellular compound. Consequently, the required tight spatial and temporal control of intracellular Ca^{2+} levels provided the basis for the emergence of calcium signaling. It is this apparent antagonism between the obvious cellular abundance of Ca^{2+} as a structural important ion in the plant and its required rareness in the cytoplasm as well as the evident question how this simple ion can specifically function in such a myriad of distinct process that has sparked considerable interest and research. Here we will discuss new insights into the signaling function of Ca^{2+} in the context of its diverse cell biological roles.

1 Introduction

Calcium (Ca^{2+}) represents very likely the most versatile ion in living organisms. It is involved in nearly all aspects of plant development and participates in a plethora of regulatory processes. Calcium is an important signaling compound, regulates cellular metabolism and is important for endocytosis and exocytosis. Moreover, it is important for energy production within mitochondria and chloroplasts. Ca^{2+} is an ion that due to its specific properties can easily be dehydrated. Because of its

O. Batistič and J. Kudla (✉)
Institut für Botanik und Botanischer Garten, Universität Münster, Schlossplatz 4, Münster 48149, Germany
e-mail: oliver.batistic@uni-muenster.de; jkudla@uni-muenster.de

R. Hell and R.-R. Mendel (eds.), *Cell Biology of Metals and Nutrients*, Plant Cell Monographs 17,
DOI 10.1007/978-3-642-10613-2_2,

flexibility in forming different coordination numbers and complex geometries, Ca^{2+} can easily form complexes with proteins, membranes and other organic acids like citrate and oxalate. This unique feature renders Ca^{2+} a toxic cellular compound because it can easily form insoluble complexes with phosphate (and consequently ATP), and also with DNA and RNA. However, on the other hand this property enabled the evolution of Ca^{2+} as an important signaling molecule because the required tight spatial and temporal control of intracellular Ca^{2+} levels provided the basis for the emergence of calcium signaling. It is this apparent antagonism between the obvious cellular abundance of Ca^{2+} as a structural important ion in certain organelles and cell structures and its required rareness in the cytoplasm as well as the yet not solved question how this simple ion can specifically function in such a myriad of distinct process that has sparked considerable interest and research. Several excellent reviews have in the past addressed the nutritional and signaling functions of Ca^{2+} (Hepler and Wayne 1985; Sanders et al. 2002; White and Broadley 2003; Hetherington and Brownlee 2004; Hirschi 2004). Here we will discuss new insights into the signaling function of Ca^{2+} in the context of its diverse cell biological roles.

2 Nutritional and Structural Functions of Ca^{2+}

2.1 Nutritional Functions of Ca^{2+}

Work in the late forties of the last century established that Ca^{2+} constitutes an important macronutrient for plant germination, growth and development (Mulder 1950; Helms 1971). Besides Mg^{2+}, Ca^{2+} is the most abundant group two element in plants and in different plant species Ca^{2+} can reach concentrations between 0.02% (e.g., Poacea species) and ~5% of the plant dry weight, with higher concentrations generally found in dicotylous plants (Broadley et al. 2003). In general, plants are classified into calcifuges that grow on soil with low Ca^{2+} content, and calcicoles, growing on high calcareous soils that can tolerate a high soluble Ca^{2+} concentration like Brassicaceae (White and Broadley 2003). Some calcicoles plants are able to deposit large amounts of Ca^{2+} in trichomes to protect the stomata from excessive Ca^{2+} (De Silva et al. 1996). Within the plant, Ca^{2+} is unequally distributed, and reaches higher concentrations in leaves and stems than in underground tissues (Demarty et al. 1984). In aerial parts, young leaves and fruits contain lower Ca^{2+} concentrations than older leaves (Kirkby and Pilbeam 1984). Calcium uptake is confined to the root tip or to nodes of newly emerging lateral roots, indicating that Ca^{2+} does not cross the "casparian strip" and follows the apoplastic pathway to reach the xylem of the central cylinder (Kirkby and Pilbeam 1984).

Calcium deficiency affects every tissue and plant organ. Generally, it causes stunted plant growth and appearance of brown spots associated with polyphenol oxidation, necrosis of primarily meristematic cells and later on in older tissues, cracking of fruits and curling of the leaves (Bussler 1962; Simon 1978). Characteristic symptoms are "tipburn" mostly described in agricultural vegetables like lettuce, or "blossom-end rot" in fruits of tomato (Simon 1978; Ho and White 2005). Especially fast growing young tissues with a low transpiration rate and therefore insufficient supply of Ca^{2+} from the xylem are prone to deficiencies and exhibit symptoms early. This effect is further enhanced by the relative immobility of Ca^{2+} within the plant because Ca^{2+} is insufficiently redistributed from older leaves, which harbor a higher Ca^{2+} content (van Goor and Wiersma 1974; White and Broadley 2003).

Root growth and the development of the root system is also affected under Ca^{2+} deficiency leading to growth arrest and finally stalling of the root (Simon 1978). Ca^{2+} also affects the uptake of other ions into the root system. It improves the selective uptake of K^+ in the presence of high Na^+ concentrations (Epstein 1961,1998). Here Ca^{2+} concurrently blocks Na^+-permeable channels and reduces Na^+ uptake (Demidchik and Tester 2002) and inhibits Na^+ induced K^+ efflux (Shabala et al. 2006). In addition, it was suggested that Ca^{2+} reduces Na^+ induced depolarization of the plasma membrane thereby minimizing K^+ efflux by depolarization-activated channels (Shabala et al. 2006).

2.2 *Structural Functions of Ca^{2+}*

The large ion radius of Ca^{2+} facilitates straightforward dehydration of the ion thereby enabling the binding of Ca^{2+} to several anionic headgroups of membrane-lipids like phosphatidylserine, phosphatidic acid or glycolipids and additionally to membrane proteins (Hauser et al. 1976; Jaiswal 2001). Consequently, Ca^{2+} represents an important regulator of different dynamic membrane processes. Within the cell, Ca^{2+} promotes the fusion of vesicles to their target membranes by cross-linking the lipids from the two different membranes (Williams 1970; Hauser et al. 1976). On the other hand, Ca^{2+} is a structure forming cation. It reduces the fluidity and therefore enables a tighter packaging of the membrane lipid bilayer, thereby reducing passive ion-fluxes of monovalent cations like H^+, Na^+ and K^+ (Williams 1970; Jaiswal 2001; White and Broadley 2003; Plieth 2005). Therefore, Ca^{2+} deficiency induces membrane leakyness for monovalent cations rendering plants susceptible to damage by salt or low pH (Plieth 2005).

Calcium is also an abundant cell wall component critically regulating the strength and pH of the cell wall (Demarty et al. 1984). Here, Ca^{2+} bridges cell wall polymers like negatively charged galacturonanes or xylans (Brett and Waldron 1996) and influences the formation of the 1,3-β-glucan "callose", which is an essential cell wall component in the cell plate of dividing cells, in growing pollen tubes, or is produced after wounding. Remarkably, the enzyme β-1,3 glucan synthase, which forms callose is directly activated by Ca^{2+} (Kauss 1987; Brett and Waldron 1996).

3 The Evolution of Ca^{2+} as a Signaling Molecule

Several physico-chemical features are unique to Ca^{2+} and predestinate this ion as potential signaling molecule. Due to the large ionic radius of a Ca^{2+} ion (99 pm) water molecules are less tightly bound to Ca^{2+} than to the smaller Mg^{2+} ion (65 pm). Therefore, less energy is required to remove the bound water to enable Ca^{2+} ligand interaction (Hepler and Wayne 1985). Consequently, Ca^{2+} ions are easily dehydrated, and can form complexes with high and variable coordination numbers (6–8, but also 5–10) and are flexible in coordination geometry and Ca^{2+}-ligand distances. Ca^{2+} binds favorably to carboxylate oxygen, while Mg^{2+} or Zn^{2+} have higher affinity toward nitrogen-based ligands, associated with amino acids, which are not as common in proteins than glutamate and aspartate (Williams 1970; Ochiai 1991). Although other ions like Ba^{2+}, Sr^{2+} and Zn^{2+} could substitute for Ca^{2+} function, they are less abundant than Ca^{2+} and other divalent cations like Cd^{2+} and Pb^{2+} are highly toxic (Jaiswal 2001).

On the other hand, these features of Ca^{2+} are also the reason why elevated levels of Ca^{2+} will react with inorganic phosphate, forming an insoluble precipitate. Therefore, high Ca^{2+} acts as a cytotoxin inhibiting the phosphate-based energy system (Hepler and Wayne 1985). Moreover, excess Ca^{2+} would compete with Mg^{2+} for binding sites on various proteins. Thus, during evolution, organisms were forced to evolve de-toxifying mechanisms that are effective in keeping Ca^{2+} at low levels in the cytoplasm. Importantly, Ca^{2+} pumps, which are responsible to extrude Ca^{2+} out of the cytoplasm, are themselves Ca^{2+}-activated resulting in a time lag of efflux activation that follows the entry of Ca^{2+} into the cytoplasm. This facet of Ca^{2+} regulation mandatory leads to a transient Ca^{2+} elevation within the cell when Ca^{2+} permeable channels open (Ochiai 1991) thereby providing a unique system to evolve the extant Ca^{2+}-regulated circuits in which Ca^{2+} itself activates Ca^{2+} channels within the cell. This Ca^{2+}-dependent release of Ca^{2+} from external or internal stores now is responsible for a rapid, amplified (100 fold) but still transient Ca^{2+} increase and forms the mechanistic basis of many Ca^{2+} signaling events (Williams 2004).

Since the concentration of Ca^{2+} in the cytoplasm is kept at low levels, the concentration can be, rapidly and efficiently modulated. In contrast, due to physiological requirements, the cytosolic concentrations of Mg^{2+} or K^{+} are relatively high. Therefore, a 100 fold change of these ions in the cytosol would be more difficult to achieve: firstly the amount of ions to be transported would be much higher and secondly, the resulting dramatic change in cellular ion homeostasis would be detrimental while a 100 fold change of the Ca^{2+} concentration only marginally effects the osmotic balance of the cytoplasm (Hepler and Wayne 1985; Ochiai 1991). Finally, due to the unique properties of Ca^{2+}, this ion can bind to and dissociate from proteins much faster than other ions (Ochiai 1991). This aspect of Ca^{2+}, allows for transient interaction with calcium binding proteins and has enabled the evolution of a signaling system that can control fast and diverse reactions (Hepler and Wayne 1985).

4 Calcium Release in Response to Signals and Stimuli

4.1 Calcium Responses to Abiotic, Biotic Factors and Development

Cytoplasmic changes of Ca^{2+} concentration from resting concentrations of 100–300 nM up to 1 μM are observed after various signals or stimuli. Abiotic cues like salt (Lynch et al. 1989), osmotic stress (Takahashi et al. 1997; Cessna et al. 1998), drought (Knight et al. 1998), ozone (Clayton et al. 1999), anoxia (Subbaiah et al. 1994), CO_2 (Webb et al. 1996), gravitation (Lee et al. 1983; Gehring et al. 1990; Fasano et al. 2002), mechanical injury and touch (Haley et al. 1995; Legue et al. 1997) all cause transient elevations in Ca^{2+} concentration. Different temperature regimes can also induce Ca^{2+} responses. Cold stress, especially the cooling rate, is reflected by specific increases of intracellular Ca^{2+} (Knight et al. 1991; Plieth et al. 1999). Development of freezing tolerance requires influx of extracellular Ca^{2+} and enhanced Ca^{2+}-dependent exocytosis to enable resealing of the membrane after mechanical disruption provoked by cold (Schapire et al. 2008; Yamazaki et al. 2008). These membrane fusions are mediated by Ca^{2+} interaction with the sensor protein synaptotagmin to promote interaction with SNARE proteins (Kesavan et al. 2007; Schapire et al. 2008; Yamazaki et al. 2008). Furthermore, tobacco plants also respond with Ca^{2+} release to heat shock (Gong et al. 1998), while Ca^{2+} transients were recorded in Arabidopsis during the recovery from heat exposure (Larkindale and Knight 2002).

Also light responses in plants are accompanied by complex and specific patterns of Ca^{2+} transients. Changes in cytoplasmic Ca^{2+} concentration were observed after plant exposure to red but not far-red light implicating that this reaction is mediated by the photoreceptor phytochrome B (Shacklock et al. 1992; Neuhaus et al. 1993). Blue light Ca^{2+} responses depend on the phototropin photoreceptors but not on cryptochromes (Baum et al. 1999; Harada et al. 2003; Stoelzle et al. 2003). However, cryptochromes could fine tune a Ca^{2+} response (Long and Jenkins 1998), and cryptochrome signaling is mediated via the Ca^{2+}- binding protein SUB1 (short under blue light), which in turn modulates phyotochrome function (Guo et al. 2001a). These observations exemplify the complexity and interconnection of Ca^{2+} responses that can occur in reaction to a single environmental cue like light. In addition, Ca^{2+} is also released in a circadian manner and responds distinctly to light intensity (Love et al. 2004).

Similar to abiotic stimuli, biotic factors like bacterial pathogens (Atkinson et al. 1990; Xu and Heath 1998; Blume et al. 2000), fungal elicitors (Knight et al. 1991), attacks by herbivores or symbiotic interactions with nitrogen-fixing bacteria or mycorrhizal fungi cause different and specific Ca^{2+} responses (Ehrhardt et al. 1996; Kosuta et al. 2008). Moreover, cell apoptosis during the hypersensitive response induced by a pathogen is mediated by a complex regulation of cellular Ca^{2+} dynamics. An initial transient increase of Ca^{2+} occurs after pathogen infection

(Levine et al. 1996) and is then followed by a silent phase and a second, but sustained increase of cytoplasmic Ca^{2+} (Grant et al. 2000). Finally, a massive efflux of Ca^{2+} to the apoplast leads to cell collapse and death (Nemchinov et al. 2008).

Transient changes in cellular Ca^{2+} concentration have also extensively been observed during tightly regulated developmental processes. Oscillatory, tip-localized gradients of Ca^{2+} are important for proper growth of pollen tubes and expansion of root hairs by enabling high exocytotic turnover at the growing tip (Rathore et al. 1991; Miller et al. 1992; Rudd and Franklin-Tong 1999). In contrast, the self-incompatibility response during pollination is mediated by a rise of Ca^{2+} originating from the nuclear region, potentially regulating gene expression to inhibit pollen tube growth (Franklin-Tong et al. 1993). Ca^{2+} is also essential for fertilization (Faure et al. 1994), and a Ca^{2+} transient lasting for several minutes has been observed during the sperm-egg fusion event (Digonnet et al. 1997).

4.2 Calcium Responses to Hormones

Calcium release and subsequent signal transduction events are important after perception of phytohormones like cytokinin (Hahm and Saunders 1991), salicylic acid (Kawano et al. 1998) and ethylene (Raz and Fluhr 1992). Jasmonic acid (Sun et al. 2006) and its precursor 12-oxophytodienoic acid (OPDA) induce a large increase of cytoplasmic Ca^{2+} (Walter et al. 2007). Giberellic acid (GA) causes a long sustained increase of cytosolic Ca^{2+} at the cell periphery by influx of extracellular Ca^{2+} (Gilroy and Jones 1992), but the release of intracellular Ca^{2+} is also important for induction of α-amylase transcription (Chen et al. 1997). GA mediated calcium influx also precedes and is important for the secretion of α-amylase (Bush 1996) and can be accounted to the effect that Ca^{2+} enhances exocytosis (Homann and Tester 1997) by promoting membrane fusion (Bhalla et al. 2006; Martens et al. 2007). Additionally, GA also increases Ca^{2+} within the endoplasmic reticulum (ER), which is important for the maturation of α-amylase (Bush et al. 1989a, b). On the other hand, GA induced Ca^{2+} release in aleurone cells can be reversed by ABA (Gilroy and Jones 1992), and Ca^{2+} uptake into the ER is also inhibited by ABA (Bush et al. 1993).

Auxin induces an oscillatory Ca^{2+} release and thereby promotes stomatal opening (Felle 1988; Irving et al. 1992). Abscisic acid (ABA) also induces oscillatory Ca^{2+} transients resulting in closure of the stomata and being essential to keep them closed in the long term (McAinsh et al. 1990; Allen et al. 1999, 2000, 2001; Staxen et al. 1999). In another study, spontaneous Ca^{2+} oscillations were terminated by ABA in some guard cells (Klusener et al. 2002). These findings suggest that ABA can have positive and negative effects on the cellular Ca^{2+} level. Moreover, since Ca^{2+} levels in the cytosol exhibit circadian oscillation, these different levels of cytoplasmic calcium could also differentially prime the guard cells and other cell types to the effect of ABA, which then leads to a different final outcome (Dodd et al. 2005). Indeed, MacRobbie (1989) found out that the stimulatory effect of ABA is stronger in the afternoon than in the morning (MacRobbie 1989). Moreover, the parameters of Ca^{2+} oscillations which lead to stomatal closure could depend on the physiological condition of the plant (Klusener et al. 2002).

These examples clearly illustrate the universality and complexity of Ca^{2+} responses as well as their intricate and often confusing interconnection with various cellular or hormonal processes. They also highlight a continuing dilemma of plant Ca^{2+} research in which a further accumulation of descriptive data linking Ca^{2+} responses to certain biological processes will not likely advance our understanding of the underlying functional principles and causalities.

4.3 Interconnection of Ca^{2+} Dynamics with other Second Messengers

A similar complex situation applies to the interconnection of Ca^{2+} with other second messenger components. Various second messengers induce Ca^{2+} release, and their generation itself is often regulated by Ca^{2+}. Diacylglycerol (DAG), Cyclic nucleotides like cAMP and cGMP (Volotovski et al. 1998), cyclic Adenosine-diphospho-Ribose (cADPR) (Allen et al. 1995), Inositol-3-phosphate (InsP3) (Alexandre 1990; Gilroy et al. 1990) or its derivate myo-Inositol-Hexakisphosphate (InsP6) (Lemtiri-Chlieh et al. 2003), Nicotinic Acid-Adenine Dinucleotidphosphate (NAADP) (Navazio et al. 2000), Sphingosine-1-phosphate (S1P) (Ng et al. 2001), extracellular Glutamate (Dennison and Spalding 2000) and extracellular ATP (Demidchik et al. 2003, 2009; Jeter et al. 2004) can mobilize Ca^{2+} either from intracellular or extracellular stores, respectively. Reactive oxygen species (ROS), generated by plasma membrane NADPH oxidases, cause a Ca^{2+} influx into the cytosol (Price et al. 1994; Pei et al. 2000; Foreman et al. 2003). Moreover, different types of ROS can differentially activate Ca^{2+} permeable channels in different root tissues (Demidchik et al. 2007). Like ROS, the Ca^{2+} releasing second messenger cADPR is also produced after application of ABA (Wu et al. 1997; Leckie et al. 1998), and is also synthesized in a circadian manner which could be important for the observed circadian Ca^{2+} oscillations (Dodd et al. 2007). These observations point to a further level of complexity interconnecting Ca^{2+} simultaneously with second messengers, hormonal responses and reactions to environmental cues.

5 Organelles and Ca^{2+}

Calcium can rapidly enter the cell (10^6 molecules/sec. per channel) but diffusion within the cytoplasm is very limited (up to 0.5 μm) and Ca^{2+} is rapidly bound by Ca^{2+} binding proteins or buffered by cell organelles (within 50 μsec) (Clapham 1995; Lecourieux et al. 2002). Indeed, near membrane Ca^{2+} concentrations in the

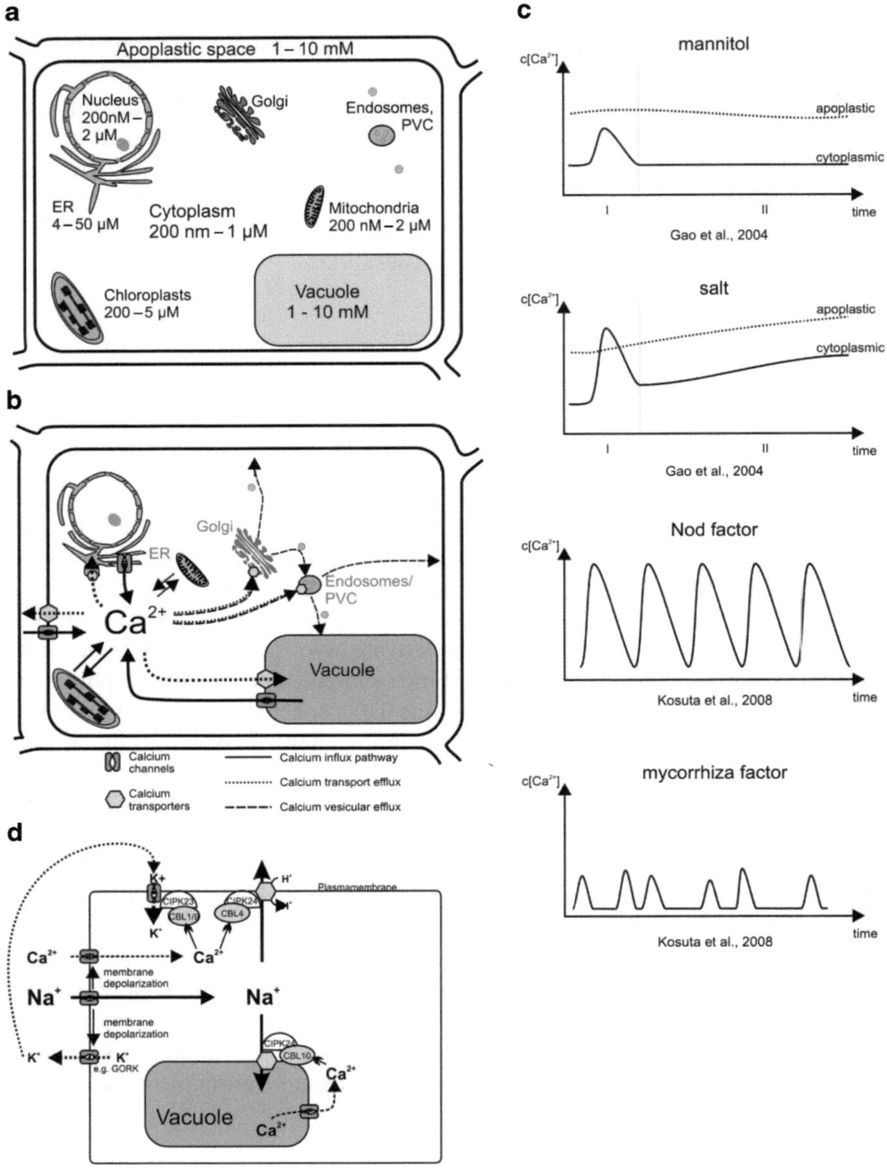

Fig. 1 (**a**) Distribution and concentration of free Ca^{2+} generally found in different compartments of the cell. Highest concentrations of calcium are found in the apoplastic space and within the vacuole (up to 10 mM). Also the ER can take up large amounts of Ca^{2+} (up to 50 μM). Within mitochondria and chloroplasts, Ca^{2+} concentrations can reach up to 2 and 5 μM. In the nucleoplasma and cytoplasma, the resting Ca^{2+} concentrations are maintained around 200 nM. The concentrations of Ca^{2+} within the Golgi and Endosomes are not known. (**b**) Different cellular compartments contribute to the cytoplasmic rise of Calcium. Influx is mainly regulated via influx

vicinity of Ca^{2+} channels are estimated to be 10–100 fold higher than the measured cytoplasmic Ca^{2+} concentrations (Etter et al. 1996; Demuro and Parker 2006). Therefore, local Ca^{2+} signals at specific microdomains are assumed to be the basis of differential Ca^{2+} signals, which can promote different responses to various signals (Berridge 2006).

Various organelles and compartments are implicated in different Ca^{2+} responses, and moreover, several compartments can act in concert to shape a Ca^{2+} signal and to establish a correct response to the signal. The major Ca^{2+} store of the cell is the apoplast which supplies extracellular Ca^{2+} (1–10 mM). From the apoplast, Ca^{2+} is released by Ca^{2+} channels which are activated by changes of the plasma membrane voltage, or by ligand activated channels (Fig. 1a).

Intracellular Ca^{2+} is stored mainly in the vacuole and ER, and released by voltage dependent and ligand gated channels. Within the vacuole, Ca^{2+} functions as counterion of inorganic and organic anions (White and Broadley 2003). Additionally, different types of vesicles, like Golgi or endosomal vesicles are potential mobile calcium stores (Wagner and Rossbacher 1980; Wick and Hepler 1980; Sakai-Wada and Yagi 1993; Trewavas 1999; Li et al. 2008), or transport excess calcium and other cations out of the cytoplasm either to the apoplast or to the vacuole (Fig. 1b) (Menteyne et al. 2006). Other organelles, like chloroplast and mitochondria, store Ca^{2+} also and are important to maintain the cellular Ca^{2+} and ATP homeostasis (Plieth 2005).

5.1 Calcium Signaling within the Nucleus

High levels of calcium are observed in the nucleus and the nuclear envelope (Wick and Hepler 1980). The nucleus can respond specifically and independently of the cytoplasm to a signal by specific temporal changes in Ca^{2+} concentration

Fig. 1 (continued) from the apoplastic space or vacuole. Also the ER, and organelles like chloroplasts and mitochondria contribute to the influx of cytosolic Ca^{2+}. The contribution of Golgi and Endosomes to Ca^{2+} influx is currently not known, but they can contribute to the uptake of Ca^{2+} out of the cytoplasma. In addition, Golgi and Endosomes can mediate the export of Ca^{2+} to the apoplast or vacuole. (**c**) Different stress stimuli provoke specific calcium signatures. Application of mannitol or sodium provokes a transient initial (I) rise of cytoplasmic Ca^{2+} concentrations, which is lower in the case of mannitol compared sodium induced responses. A second phase (II) of sustained Ca^{2+} increase is observed after sodium application, but not after mannitol treatment. Moreover, apoplastic Ca^{2+} concentrations rise also under sodium stress, but not in response to mannitol stress. Additionally, different symbiotic factors induce distinct Ca^{2+} spiking pattern. While Nod factors mediate a repetitive, constant Ca^{2+} spiking, mycorrhiza provokes a non-uniform, chaotic Ca^{2+} spiking, which exhibits lower maximal amplitudes. (**d**) Model of CBL/CIPK function in salt stress responses. During Na^+ stress, sodium mediates influx of Ca^{2+} from the apoplastic space. While apoplastic Ca^{2+} triggers CBL4/CIPK24 to activate the sodium extrusion exchanger SOS1 to directly transport sodium out of the cell, CBL10 together with CIPK24 could be activated by vacuolar Ca^{2+} to transport excess Na^+ into the vacuole. Additionally, K^+ ions that are released due to depolarisation of the plasma membrane by Na^+ could be transported back into the cell by activation of CBL1/CBL9/CIPK23 which in turn activate K^+ inward rectifying channels (van Der Luit et al. 1999; Pauly et al. 2001). Bacterial elicitors induce a Ca^{2+} release in the cytosol and nucleus. However, the nitric oxide (NO) signal following elicitor application is important for the Ca^{2+} release in the cytosol, but it does not trigger a nuclear Ca^{2+} response (Lamotte et al. 2004). The nucleus also exhibits specific calcium signals in response to different elicitors. Harpin and flagellin resulted in a different Ca^{2+} release than observed in response to carbohydrate elicitors like oligogalacturonides. Remarkably, the cytosolic Ca^{2+} response to these elicitors was comparable (Lecourieux et al. 2005).

These observations suggest that the nucleus does indeed harbor an independent Ca^{2+} machinery which could involve P-ATPases and nucleotide gated channels located at the inner membrane of the nucleus to regulate the nuclear Ca^{2+} reservoir (Mazars et al. 2009). Although it was reported that nuclei are Ca^{2+} impermeable (Xiong et al. 2004), inhibition of Ca^{2+} entry from the extracellular milieu prevented nuclear Ca^{2+} rises (Pauly et al. 2001; Mazars et al. 2009). In additon, it was reported that the nuclear membrane can take up Ca^{2+} in an ATP-dependent manner, which then can be released into the nucleus, implicating that nuclei harbor autonomous Ca^{2+} machineries (Bunney et al. 2000).

In legumes, Nod factor-mediated perinuclear oscillations of Ca^{2+} that occur 10-30 minutes after the initial Ca^{2+} rise in the cytosol (Ehrhardt et al. 1996; Felle et al. 1999), are mediated by the proteins CASTOR and POLLUX, which resemble bacterial potassium channels. However, the exact function of these proteins has remained unclear (Charpentier et al. 2008).

5.2 Calcium Regulation by the ER

Especially due to its large surface within the cell the ER is likely to represent an important intracellular Ca^{2+} store. The ER contains different types of Ca^{2+} channels and transporters, to regulate cytoplasmic and luminal Ca^{2+} levels. However, no ER-localized Ca^{2+} channel from plants has been identified at the molecular level and our understanding of the contribution of the ER to cellular dynamics of Ca^{2+} is much less advanced than in animal systems. Considering the absence of channels that exhibit recognizable similarity to animal InsP3 and Ryanodine-receptors in plants (with the exception of Chlamydomonas and Volvox (Wheeler and Brownlee 2008)) the interconnection of the ER with the cellular Ca^{2+} homeostasis in plants may be fundamentally different than in animals cells.

The ER is loosely associated with the mitotic apparatus, and Ca^{2+} levels regulated by the ER could be therefore important to control cell division (Hepler 2005). Within the lumen of the ER, calcium is important for the maturation of proteins (Bush et al. 1989b). Ca^{2+} concentration within the lumen of the ER is tightly controlled by Calreticulin (CRT), a high capacity calcium binding protein (25 calcium ions/protein) (Persson et al. 2001). Overexpression or reducing the expression either enhances resistance to Ca^{2+} depletion or leads to increased sensitivity to low Ca^{2+}, respectively (Persson et al. 2001). Moreover, increasing the Ca^{2+} buffer capacity by overexpressing CRT also enhances the stimulus induced Ca^{2+} release from the ER (Persson et al. 2001).

5.3 Mitochondrial Calcium Dynamics

Mitochondria have to retain high concentrations of Ca^{2+} to maintain the activity of enzymes like the NADH dehydrogenase (Moore and Åkerman 1984). Moreover, mitochondria are able to take up enormous amounts of Ca^{2+} from the cytosol (Dieter and Marme 1980). The resting concentration of Ca^{2+} in mitochondria has been estimated to be around 200 nM (Subbaiah et al. 1998; Logan and Knight 2003), and the mitochondrial Ca^{2+} content increases dramatically in response to cytoplasmic Ca^{2+} rises (Logan and Knight 2003). Therefore, mitochondria represent important cellular Ca^{2+} sinks that can contribute to the reduction of cytosolic Ca^{2+} levels after a stimulus induced elevation of cytoplasmic Ca^{2+} concentration (Bygrave 1978; Dieter and Marme 1980, 1983).

In addition, specific stimuli like touch or hydrogen peroxide induce a transient increase of mitochondrial Ca^{2+} concentration that was suggested to occur independently of the cytosolic rise of Ca^{2+}, pointing to a semi-autonomous mitochondrial calcium signaling pathway (Logan and Knight 2003). Moreover, mitochondria can also contribute to the cytosolic rise of Ca^{2+} by release of Ca^{2+} during anoxic conditions (Subbaiah et al. 1998).

5.4 The Role of Chloroplasts in Cellular Calcium Homeostasis

Chloroplasts can accumulate Ca^{2+} in the millimolar range and thereby can contribute to the cellular Ca^{2+} homeostasis (Portis and Heldt 1976). The level of Ca^{2+} in chloroplasts can rise upon illumination by light (Moore and Åkerman 1984; Miller and Sanders 1987; Kreimer et al. 1988) or after dark transition (Sai and Johnson 2002), and can follow a circadian rhythm (Johnson et al. 1995), but does not respond to mechanical stress or cold (van Der Luit et al. 1999). Within the chloroplast, Ca^{2+} is required for the electron flow at photosystem II (Kauss 1987), stabilizes the high redox potential form of cytochrome b-559 (McNamara and Gounaris 1995) and is an important co-factor and activity-regulator of enzymes like NAD kinase (Moore and Åkerman 1984). Indeed, the differential distribution of Ca^{2+} and Mg^{2+} between the stroma and thylakoid lumen during the dark and light phase has been shown to contribute to the regulation of the "on-off" state of chloroplasts (Ettinger et al. 1999).

A direct influence of the chloroplast on the cytoplasmic Ca^{2+} dynamics was revealed by the analysis of the Ca^{2+} -binding protein CAS (Ca^{2+} sensing receptor) (Han et al. 2003; Nomura et al. 2008; Weinl et al. 2008). CAS has originally been reported as an extracellular Ca^{2+}-sensing receptor, exhibting a high capacity to bind calcium (10–12 calcium ions per molecule) (Han et al. 2003). However, several studies revealed that the protein is exlusively localized in chloroplasts (Nomura et al. 2008; Vainonen et al. 2008; Weinl et al. 2008). Within the chloroplast, CAS is targeted to the thylakoid membrane, and is light-dependent phosphorylated (Vainonen et al. 2008; Weinl et al. 2008). CAS knock-out plants exhibit retarded growth, however, activity of the photosystem is not affected in these plants (Vainonen et al. 2008). When grown under low Ca^{2+} conditions, CAS knock-down plants display delayed bolting and are not able to induce flowering (Han et al. 2003). Ca^{2+}-induced cytoplasmic Ca^{2+} release is impaired in CAS knock-down and knock-out plants and these lines are impaired in Ca^{2+}-induced stomatal closure responses (Han et al. 2003; Nomura et al. 2008; Weinl et al. 2008). However, CAS knock-out plants can respond to externally imposed Ca^{2+} oscillations and then display normal stomatal closure reactions, indicating that the ability to respond to cytoplasmic Ca^{2+} elevations in mutant plants is not affected. This points to a function of CAS in the generation of cytoplasmic Ca^{2+} transients that are required for stomatal closure (Weinl et al. 2008) and indicates that the chloroplast targeted Ca^{2+} sensor protein CAS somehow connects cytoplasmic and chloroplast Ca^{2+} dynamics. This function somewhat resembles that of the Ca^{2+} buffer protein CRT in the ER. Similarly, loss of CAS could lead to a reduced buffer capacity of the chloroplasts, implicating that less Ca^{2+} can be allocated from the chloroplasts to the transient cytoplasmic increase of Ca^{2+}. These recent findings surprisingly highlighted the interconnection and importance of chloroplasts for the cellular Ca^{2+} machinery.

6 Channels and Transporters shaping Ca^{2+} Signals

6.1 *Influx of Ca^{2+}*

Several different Ca^{2+} permeable channel activities were reported to exist at the plasma membrane of plants that can mediate the influx of Ca^{2+} into the cytosol and have the potential to modulate the cellular Ca^{2+} signature depending on their specific activation properties (White et al. 2002; Demidchik and Maathuis 2007). In general, Ca^{2+} permeable channels can be classified as voltage dependent and voltage independent/ligand dependent channels (White et al. 2002). Additionally, stretch activated calcium channels exist at the plasma membrane (Cosgrove and Hedrich 1991; Dutta and Robinson 2004; Nakagawa et al. 2007). These different channel types can co-exist in certain cell types, allowing the cells to respond to a wide range of signals and to differentiate between nutrient and signaling requirements (Miedema et al. 2001; Miedema et al. 2008). Variability in the specific abundance of the different channel types could contribute to the specific needs of a cell type or tissue (Demidchik et al. 2002). However, it should be noted that the molecular identification and characterization of true Ca^{2+} – specific channels from plants has still not been reported.

6.1.1 Voltage Dependent Channels

Voltage dependent channels are separated in to depolarization activated Ca^{2+} permeable channels (DACCs) and hyperpolarization activated Ca^{2+} permeable channels (HACCs) (White et al. 2002). Depolarization activated channels contribute to the short transient influx of Ca^{2+} during signal responses, since they enter a quiescent state at constant depolarized membrane voltage (Thion et al. 1998). Hyperpolarization activated channels exhibit a large Ca^{2+} conductance and could contribute to a sustained Ca^{2+} influx regulating signaling and nutrition in fast growing tissues or cell types (Miedema et al. 2001, 2008), but are inactivated by increased levels of intracellular Ca^{2+} (Hamilton et al. 2000). Consequently, both channel types could function interdependently. Hyperpolarization of the membrane could activate HAC channels. Influx of Ca^{2+} would inhibit HACC activity and depolarize the membrane which then would activate DAC channels (Hamilton et al. 2000; Miedema et al. 2001). In addition, voltage dependent K^+ channels could also contribute to the influx of Ca^{2+} (Fairley-Grenot and Assmann 1992; Wegner and De Boer 1997; White et al. 2002).

Plant annexins are small proteins capable of Ca^{2+}-dependent membrane binding and insertion and appear to create Ca^{2+} influx pathways especially during stress responses involving acidosis (Mortimer et al. 2008). Some annexins were recently shown to assemble into Ca^{2+}-permeable channels in the plasma membrane and endomembranes of plant cells, which could be activated by Ca^{2+}, hyperpolarization and ROS (Demidchik and Maathuis 2007; Mortimer et al. 2008). In addition, annexins from *Zea mays* were reported to create Ca^{2+} permeable transport pathways and to regulate cytoplasmic Ca^{2+} concentration (Laohavisit et al. 2009).

6.1.2 Ligand Gated Channels

While the molecular identity of voltage dependent channels is still of uncertainty, ligand gated channels are nonselective cation channels represented by Cyclic nucleotide gated channels (CNGCs) and Glutamate receptors, which are important for ion homeostasis of Ca^{2+} and different other cations like K^+, Na^+ and others (Hua et al. 2003a; Ali et al. 2006). Individual CNGCs harbor different selectivity filters, indicating that certain CNGCs exhibit distinct selectivity for cations (Kaplan et al. 2007). Indeed, tobacco plants overexpressing a CNGC gene were hypersensitive to Pb^{2+}, while Arabidopsis plants harboring a T-DNA insertion within the respective CNGC1 gene were more tolerant to Pb^{2+} (Arazi et al. 1999; Sunkar et al. 2000). In Arabidopsis 20 CNGC were identified which are expressed in different tissues of the plant (White et al. 2002). In general, CNGCs are activated by cAMP and cGMP and harbor a binding site for Calmodulin which partially overlaps with the binding domain for cNMPs. Therefore, binding of Ca^{2+}/Calmodulin results in inactivation of CNGCs, due to blocking of the cNMP binding domain (Hua et al. 2003b; Ali et al. 2006). Moreover, for CNGC2 it was also suggested that the channel is blocked by high external Ca^{2+} concentrations (Hua et al. 2003a). CNGC2 was originally identified as the "defence no death" (dnd) 1 mutant, which fails to induce the Ca^{2+} mediated hypersensitive response to an avirulent strain of the pathogen *Pseudomonas syringae* and exhibits enhanced resistance to pathogens (Yu et al. 1998; Clough et al. 2000). Interestingly, mutants of CNGC2 are specifically hypersensitive to external Ca^{2+}, but have a normal Ca^{2+} content. Therefore, it was suggested that CNGC2 contributes to calcium signaling (Chan et al. 2003). In addition to CNGC2, CNGC4 (Balague et al. 2003), CNGC11 and CNGC12 (Yoshioka et al. 2006; Urquhart et al. 2007) were also implicated in mediating pathogen responses, while CNGC18 mediates tip growth of pollen (Frietsch et al. 2007).

Similar to CNGCs, Glutamate receptors (GLR) are also non-selective cation channels and 20 genes encoding GLRs were identified in Arabidopsis (White et al. 2002). GLRs are activated by Glutamate and Glycine as well as by other amino acids and mediate increases of cytosolic Ca^{2+} concentration (Qi et al. 2006). It is assumed that GLRs are important for plant Ca^{2+} nutrition (Demidchik and Maathuis 2007) but also have a role in calcium dependent photomorphogenesis. Application of GLR antagonists impaired plant light-signal transduction and resulted in enlarged hypocotyls and reduced chlorophyll accumulation (Lam et al. 1998; Brenner et al. 2000).

6.1.3 Vacuolar and ER Ca^{2+} Channels

Electrophysiological analyzes of the vacuolar membrane identified currents that are indicative for the function of voltage dependent channels and ligand gated channels. Among these, the "Two-pore channel" 1 (TPC1) appears to encode the depolarization activated slow vacuolar (SV) channel of the tonoplast (Hedrich and Neher 1987; Peiter et al. 2005). Plants lacking TPC1 are deficient in SV channel activity (Peiter et al. 2005). Remarkably, although the SV channel is the most abundant vacuolar channel, loss of SV channel function does not or only marginally impair calcium signaling events mediated by ABA or different biotic and abiotic factors which partially rely on influx of Ca^{2+} from intracellular stores (Peiter et al. 2005; Ranf et al. 2008). These observations suggest that SV-channels contribute only modestly to the modulation of cytoplasmic Ca^{2+} concentration by Ca^{2+} influx from the vacuole (Perez et al. 2008). Therefore, the exact functional role of TPC1 is still uncertain (Pottosin and Schonknecht 2007). Moreover, it needs to be considered that further not well characterized voltage dependent channels, that have been described as a fast vacuolar channel (FV) (Hedrich and Neher 1987) and as a Ca^{2+} insensitive vacuolar channel (CIVC) (Ranf et al. 2008), are likely to contribute to Ca^{2+} fluxes across the vacuolar membrane (Allen and Sanders 1994).

The identity and characterization of plant vacuolar ligand-gated Ca^{2+} channels is even less advanced than that of ligand-gated channels from the plasma membrane. By applying either caged compounds, or by direct patch clamp techniques, it was revealed that InsP3/InsP6 and cADPR mediated Ca^{2+} release, suggesting the existence of ligand-gated channels (Schumaker and Sze 1987; Alexandre 1990; Gilroy et al. 1990; Allen et al. 1995; Lemtiri-Chlieh et al. 2003). However, considering the paucity of molecular data confirming the existence of such channels, it remains possible that these compounds may indirectly activate channels by binding to receptors which subsequently activate voltage dependent channels like SV and FV channels (Lemtiri-Chlieh et al. 2003).

Similarly, it was also suggested that the ER contributes to InsP3/InsP6 and cADPR mediated Ca^{2+} release (Muir and Sanders 1997; Martinec et al. 2000; Navazio et al. 2001). Besides these two types of channels, a unique ligand gated channel appears to exist at the ER which is activated by NAADP (Navazio et al. 2000).

6.2 *Efflux of Calcium*

After release of Ca^{2+} into the cytosol, Ca^{2+} is actively transported out of the cytoplasma against the electro-chemical gradient to restore the normal cytoplasmic Ca^{2+} level. This finally leads to the observed Ca^{2+} transient and it should be emphasized that a tight regulation of Ca^{2+} efflux is as equally important for Ca^{2+} signaling as the more intensively studied influx mechanisms.

Extrusion of calcium is achieved by P-type calcium-ATPases and by Ca^{2+}/H^+ antiporter systems. While pumps mediate high-affinity low-turnover Ca^{2+} export, antiporter provoke low-affinity high-capacity export. Therefore, antiporter reduce the Ca^{2+} cytoplasmic level to a few micromolar after signal mediated influx of Ca^{2+}, while calcium-ATPases further lessen the cytoplasmic Ca^{2+} concentration to the resting level and maintain the Ca^{2+} homeostasis (Bush 1993; Hirschi 1999).

Ca^{2+} efflux transport activity appears to be coordinatively regulated with the influx of Ca^{2+} and specific regulation in response to defined stimuli has been reported (Bush et al. 1993; Gao et al. 2004). However, the underlying principles of Ca^{2+} efflux regulation are still poorly understood. Specific hormones can differentially activate the transporter systems of the ER or tonoplast. After a Ca^{2+} transient, Ca^{2+} released by GA seems to be mainly transported out of the cytoplasma via the ER transporters. In contrast, ABA activates transport activity at the ER and the tonoplast (Bush et al. 1993).

6.2.1 Calcium-Proton Antiporter

In the Arabidopsis genome 6 genes encode for putative Ca^{2+}/H^+ antiporters, designated as cation exchangers (CAX) (Maser et al. 2001; Shigaki et al. 2006) that contribute to the regulation of Ca^{2+} (Catala et al. 2003; Cheng et al. 2003; Zhao et al. 2008). In addition, five cation calcium exchanger (CCX) proteins (also termed CAX7-11), related to K^+ dependent Na^+/Ca^{2+} antiporters are encoded in the Arabidopsis genome (Shigaki et al. 2006). Moreover, four putative antiporters are encoded in the genome of Arabidopsis, which exhibit EF hand Ca^{2+} binding motifs suggesting that they are directly regulated by Ca^{2+} (Shigaki et al. 2006).

CAX proteins harbor a *N*-terminal regulatory/autoinhibitory domain, which binds to an adjacent region within the *N*-terminus (Pittman et al. 2002a; Mei et al. 2007). It has been observed that individual CAX proteins can have different transport capacities, metal selectivity and transcriptional regulation (Hirschi et al. 2000; Pittman et al. 2002b). Although individual CAX proteins can function specifically in distinct responses to definite stimuli (Zhao et al. 2008), CAX1 and CAX3 could also form functional heteromers (Cheng et al. 2005; Zhao et al. 2009). Additionally, different regulatory proteins could interact with CAX proteins to modulate their transport activity (Cheng and Hirschi 2003; Cheng et al. 2004a, b). CAX1-CAX4 are localized to the vacuole (Hirschi et al. 2000; Cheng et al. 2002a, 2003, 2005), but anti-porter activity was also reported to reside at the plasma membrane (Kasai and Muto 1990; Luo et al. 2005).

Several attempts were performed to change the cellular calcium levels of plants by overexpressing CAX proteins, either to improve plant tolerance against various stress regimes or to improve the availability of calcium for human nutrition. Overexpression of the truncated version (lacking the regulatory domain) of the vacuolar antiporter CAX1 from Arabidopsis in tobacco leads to an altered Ca^{2+} homeostasis. Although plants contained more total Ca^{2+}, plants showed Ca^{2+} deficiency symptoms. In accordance with this, plants also displayed hypersensitivity to Mg^{2+}, Na^{+} and to cold shock (Hirschi 1999). It was discussed that overexpression of AtCAX1 resulted in over-accumulation of Ca^{2+} in the vacuole and, therefore, by simultaneous reduction of the cytoplasmic Ca^{2+} concentration caused the deficiency symptoms.

6.2.2 Phosphorylated-type ATPases

Classical Ca^{2+} P-ATPases belong to the second subclass (II) of Phosphorylated(P)-type ATPases. Two different types of P_{II} ATPases are found in plants. P_{IIB} ATPases contain an autoinhibtory *N*-terminal region (autoinhibited calcium ATPases, ACAs; 10 members), which is absent in P_{IIA} type proteins (ER type calcium ATPases, ECA, 4 members) (Sze et al. 2000). The autoinhibitory domain in P_{IIB} type proteins can be relieved by the binding of Calmodulin, which results in activation of the pump (Harper et al. 1998). On the other hand, the activity of the P_{IIB} type Ca^{2+}-ATPase ACA2, is inhibited by phosphorylation within the *N*-terminal regulatory domain. Interestingly, this regulatory function is mediated by an another Ca^{2+} binding protein, a CDPK (Hwang et al. 2000). P_{IIA} type ATPases are found at the ER (ECA1) (Liang et al. 1997), the Golgi (ECA3) (Mills et al. 2008) and endosomes (also ECA 3) (Li et al. 2008). Besides being Ca^{2+} transporters, ECAs are also important for regulating the Mn^{2+} homeostasis of plants, transporting excess Mn^{2+} out of the cytoplasm (Wu et al. 2002; Li et al. 2008; Mills et al. 2008). The existence of ECAs at the Golgi and/or Endosomes could also be important for exocytotic processes as, for example, the vacuolar sorting receptor PV72 interacts with target proteins in a Ca^{2+} dependent manner (Shimada et al. 2002; Watanabe et al. 2002).

P_{IIB} type ATPases are found at the ER (ACA2) (Harper et al. 1998), the vacuole (ACA4, ACA11) (Geisler et al. 2000; Lee et al. 2007), the plasma membrane (ACA8, ACA9, ACA10) (Bonza et al. 2000; Schiott et al. 2004; George et al. 2008) and also at the plastid envelope (ACA1) (Huang et al. 1993). Transcript levels of ACAs are stress regulated (Carena et al. 2006). The importance of a P_{IIa} type Ca^{2+}-ATPase activity in regulating the cytoplasmic Ca^{2+} level is exemplified by an analysis of a Ca^{2+}-ATPase loss-of-function mutant in the moss *Physcomitrella patens*. While wildtype plants exhibit a transient Ca^{2+} release after applying Na^{+} stress, loss of function mutant lines exhibit a sustained elevation of Ca^{2+} (Qudeimat et al. 2008). Interestingly, the sustained increase of Ca^{2+} concentration impaired the expression of salt stress-induced genes and rendered mutant plants less tolerant to Na^{+} stress (Qudeimat et al. 2008), implicating a direct causal relation between the proper formation of a Ca^{2+} signature and stress tolerance. In Arabidopsis, analyses of loss-of-function mutations of ACA9 and ACA10 implicate these pumps in specific functions, like in pollen tube growth and in inflorescence development of plants, respectively (Schiott et al. 2004; George et al. 2008)

Moreover, several P_I type proteins, which are mainly heavy metal transporters are also implicated in Ca^{2+} transport. AtHMA1 (heavy metal ATPase), a heavy metal transporter supposed to function in detoxification processes for heavy metals is a P_I-ATPase that localizes to the chloroplast envelope. In addition to heavy metals like Cu^{2+}, AtHMA1 transports Ca^{2+} with high affinity and is specifically inhibited by thapsigargin like SERCA (sarcoplasmic/endoplasmic reticulum Ca^{2+}ATPase)-pumps from animals (Seigneurin-Berny et al. 2006; Moreno et al. 2008).

7 Signal Response Coupling of Calcium

Diverse stimuli specifically induce changes in cellular and apoplastic Ca^{2+} concentration (Cessna et al. 1998; Pauly et al. 2001; Lecourieux et al. 2002, 2005; Rentel and Knight 2004; Kosuta et al. 2008). The specific signatures of such Ca^{2+} transients can be unique to a defined cue but different stimuli can also induce similar Ca^{2+} responses. Typical examples of such Ca^{2+} signature are presented in Fig. 1c, and below we discuss only selected representative instances to illustrate this facet of Ca^{2+} signaling.

7.1 Differences in Salt and Mannitol Responses

Several reports showed that specific Ca^{2+} transients can be detected after salt or drought stress (imposed by mannitol application) in different cell types of the root, which, however, exhibited similar response signatures (Knight et al.

1997; Kiegle et al. 2000). Therefore, the existence of a Ca^{2+} independent factor was suggested that would allow to discriminate between the Ca^{2+} signatures evoked by drought and salt stress (Knight et al. 1997). In contrast, aleurone cells of barley responded differentially to either mannitol or sodium, leading to decrease and increase of cytosolic Ca^{2+}, respectively (Bush 1996). A detailed study by Gao et al. (2004) addressing the interconnection of cytoplasmic and apoplastic dynamics of Ca^{2+}concentration in roots of Arabidopsis provides new insights how Ca^{2+} could function as a specific signal in these different stress responses (Gao et al. 2004). Here, the Ca^{2+} elevation within the cytosol was simultaneously recorded to the changes of apoplastic Ca^{2+}. The response to mannitol resulted only in a minor transient cytoplasmic Ca^{2+} elevation. In contrast, sodium stress resulted in a transient cytoplasmic Ca^{2+} spike which was more pronounced than after mannitol application (Pauly et al. 2001; Gao et al. 2004). However, after this initial spike a slow but constant rise of cytoplasmic Ca^{2+} concentration was observed (see Fig. 1c) (Gao et al. 2004). Additionally, apoplastic Ca^{2+} elevation was also observed under sodium stress (Gao et al. 2004), which could result from enhanced exocytosis (Homann and Tester 1997; Belan et al. 1998). As sodium ions can exchange Ca^{2+} from the membrane or from cell wall components this apoplastic rise in Ca^{2+} concentration may contribute to membrane protection (Williams 1970; Hauser et al. 1976; Brett and Waldron 1996). An apoplastic calcium rise was not observed under drought stress, most likely since mannitol does not exchange calcium from membrane and cell wall components (Gao et al. 2004).

7.2 Differences in Symbiotic Calcium Responses

Symbiotic plants can interact with nitrogen-fixing rhizobial bacteria or with arbuscular mycorrhizal fungi that aid nutrient uptake of the plant. Although both signals (Nod factor, or mycorrhizal factors) require the same machinery for Ca^{2+} signaling, the outcome is quite different. While Nod factors induce nodulation, interaction with symbiotic fungi does not induce nodulation (Kosuta et al. 2008). Although both factors induce Ca^{2+} oscillations, their signatures are remarkably different. Initial, Ca^{2+} spiking triggered by Nod factor perception induces a subsequent large rise of cytoplasmic Ca^{2+} in regard of duration and amplitude. After a descent of cytoplasmic Ca^{2+} to basal levels, immediate and recurrent increases of calcium occur which finally result in repetitive and periodic spiking. This is in contrast to the oscillations induced by the mycorrhizal factor. Here, the transient is shorter and lower in amplitude (only 17% of the Nod factor spike). After the drop to basal level, there is a gap of different length of low Ca^{2+} before the next spike is recorded which can differ in duration and amplitude from the previous spike (see Fig. 1c) (Kosuta et al. 2008). These differences could result in the different observed symbiotic outcomes.

8 Calcium Binding Proteins

Different classes of Ca^{2+} binding proteins represent the cellular "currency" to translate Ca^{2+} signals into defined downstream response reactions. Here we will focus on three main classes of Ca^{2+} binding proteins, which all harbor EF hands for Ca^{2+} binding. These are the Calmodulins (CaMs) and their related Calmodulin like proteins (CMLs), the Calcium-dependent protein kinases (CDPKs) and the calcineurin-B like (CBL) proteins which form a network with CBL-interacting protein kinases (CIPKs).

8.1 Calmodulin

Calmodulins are small proteins of about 148 amino acids, composed of four Ca^{2+}-binding EF hands which are arranged in a dumbbell structure in that EF hands 1/2 and 3/4 are separated by a alpha-helical handle (Strynadka and James 1989). Upon Ca^{2+} binding, CaMs undergo a structural change, from a closed to an open conformation which enables binding to target proteins (Chin and Means 2000).

In Arabidopsis, 7 genes encode for only 4 Calmodulin isoforms because 3 Calmodulins share an identical amino acid sequence (CaMs 2, 3 and 5) (McCormack et al. 2005). Additionally, Arabidopsis contains 50 CaM-like proteins, which harbor variable numbers (between 2 and 6) of EF hands that could account for different Ca^{2+} binding affinities (McCormack and Braam 2003).

CaMs are generally cytoplasmic proteins, which can translocate into the nucleus or to cellular membranes upon binding to different target proteins (Deisseroth et al. 1998; Chung et al. 2000). Additionally, Ca^{2+} binding proteins related to Calmodulin can be secreted into the apoplastic space, potentially regulating cell wall regeneration (Sun et al. 1995) and the growth of pollen by affecting activity of hyperpolarization activated channels (Ma et al. 1999; Shang et al. 2005). CML18 has been reported to localize within the vacuole here regulating the function of the Na^+/H^+ antiporter NHX1 (Yamaguchi et al. 2005).

About 170 different target proteins of CaMs were identified up to now (Jun et al. 1996; Reddy et al. 2002; Yang and Poovaiah 2003; Popescu et al. 2007). CaMs/CMLs are important regulators of metabolism, cell structure proteins, proteins of the phyotohormone signal network, ion transporters, heat shock proteins and proteins regulating translation and transcription (Reddy et al. 2002; Du and Poovaiah 2005; Popescu et al. 2007; Du et al. 2009). CaM or CMLs are directly implicated in fine tuning the hypersensitivity response after pathogen infection. While NO production is mediated by CaMs or CMLs, like CML24 (Ma et al. 2008), the channel which is important for Ca^{2+} release, which then results in NO production, is also inhibited by CaMs (Ali et al. 2007). This could be important to prevent excess accumulation of Ca^{2+}, and would lead to the observed transient increase of Ca^{2+}. CaMs are also supposed to have an important role during photomorphogenesis (Neuhaus et al. 1993) and a specific function in light dependent development of seedlings was assessed for CaM7 that can act as a transcriptional regulator by binding to Z- and G-box light responsive promoter elements (Kushwaha et al. 2008).

Different mechanisms contribute to the specificity of responses mediated by individual CaMs and CMLs. In the tobacco plant *Nicotiana plumbaginifolia*, individual CaMs are differentially transcriptionally regulated. While NpCaM1 is upregulated by wind and cold, the expression of NpCaM2 is unaffected (van Der Luit et al. 1999). In Arabidopsis, one CaM and different CMLs are induced by touch (Braam et al. 1997). Additionally, CML expression can be tissue specific and can be modulated by stress, hormones and light (McCormack et al. 2005; Magnan et al. 2008). Post-transcriptional regulation of protein stability by methylation or phosphorylation could also play an important role (Oh and Roberts 1990; Leclerc et al. 1999; Kushwaha et al. 2008). An additional layer of functional diversification results from differential target protein regulation by CaM/CMLs (Lee et al. 2000; Popescu et al. 2007). Remarkably, it has been reported that targets can be activated by one CaM/CML and reciprocally inactivated by another CaM/CML (Lee et al. 1997; Leclerc et al. 1999; Kushwaha et al. 2008).

8.2 CDPKs

Calcium dependent protein kinases (CDPKs) evolutionary arose by fusion of a *N*-terminal serine/threonine kinase with a *C*-terminal Calmodulin EF hand Ca^{2+} binding domain, separated by an autoinhibitory junction domain. Classical CDPKs contain four EF hands, while "CDPK-related kinases" (CRKs) harbor three or less functional EF hands (Harper et al. 2004).

The junction domain functions as a pseudosubstrate. When no Ca^{2+} is bound to the Calmodulin domain, the junction domain interacts with the kinase domain and blocks phosphorylation of target proteins. Upon Ca^{2+} binding, the Calmodulin domain binds the junction domain displacing it from the kinase domain and thereby leading to activation of the kinase (Hrabak et al. 2003).

In Arabidopsis 34 CDPKs are encoded in the genome. All CDPKs harbor a *N*-terminal "variable domain" upstream of the kinase domain, which can differ in sequence and length (Cheng et al. 2002b). The "variable domain" can determine CDPK localization and can represent a docking site for regulatory 14-3-3 proteins (Lu and Hrabak 2002). CDPKs are differentially localized, and have been found to reside in the cytoplasma and nucleoplasma, but are also associated with the cytoskeleton, plasma membrane, ER or peroxisomes (Putnam-Evans et al. 1989; Martin and Busconi 2000; Lu and Hrabak 2002; Dammann et al. 2003; Choi et al. 2005).

Regulation of CDPK kinase activity is achieved by a complex interplay of membrane translocation, phosho-lipid binding and auto- as well as trans-phosphorylation (Schaller et al. 1992; Farmer and Choi 1999; Szczegielniak et al. 2000,

2005). Some CDPKs are transcriptionally regulated (Hrabak et al. 2003) and could be regulated by their protein stability (Satterlee and Sussman 1998; Zhu et al. 2007). Since each CDPKs display distinct composition of various EF hands with different affinities to calcium, differential activation of CDPKs depending on the respective cellular Ca^{2+} concentration may occur. Certain CDPKs display a very low activation requirement suggesting that these kinases are already constitutively active at resting cytoplasmic Ca^{2+} levels (Lee et al. 1998).

Functional analysis of two CDPKs from Arabidopsis that are strongly expressed in guard cells (CPK3 and CPK6), revealed that loss-of-function mutants of CPK3 and/or CPK6 are impaired in Ca^{2+} and ABA-dependent activation of S-type anion channels (Schroeder and Hagiwara 1989; Mori et al. 2006) and in the ABA-dependent activation of Ca^{2+} channels (Hamilton et al. 2000; Pei et al. 2000; Mori et al. 2006). Consequently, these mutant plants exhibited reduced stomatal closure after application of ABA or after artificially imposing Ca^{2+} oscillations (Mori et al. 2006). However, long term stomatal closure in these mutants as well as ABA-mediated inhibition of seed germination were not affected, implicating that CPK3/CPK6 are specifically regulating the rapid stomatal closure (Mori et al. 2006). In contrast, loss of function of CPK4/CPK11, two highly related cytoplasmic/nucleoplasmic localized CDPKs, reduces the sensitivity of mutant plants to ABA in seedling germination and growth, and stomatal closure is partially impaired in response to ABA (Zhu et al. 2007). CPK4 and CPK11 interact and phosphorylate ABA-responsive transcription factors ABF1 and ABF4 in the presence of ABA (Uno et al. 2000; Zhu et al. 2007). However, ABFs also interact and are phosphorylated by other CDPKs, implicating further kinases in regulating ABA responses in Arabidopsis (Choi et al. 2005).

Phosphorylation of the transcriptional activator "Repression of shoot growth" (RSG) by CDPK1 from tobacco enables binding of 14-3-3 proteins and represses the function of RSG during GA responses (Ishida et al. 2008). CDPK2 from tobacco is transiently activated by phosphorylation specifically after pathogen infection (Romeis et al. 2000, 2001), and is triggering different stress response pathways within the cell (Ludwig et al. 2005). Furthermore, CDPKs are implicated in root development (Ivashuta et al. 2005), wound response (Szczegielniak et al. 2005), secretion and vacuolar function in response to GA (McCubbin et al. 2004), pollen tube growth and pollen tube polarity (Estruch et al. 1994; Yoon et al. 2006), response to salt, drought stress and potassium homeostasis (Saijo et al. 2000; Ma and Wu 2007). These findings illustrate the functional diversity of CDPKs in various biological processes.

8.3 CBLs and CIPKs

The third class of Ca^{2+} binding proteins is represented by the group of Calcineurin-B like (CBL) proteins (Kudla et al. 1999; Batistic and Kudla 2004). Similar to CaMs, CBLs contain four EF hands to bind Ca^{2+} (Nagae et al. 2003; Kolukisaoglu et al. 2004; Sanchez-Barrena et al. 2005), but contain an unconventional first EF hand, which encompasses 14 aminoacids instead of 12 aminoacids typical for a canonical EF hand (Nagae et al. 2003). Nevertheless, this unique EF hand is still able to bind Ca^{2+} (Nagae et al. 2003; Sanchez-Barrena et al. 2005, 2007), and could play an important role in the selective interaction with CBL-partner proteins. In Arabidopsis 10 genes encode for CBL proteins (Kolukisaoglu et al. 2004). CBL1, CBL4, CBL5 and CBL9 are *N*-terminal myristoylated proteins (Ishitani et al. 2000; Batistic et al. 2008) that, in addition, harbor further cysteine residues in the vicinity of the myristoylated glycine suggesting further acylation. Indeed, CBL1 has been shown to undergo modification by palmitate or stearate, and together with the myristoyl modification, these lipid modifications are important for correct plasma membrane targeting and function of the CBL1 protein in stress response (Batistic et al. 2008). CBL2, CBL3, CBL6 and CBL10 lack a classical N-myristoylation site. Instead, these proteins harbor an extended *N*-terminal region that is important for correct sub-cellular targeting of the proteins (Batistic et al. 2009).

CBL proteins interact and regulate the activity of a certain class of protein kinases, designated as CBL-interacting protein kinases (CIPKs) (Shi et al. 1999). In Arabidopsis, 26 genes encode for CIPKs, which belong to the third subgroup of SNF-related protein kinases (SnRK3) (Hrabak et al. 2003; Batistic and Kudla 2009). It has been suggested that binding of CBLs to the CIPKs via the conserved NAF domain of these kinases (Albrecht et al. 2001) relieves autoinhibition of the kinase, which then results in kinase activation and target phosphorylation (Guo et al. 2001b; Fujii and Zhu 2009). Moreover, CIPKs can interact with PP2Cs (Ohta et al. 2003) and crystallization studies implicate that CIPK24 either interacts with CBLs or PP2Cs, excluding the formation of a trimeric complex (Sanchez-Barrena et al. 2007). Therefore, the on-off state of the CIPKs may be regulated by the interaction with CBLs (on state) or type 2C protein phosphatases (off state).

Several mechanisms contribute to generating signaling specificity within the CBL-CIPK network. Preferential complex formation between certain CBLs and CIPKs enable a focused signal transmission of signals from the calcium sensor proteins to the kinases (Albrecht et al. 2001). Additionally, certain pairs of CBL-CIPK complexes are localized at different cellular compartments, and are differentially expressed in different tissues or in response to stresses, thereby enabling spatial and temporal regulation of the network. For example, CBL4 is mainly expressed in roots, while CBL10 is mainly expressed in leaves (Kim et al. 2007). Both calcium sensor proteins can interact with CIPK24, which is expressed in both tissues. However, CBL4/CIPK24 complexes are localized at the plasma membrane while CBL10/CIPK24 complexes accumulate at the tonoplast thereby creating a dual functioning kinase (Fig. 1d). The alternative formation of CBL/CIPK24 complexes may enable simulatenous Ca^{2+}-dependent regulation of Na^{+} extrusion in the root and Na^{+} sequestration into the vacuole in the shoot of salt stressed plants (Kim et al. 2007). In general, CBL and CIPK proteins are critical for controlling the response to different stress situations like salt and osmotic stress (Albrecht et al. 2003; Cheong et al. 2003; D'Angelo et al. 2006; Tripathi et al. 2009), response to and regulation of ABA synthesis (Kim et al. 2003; Pandey et al. 2004), nitrate homeostasis (Hu et al. 2009) root development (Tripathi et al. 2009) and stomatal movement (Cheong et al. 2007). CIPK11 negatively regulates the plasma membrane Arabidopsis H^+-ATPase 2 (AHA2), which mediates hyperpolarization of the plasma membrane (Fuglsang et al. 2007). CBL4 together with its interacting protein kinase CIPK24 form the specific "Salt overly sensitive" (SOS) pathway, which regulate the sodium/proton antiporter SOS1 at the plasma membrane (Fig. 1d). During salt stress, calcium influx is detected by CBL4, which activates CIPK24 and subsequently activates SOS1, to extrude excess sodium out of the cell (Halfter et al. 2000; Qiu et al. 2002; Quintero et al. 2002). CBL1 and CBL9 target CIPK23 to the plasma membrane to activate the potassium channel AKT1 to maintain K^+ homeostasis under low potassium conditions (Li et al. 2006; Xu et al. 2006; Cheong et al. 2007) (Fig. 1d). The identification of further targets fo CBL/CIPK complexes currently remains one of the main challenges to further our understanding of this complex signaling network.

9 Conclusions

Beginning in the middle of the last century plant biologists uncovered the crucial nutritional and structural role of Ca^{2+} for plants. However, in this regard Ca^{2+} never attracted as much attention as for example K^+. It was the surprising notion, that only a tiny fraction of the bio-available calcium, namely the free cytoplasmic Ca^{2+} pool and its regulated dynamics, modulates a plethora of biological processes that sparked an immense interest in this ion. Consequently, during the following decades of the last century an immense amount of observations accumulated that linked changes in cellular Ca^{2+} concentration and distribution to the regulation of many diverse processes of plant growth and development.

The extensive involvement of Ca^{2+} frequently leads to the vexing question: how can one ion specifically control so many events? Current research is beginning to provide answers. Ca^{2+} regulation in plants involves many facets that can define and adjust responses in both time and space. The unequal distribution of Ca^{2+} in the cell provides the basis for rapid Ca^{2+} fluxes and the resulting concentration changes. Influx channels on the plasma membrane and release channels from internal stores provide several ways to generate rapid ion elevations or to create local gradients. The frequency as well as amplitude modulation, provide means of generating signals that have unique properties. Once these signals are generated, then a wide variety of Ca^{2+}-decoding components interpret and relay these signals. Complex signaling networks, prominently involving CDPKs and CIPKs translate this information into phosphorylation events thereby simultaneously amplifying and specifying response reactions. It is an emerging picture that plants possess myriad ways in which Ca^{2+} can operate as the intermediary in transducing stimuli into the appropriate responses. The challenge for the near future lies in characterizing the underlying functional principles of signal response coupling and in identifying the prime targets of Ca^{2+} regulated phosphorylation events.

References

Albrecht V, Ritz O, Linder S, Harter K, Kudla J (2001) The NAF domain defines a novel protein-protein interaction module conserved in $Ca2^{+}$-regulated kinases. EMBO J 20:1051–1063

Albrecht V, Weinl S, Blazevic D, D'Angelo C, Batistic O, Kolukisaoglu U, Bock R, Schulz B, Harter K, Kudla J (2003) The calcium sensor CBL1 integrates plant responses to abiotic stresses. Plant J 36:457–470

Alexandre J (1990) Opening of Ca^{2+} channels in isolated red beet root vacuole membrane by inositol 1, 4, 5-trisphosphate. Nature 343:567–570

Ali R, Zielinski RE, Berkowitz GA (2006) Expression of plant cyclic nucleotide-gated cation channels in yeast. J Exp Bot 57:125–138

Ali R, Ma W, Lemtiri-Chlieh F, Tsaltas D, Leng Q, von Bodman S, Berkowitz GA (2007) Death don't have no mercy and neither does calcium: Arabidopsis CYCLIC NUCLEOTIDE GATED CHANNEL2 and innate immunity. Plant Cell 19:1081–1095

Allen GJ, Sanders D (1994) Two Voltage-Gated, Calcium Release Channels Coreside in the Vacuolar Membrane of Broad Bean Guard Cells. Plant Cell 6:685–694

Allen GJ, Muir SR, Sanders D (1995) Release of Ca^{2+} from individual plant vacuoles by both InsP3 and cyclic ADP-ribose. Science 268:735–737

Allen GJ, Kwak JM, Chu SP, Llopis J, Tsien RY, Harper JF, Schroeder JI (1999) Cameleon calcium indicator reports cytoplasmic calcium dynamics in Arabidopsis guard cells. Plant J 19:735–747

Allen GJ, Chu SP, Schumacher K, Shimazaki CT, Vafeados D, Kemper A, Hawke SD, Tallman G, Tsien RY, Harper JF, Chory J, Schroeder JI (2000) Alteration of stimulus-specific guard cell calcium oscillations and stomatal closing in Arabidopsis det3 mutant. Science 289:2338–2342

Allen GJ, Chu SP, Harrington CL, Schumacher K, Hoffmann T, Tang YY, Grill E, Schroeder JI (2001) A defined range of guard cell calcium oscillation parameters encodes stomatal movements. Nature 411:1053–1057

Arazi T, Sunkar R, Kaplan B, Fromm H (1999) A tobacco plasma membrane calmodulin-binding transporter confers Ni^{2+} tolerance and Pb^{2+} hypersensitivity in transgenic plants. Plant J 20:171–182

Atkinson MM, Keppler LD, Orlandi EW, Baker CJ, Mischke CF (1990) Involvement of Plasma Membrane Calcium Influx in Bacterial Induction of the K/H and Hypersensitive Responses in Tobacco. Plant Physiol 92:215–221

Balague C, Lin B, Alcon C, Flottes G, Malmstrom S, Kohler C, Neuhaus G, Pelletier G, Gaymard F, Roby D (2003) HLM1, an essential signaling component in the hypersensitive response, is a member of the cyclic nucleotide-gated channel ion channel family. Plant Cell 15:365–379

Batistic O, Kudla J (2004) Integration and channeling of calcium signaling through the CBL calcium sensor/CIPK protein kinase network. Planta 219:915–924

Batistic O, Kudla J (2009) Plant calcineurin B-like proteins and their interacting protein kinases. Biochim Biophys Acta 1793:985–992

Batistic O, Sorek N, Schultke S, Yalovsky S, Kudla J (2008) Dual fatty acyl modification determines the localization and plasma membrane targeting of CBL/CIPK Ca^{2+} signaling complexes in Arabidopsis. Plant Cell 20:1346–1362

Batistic O, Waadt R, Steinhorst L, Held K, Kudla J (2009) CBL-mediated targeting of CIPKs facilitates the decoding of calcium signals emanating from distinct cellular stores. Plant J, doi: 10.1111/j.1365-1313X.2009.04045.x

Baum G, Long JC, Jenkins GI, Trewavas AJ (1999) Stimulation of the blue light phototropic receptor NPH1 causes a transient increase in cytosolic Ca^{2+}. Proc Natl Acad Sci USA 96:13554–13559

Belan P, Gardner J, Gerasimenko O, Gerasimenko J, Mills CL, Petersen OH, Tepikin AV (1998) Isoproterenol evokes extracellular Ca^{2+} spikes due to secretory events in salivary gland cells. J Biol Chem 273:4106–4111

Berridge MJ (2006) Calcium microdomains: organization and function. Cell Calcium 40:405–412
Bhalla A, Chicka MC, Tucker WC, Chapman ER (2006) Ca^{2+}-synaptotagmin directly regulates t-SNARE function during reconstituted membrane fusion. Nat Struct Mol Biol 13:323–330
Blume B, Nurnberger T, Nass N, Scheel D (2000) Receptor-mediated increase in cytoplasmic free calcium required for activation of pathogen defense in parsley. Plant Cell 12:1425–1440
Bonza MC, Morandini P, Luoni L, Geisler M, Palmgren MG, De Michelis MI (2000) At-ACA8 encodes a plasma membrane-localized calcium-ATPase of Arabidopsis with a calmodulin-binding domain at the N terminus. Plant Physiol 123:1495–1506
Braam J, Sistrunk ML, Polisensky DH, Xu W, Purugganan MM, Antosiewicz DM, Campbell P, Johnson KA (1997) Plant responses to environmental stress: regulation and functions of the Arabidopsis TCH genes. Planta 203:S35–S41
Brenner ED, Martinez-Barboza N, Clark AP, Liang QS, Stevenson DW, Coruzzi GM (2000) Arabidopsis mutants resistant to S(+)-beta-methyl-alpha, beta-diaminopropionic acid, a cycad-derived glutamate receptor agonist. Plant Physiol 124:1615–1624
Brett C, Waldron K (1996) Physiology and biochemistry of plant cell walls, second editionth edn. Chapman & Hall, London
Broadley MR, Bowen HC, Cotterill HL, Hammond JP, Meacham MC, Mead A, White PJ (2003) Variation in the shoot calcium content of angiosperms. J Exp Bot 54:1431–1446
Bunney TD, Shaw PJ, Watkins PA, Taylor JP, Beven AF, Wells B, Calder GM, Drobak BK (2000) ATP-dependent regulation of nuclear Ca^{2+} levels in plant cells. FEBS Lett 476:145–149
Bush DS (1993) Regulation of cytosolic calcium in plants. Plant Physiol 103:7–13
Bush DS (1996) Effects of gibberellic acid and environmental factors on cytosloic calcium in wheat aleurone cells. Planta 199:89–99
Bush DS, Biswas AK, Jones RL (1989a) Gibberellic-acid-stimulated Ca^{2+} accumulation in endoplasmic reticulum of barley aleurone: Ca^{2+} transport and steady-state levels. Planta 178:411–420
Bush DS, Sticher L, van Huystee R, Wagner D, Jones RL (1989b) The calcium requirement for stability and enzymatic activity of two isoforms of barley aleurone alpha-amylase. J Biol Chem 264:19392–19398
Bush DS, Biswas AK, Jones RL (1993) Hormonal regulation of Ca^{2+} transport in the endomembrane system of the barley aleurone. Planta 189:507–515
Bussler W (1962) Die Entwicklung von Calcium-Mangelsymptomen. Z Pflanzenernahr Dung Bodenkde 100:53–58
Bygrave FL (1978) Mitochondria and the control of intracellular calcium. Biol Rev Camb Philos Soc 53:43–79
Carena M, Bonza MC, Harris R, Sanders D, De Michelis MI (2006) Abscisic acid stimulates the expression of two isoforms of plasma membrane Ca^{2+}-ATPase in Arabidopsis thaliana seedlings. Plant Biol 8:572–578
Catala R, Santos E, Alonso JM, Ecker JR, Martinez-Zapater JM, Salinas J (2003) Mutations in the Ca^{2+}/H^{+} transporter CAX1 increase CBF/DREB1 expression and the cold-acclimation response in Arabidopsis. Plant Cell 15:2940–2951
Cessna SG, Chandra S, Low PS (1998) Hypo-osmotic shock of tobacco cells stimulates Ca^{2+} fluxes deriving first from external and then internal Ca^{2+} stores. J Biol Chem 273:27286–27291
Chan CW, Schorrak LM, Smith RK, Bent AF, Sussman MR (2003) A cyclic nucleotide-gated ion channel, CNGC2, is crucial for plant development and adaptation to calcium stress. Plant Physiol 132:728–731
Charpentier M, Bredemeier R, Wanner G, Takeda N, Schleiff E, Parniske M (2008) Lotus japonicus CASTOR and POLLUX Are Ion Channels Essential for Perinuclear Calcium Spiking in Legume Root Endosymbiosis. Plant Cell 20:3467–3479
Chen X, Chang M, Wang B, Wu B (1997) Cloning of a Ca^{2+}-ATPase gene and the role of cytosolic Ca^{2+} in the gibberellin-dependent signaling pathway in aleurone cells. Plant J 11:363–371

Cheng NH, Hirschi KD (2003) Cloning and characterization of CXIP1, a novel PICOT domain-containing Arabidopsis protein that associates with CAX1. J Biol Chem 278:6503–6509

Cheng NH, Pittman JK, Shigaki T, Hirschi KD (2002a) Characterization of CAX4, an Arabidopsis H^+/cation antiporter. Plant Physiol 128:1245–1254

Cheng SH, Willmann MR, Chen HC, Sheen J (2002b) Calcium signaling through protein kinases. The Arabidopsis calcium-dependent protein kinase gene family. Plant Physiol 129:469–485

Cheng NH, Pittman JK, Barkla BJ, Shigaki T, Hirschi KD (2003) The Arabidopsis cax1 mutant exhibits impaired ion homeostasis, development, and hormonal responses and reveals interplay among vacuolar transporters. Plant Cell 15:347–364

Cheng NH, Liu JZ, Nelson RS, Hirschi KD (2004a) Characterization of CXIP4, a novel Arabidopsis protein that activates the H^+/Ca^{2+} antiporter, CAX1. FEBS Lett 559:99–106

Cheng NH, Pittman JK, Zhu JK, Hirschi KD (2004b) The protein kinase SOS2 activates the Arabidopsis H^+/Ca^{2+} antiporter CAX1 to integrate calcium transport and salt tolerance. J Biol Chem 279:2922–2926

Cheng NH, Pittman JK, Shigaki T, Lachmansingh J, LeClere S, Lahner B, Salt DE, Hirschi KD (2005) Functional association of Arabidopsis CAX1 and CAX3 is required for normal growth and ion homeostasis. Plant Physiol 138:2048–2060

Cheong YH, Kim KN, Pandey GK, Gupta R, Grant JJ, Luan S (2003) CBL1, a calcium sensor that differentially regulates salt, drought, and cold responses in Arabidopsis. Plant Cell 15:1833–1845

Cheong YH, Pandey GK, Grant JJ, Batistic O, Li L, Kim BG, Lee SC, Kudla J, Luan S (2007) Two calcineurin B-like calcium sensors, interacting with protein kinase CIPK23, regulate leaf transpiration and root potassium uptake in Arabidopsis. Plant J 52:223–239

Chin D, Means AR (2000) Calmodulin: a prototypical calcium sensor. Trends Cell Biol 10: 322–328

Choi HI, Park HJ, Park JH, Kim S, Im MY, Seo HH, Kim YW, Hwang I, Kim SY (2005) Arabidopsis calcium-dependent protein kinase AtCPK32 interacts with ABF4, a transcriptional regulator of abscisic acid-responsive gene expression, and modulates its activity. Plant Physiol 139:1750–1761

Chung WS, Lee SH, Kim JC, Heo WD, Kim MC, Park CY, Park HC, Lim CO, Kim WB, Harper JF, Cho MJ (2000) Identification of a calmodulin-regulated soybean Ca^{2+}-ATPase (SCA1) that is located in the plasma membrane. Plant Cell 12:1393–1407

Clapham DE (1995) Calcium signaling. Cell 80:259–268

Clayton H, Knight MR, Knight H, McAinsh MR, Hetherington AM (1999) Dissection of the ozone-induced calcium signature. Plant J 17:575–579

Clough SJ, Fengler KA, Yu IC, Lippok B, Smith RK Jr, Bent AF (2000) The Arabidopsis dnd1 "defense, no death" gene encodes a mutated cyclic nucleotide-gated ion channel. Proc Natl Acad Sci USA 97:9323–9328

Cosgrove DJ, Hedrich R (1991) Stretch-activated chloride, potassium, and calcium channels coexisting in plasma membranes of guard cells of Vicia faba L. Planta 186:143–153

D'Angelo C, Weinl S, Batistic O, Pandey GK, Cheong YH, Schultke S, Albrecht V, Ehlert B, Schulz B, Harter K, Luan S, Bock R, Kudla J (2006) Alternative complex formation of the Ca^{2+}-regulated protein kinase CIPK1 controls abscisic acid-dependent and independent stress responses in Arabidopsis. Plant J 48:857–872

Dammann C, Ichida A, Hong B, Romanowsky SM, Hrabak EM, Harmon AC, Pickard BG, Harper JF (2003) Subcellular targeting of nine calcium-dependent protein kinase isoforms from Arabidopsis. Plant Physiol 132:1840–1848

De Silva DLR, Hetherington AM, Mansfield TA (1996) Where does all the calcium go? Evidence of an important regulatory role for trichomes in two calcicoles. Plant Cell Environ 19:880–886

Deisseroth K, Heist EK, Tsien RW (1998) Translocation of calmodulin to the nucleus supports CREB phosphorylation in hippocampal neurons. Nature 392:198–202

Demarty M, Morvan C, Thellier M (1984) Calcium and the cell wall. Plant Cell Environ 7:441–448

Demidchik V, Maathuis FJ (2007) Physiological roles of nonselective cation channels in plants: from salt stress to signalling and development. New Phytol 175:387–404
Demidchik V, Tester M (2002) Sodium fluxes through nonselective cation channels in the plasma membrane of protoplasts from Arabidopsis roots. Plant Physiol 128:379–387
Demidchik V, Bowen HC, Maathuis FJ, Shabala SN, Tester MA, White PJ, Davies JM (2002) Arabidopsis thaliana root non-selective cation channels mediate calcium uptake and are involved in growth. Plant J 32:799–808
Demidchik V, Nichols C, Oliynyk M, Dark A, Glover BJ, Davies JM (2003) Is ATP a signaling agent in plants? Plant Physiol 133:456–461
Demidchik V, Shabala SN, Davies JM (2007) Spatial variation in H_2O_2 response of Arabidopsis thaliana root epidermal Ca^{2+} flux and plasma membrane Ca^{2+} channels. Plant J 49:377–386
Demidchik V, Shang Z, Shin R, Thompson E, Rubio L, Laohavisit A, Mortimer JC, Chivasa S, Slabas AR, Glover BJ, Schachtman DP, Shabala SN, Davies JM (2009) Plant extracellular ATP signalling by plasma membrane NADPH oxidase and Ca^{2+} channels. Plant J 58:903–913
Demuro A, Parker I (2006) Imaging single-channel calcium microdomains. Cell Calcium 40:413–422
Dennison KL, Spalding EP (2000) Glutamate-gated calcium fluxes in Arabidopsis. Plant Physiol 124:1511–1514
Dieter P, Marme D (1980) Ca^{2+} transport in mitochondiral and microsomal fraction from higher plants. Planta 150:1–8
Dieter P, Marmé D (1983) The effect of calmodulin and far-red light on the kinetic properties of the mitochondrial and microsomal calcium-ion transport system from corn. Planta 159:277–281
Digonnet C, Aldon D, Leduc N, Dumas C, Rougier M (1997) First evidence of a calcium transient in flowering plants at fertilization. Development 124:2867–2874
Dodd AN, Love J, Webb AA (2005) The plant clock shows its metal: circadian regulation of cytosolic free Ca^{2+}. Trends Plant Sci 10:15–21
Dodd AN, Gardner MJ, Hotta CT, Hubbard KE, Dalchau N, Love J, Assie JM, Robertson FC, Jakobsen MK, Goncalves J, Sanders D, Webb AA (2007) The Arabidopsis circadian clock incorporates a cADPR-based feedback loop. Science 318:1789–1792
Du L, Poovaiah BW (2005) Ca^{2+}/calmodulin is critical for brassinosteroid biosynthesis and plant growth. Nature 437:741–745
Du L, Ali GS, Simons KA, Hou J, Yang T, Reddy AS, Poovaiah BW (2009) Ca^{2+}/calmodulin regulates salicylic-acid-mediated plant immunity. Nature 457:1154–1158
Dutta R, Robinson KR (2004) Identification and characterization of stretch-activated ion channels in pollen protoplasts. Plant Physiol 135:1398–1406
Ehrhardt DW, Wais R, Long SR (1996) Calcium spiking in plant root hairs responding to Rhizobium nodulation signals. Cell 85:673–681
Epstein E (1961) The essential role of calcium in selective cation transport by plant cells. Plant Physiol 36:437–444
Epstein E (1998) How calcium enhances plant salt tolerance. Science 280:1906–1907
Estruch JJ, Kadwell S, Merlin E, Crossland L (1994) Cloning and characterization of a maize pollen-specific calcium-dependent calmodulin-independent protein kinase. Proc Natl Acad Sci USA 91:8837–8841
Etter EF, Minta A, Poenie M, Fay FS (1996) Near-membrane Ca^{2+} transients resolved using the Ca^{2+} indicator FFP18. Proc Natl Acad Sci USA 93:5368–5373
Ettinger WF, Clear AM, Fanning KJ, Peck ML (1999) Identification of a Ca^{2+}/H^+ antiport in the plant chloroplast thylakoid membrane. Plant Physiol 119:1379–1386
Fairley-Grenot KA, Assmann SM (1992) Permeation of Ca^{2+} through K^+ channels in the plasma membrane of Vicia faba guard cells. J Membr Biol 128:103–113
Farmer PK, Choi JH (1999) Calcium and phospholipid activation of a recombinant calcium-dependent protein kinase (DcCPK1) from carrot (Daucus carota L.). Biochim Biophys Acta 1434:6–17

Fasano JM, Massa GD, Gilroy S (2002) Ionic signaling in plant responses to gravity and touch. J Plant Growth Regul 21:71–88

Faure JE, Digonnet C, Dumas C (1994) An in Vitro System for Adhesion and Fusion of Maize Gametes. Science 263:1598–1600

Felle H (1988) Auxin causes oscillation of cytosolic free calcium and pH in Zea mays coleoptiles. Planta 174:495–499

Felle HH, Kondorosi E, Kondorosi A, Schultze M (1999) Elevation of the cytosolic free Ca^{2+} is indispensable for the transduction of the Nod factor signal in alfalfa. Plant Physiol 121:273–280

Foreman J, Demidchik V, Bothwell JH, Mylona P, Miedema H, Torres MA, Linstead P, Costa S, Brownlee C, Jones JD, Davies JM, Dolan L (2003) Reactive oxygen species produced by NADPH oxidase regulate plant cell growth. Nature 422:442–446

Franklin-Tong VE, Ride JP, Read ND, Trewavas A, Franklin FCH (1993) The self-incompatibiity response in Papaver rhoeas is mediated by cytosolic free calcium. Plant J 4:163–177

Frietsch S, Wang YF, Sladek C, Poulsen LR, Romanowsky SM, Schroeder JI, Harper JF (2007) A cyclic nucleotide-gated channel is essential for polarized tip growth of pollen. Proc Natl Acad Sci USA 104:14531–14536

Fuglsang AT, Guo Y, Cuin TA, Qiu Q, Song C, Kristiansen KA, Bych K, Schulz A, Shabala S, Schumaker KS, Palmgren MG, Zhu JK (2007) Arabidopsis protein kinase PKS5 inhibits the plasma membrane H^+-ATPase by preventing interaction with 14-3-3 protein. Plant Cell 19:1617–1634

Fujii H, Zhu JK (2009) An autophosphorylation site of the protein kinase SOS2 is important for salt tolerance in Arabidopsis. Mol Plant 2:183–190

Gao D, Knight MR, Trewavas AJ, Sattelmacher B, Plieth C (2004) Self-reporting Arabidopsis expressing pH and Ca^{2+} indicators unveil ion dynamics in the cytoplasm and in the apoplast under abiotic stress. Plant Physiol 134:898–908

Gehring CA, Williams DA, Cody SH, Parish RW (1990) Phototropism and geotropism in maize coleoptiles are spatially correlated with increases in cytosolic free calcium. Nature 345:528–530

Geisler M, Frangne N, Gomes E, Martinoia E, Palmgren MG (2000) The ACA4 gene of Arabidopsis encodes a vacuolar membrane calcium pump that improves salt tolerance in yeast. Plant Physiol 124:1814–1827

George L, Romanowsky SM, Harper JF, Sharrock RA (2008) The ACA10 Ca^{2+}-ATPase regulates adult vegetative development and inflorescence architecture in Arabidopsis. Plant Physiol 146:716–728

Gilroy S, Jones RL (1992) Gibberellic acid and abscisic acid coordinately regulate cytoplasmic calcium and secretory activity in barley aleurone protoplasts. Proc Natl Acad Sci USA 89:3591–3595

Gilroy S, Read ND, Trewavas AJ (1990) Elevation of cytoplasmic calcium by caged calcium or caged inositol triphosphate initiates stomatal closure. Nature 346:769–771

Gong M, van der Luit AH, Knight MR, Trewavas AJ (1998) Heat-shock-induced changes in intracellular Ca^{2+} level in tobacco seedlings in relation to thermotolerance. Plant Physiol 116:429–437

Grant M, Brown I, Adams S, Knight M, Ainslie A, Mansfield J (2000) The RPM1 plant disease resistance gene facilitates a rapid and sustained increase in cytosolic calcium that is necessary for the oxidative burst and hypersensitive cell death. Plant J 23:441–450

Guo H, Mockler T, Duong H, Lin C (2001a) SUB1, an Arabidopsis Ca^{2+}-binding protein involved in cryptochrome and phytochrome coaction. Science 291:487–490

Guo Y, Halfter U, Ishitani M, Zhu JK (2001b) Molecular characterization of functional domains in the protein kinase SOS2 that is required for plant salt tolerance. Plant Cell 13:1383–1400

Hahm SH, Saunders MJ (1991) Cytokinin increases intracellular Ca^{2+} in Funaria: detection with Indo-1. Cell Calcium 12:675–681

Haley A, Russell AJ, Wood N, Allan AC, Knight M, Campbell AK, Trewavas AJ (1995) Effects of mechanical signaling on plant cell cytosolic calcium. Proc Natl Acad Sci USA 92:4124–4128
Halfter U, Ishitani M, Zhu JK (2000) The Arabidopsis SOS2 protein kinase physically interacts with and is activated by the calcium-binding protein SOS3. Proc Natl Acad Sci USA 97:3735–3740
Hamilton DW, Hills A, Kohler B, Blatt MR (2000) Ca^{2+} channels at the plasma membrane of stomatal guard cells are activated by hyperpolarization and abscisic acid. Proc Natl Acad Sci USA 97:4967–4972
Han S, Tang R, Anderson LK, Woerner TE, Pei ZM (2003) A cell surface receptor mediates extracellular Ca^{2+} sensing in guard cells. Nature 425:196–200
Harada A, Sakai T, Okada K (2003) Phot1 and phot2 mediate blue light-induced transient increases in cytosolic Ca^{2+} differently in Arabidopsis leaves. Proc Natl Acad Sci USA 100:8583–8588
Harper JF, Hong B, Hwang I, Guo HQ, Stoddard R, Huang JF, Palmgren MG, Sze H (1998) A novel calmodulin-regulated Ca^{2+}-ATPase (ACA2) from Arabidopsis with an N-terminal autoinhibitory domain. J Biol Chem 273:1099–1106
Harper JF, Breton G, Harmon A (2004) Decoding Ca^{2+} signals through plant protein kinases. Annu Rev Plant Biol 55:263–288
Hauser H, Levine BA, Williams RJP (1976) Interaction of ions with membranes. Trends Biochem Sci 1:278–281
Hedrich R, Neher E (1987) Cytoplasmic calcium regulates voltage-dependent ion channels in plant vacuoles. Nature 329:833–836
Helms K (1971) Calcium Deficiency of Dark-grown Seedlings of Phaseolus vulgaris L. Plant Physiol 47:799–804
Hepler PK (2005) Calcium: a central regulator of plant growth and development. Plant Cell 17:2142–2155
Hepler PK, Wayne RO (1985) Calcium and plant development. Ann Rev Plant Physiol 36: 397–439
Hetherington AM, Brownlee C (2004) The generation of Ca^{2+} signals in plants. Annu Rev Plant Biol 55:401–427
Hirschi KD (1999) Expression of Arabidopsis CAX1 in tobacco: altered calcium homeostasis and increased stress sensitivity. Plant Cell 11:2113–2122
Hirschi KD (2004) The calcium conundrum. Both versatile nutrient and specific signal. Plant Physiol 136:2438–2442
Hirschi KD, Korenkov VD, Wilganowski NL, Wagner GJ (2000) Expression of arabidopsis CAX2 in tobacco. Altered metal accumulation and increased manganese tolerance. Plant Physiol 124:125–133
Ho LC, White PJ (2005) A cellular hypothesis for the induction of blossom-end rot in tomato fruit. Ann Bot (Lond) 95:571–581
Homann U, Tester M (1997) Ca^{2+}-independent and Ca^{2+}/GTP-binding protein-controlled exocytosis in a plant cell. Proc Natl Acad Sci USA 94:6565–6570
Hrabak EM, Chan CW, Gribskov M, Harper JF, Choi JH, Halford N, Kudla J, Luan S, Nimmo HG, Sussman MR, Thomas M, Walker-Simmons K, Zhu JK, Harmon AC (2003) The Arabidopsis CDPK-SnRK superfamily of protein kinases. Plant Physiol 132:666–680
Hu HC, Wang YY, Tsay YF (2009) AtCIPK8, a CBL-interacting protein kinase, regulates the low-affinity phase of the primary nitrate response. Plant J 57:264–278
Hua BG, Mercier RW, Leng Q, Berkowitz GA (2003a) Plants do it differently. A new basis for potassium/sodium selectivity in the pore of an ion channel. Plant Physiol 132:1353–1361
Hua BG, Mercier RW, Zielinski RE, Berkowitz GA (2003b) Functional interaction of calmodulin with a plant cyclic nucleotide gated cation channel. Plant Physiol Biochem 41:945–954
Huang L, Berkelman T, Franklin AE, Hoffman NE (1993) Characterization of a gene encoding a Ca^{2+}-ATPase-like protein in the plastid envelope. Proc Natl Acad Sci USA 90:10066–10070

Hwang I, Sze H, Harper JF (2000) A calcium-dependent protein kinase can inhibit a calmodulin-stimulated Ca^{2+} pump (ACA2) located in the endoplasmic reticulum of Arabidopsis. Proc Natl Acad Sci USA 97:6224–6229

Irving HR, Gehring CA, Parish RW (1992) Changes in cytosolic pH and calcium of guard cells precede stomatal movements. Proc Natl Acad Sci USA 89:1790–1794

Ishida S, Yuasa T, Nakata M, Takahashi Y (2008) A tobacco Calcium-dependent protein kinase, CDPK1, regulates the transcription factor REPRESSION OF SHOOT GROWTH in response to gibberellins. Plant Cell 20:3273–3288

Ishitani M, Liu J, Halfter U, Kim CS, Shi W, Zhu JK (2000) SOS3 function in plant salt tolerance requires N-myristoylation and calcium binding. Plant Cell 12:1667–1678

Ivashuta S, Liu J, Lohar DP, Haridas S, Bucciarelli B, VandenBosch KA, Vance CP, Harrison MJ, Gantt JS (2005) RNA interference identifies a calcium-dependent protein kinase involved in Medicago truncatula root development. Plant Cell 17:2911–2921

Jaiswal JK (2001) Calcium - how and why? J Biosci 26:357–363

Jeter CR, Tang W, Henaff E, Butterfield T, Roux SJ (2004) Evidence of a novel cell signaling role for extracellular adenosine triphosphates and diphosphates in Arabidopsis. Plant Cell 16:2652–2664

Johnson CH, Knight MR, Kondo T, Masson P, Sedbrook J, Haley A, Trewavas A (1995) Circadian oscillations of cytosolic and chloroplastic free calcium in plants. Science 269:1863–1865

Jun T, Wu S, Bai J, Sun D (1996) Extracellular calmodulin-binding proteins in plants: purification of a 21-kDa calmodulin-binding protein. Planta 198:510–516

Kaplan B, Sherman T, Fromm H (2007) Cyclic nucleotide-gated channels in plants. FEBS Lett 581:2237–2246

Kasai M, Muto S (1990) Ca^{2+} pump and Ca^{2+}/H^{+} antiporter in plasma membrane vesicles isolated by aqueous two-phase partitioning from corn leaves. J Membr Biol 114:133–142

Kauss H (1987) Some aspects of Calcium-dependent regulation in plant metabolism. Ann Rev Plant Physiol 38:47–72

Kawano T, Sahashi N, Takahashi K, Uozumi N, Muto S (1998) Salicylic acid induces extracellular superoxide generation followed by an increase in cytosolic calcium ion in tobacco suspension culture: the earliest events in salicylic acid signal transduction. Plant Cell Physiol 39:721–730

Kesavan J, Borisovska M, Bruns D (2007) v-SNARE actions during Ca^{2+}-triggered exocytosis. Cell 131:351–363

Kiegle E, Moore CA, Haseloff J, Tester MA, Knight MR (2000) Cell-type-specific calcium responses to drought, salt and cold in the Arabidopsis root. Plant J 23:267–278

Kim KN, Cheong YH, Grant JJ, Pandey GK, Luan S (2003) CIPK3, a calcium sensor-associated protein kinase that regulates abscisic acid and cold signal transduction in Arabidopsis. Plant Cell 15:411–423

Kim BG, Waadt R, Cheong YH, Pandey GK, Dominguez-Solis JR, Schultke S, Lee SC, Kudla J, Luan S (2007) The calcium sensor CBL10 mediates salt tolerance by regulating ion homeostasis in Arabidopsis. Plant J 52:473–484

Kirkby EA, Pilbeam DJ (1984) Calcium as a plant nutrient. Plant Cell Environ 7:397–405

Klusener B, Young JJ, Murata Y, Allen GJ, Mori IC, Hugouvieux V, Schroeder JI (2002) Convergence of calcium signaling pathways of pathogenic elicitors and abscisic acid in Arabidopsis guard cells. Plant Physiol 130:2152–2163

Knight MR, Campbell AK, Smith SM, Trewavas AJ (1991) Transgenic plant aequorin reports the effects of touch and cold-shock and elicitors on cytoplasmic calcium. Nature 352: 524–526

Knight H, Trewavas AJ, Knight MR (1997) Calcium signalling in Arabidopsis thaliana responding to drought and salinity. Plant J 12:1067–1078

Knight H, Brandt S, Knight MR (1998) A history of stress alters drought calcium signalling pathways in Arabidopsis. Plant J 16:681–687

Kolukisaoglu U, Weinl S, Blazevic D, Batistic O, Kudla J (2004) Calcium sensors and their interacting protein kinases: genomics of the Arabidopsis and rice CBL-CIPK signaling networks. Plant Physiol 134:43–58

Kosuta S, Hazledine S, Sun J, Miwa H, Morris RJ, Downie JA, Oldroyd GE (2008) Differential and chaotic calcium signatures in the symbiosis signaling pathway of legumes. Proc Natl Acad Sci USA 105:9823–9828

Kreimer G, Melkonian M, Holtum JA, Latzko E (1988) Stromal Free Calcium Concentration and Light-Mediated Activation of Chloroplast Fructose-1, 6-Bisphosphatase. Plant Physiol 86:423–428

Kudla J, Xu Q, Harter K, Gruissem W, Luan S (1999) Genes for calcineurin B-like proteins in Arabidopsis are differentially regulated by stress signals. Proc Natl Acad Sci USA 96:4718–4723

Kushwaha R, Singh A, Chattopadhyay S (2008) Calmodulin7 plays an important role as transcriptional regulator in Arabidopsis seedling development. Plant Cell 20:1747–1759

Lam HM, Chiu J, Hsieh MH, Meisel L, Oliveira IC, Shin M, Coruzzi G (1998) Glutamate-receptor genes in plants. Nature 396:125–126

Lamotte O, Gould K, Lecourieux D, Sequeira-Legrand A, Lebrun-Garcia A, Durner J, Pugin A, Wendehenne D (2004) Analysis of nitric oxide signaling functions in tobacco cells challenged by the elicitor cryptogein. Plant Physiol 135:516–529

Laohavisit A, Mortimer JC, Demidchik V, Coxon KM, Stancombe MA, Macpherson N, Brownlee C, Hofmann A, Webb AA, Miedema H, Battey NH, Davies JM (2009) Zea mays Annexins Modulate Cytosolic Free Ca^{2+} and Generate a Ca^{2+}-Permeable Conductance. Plant Cell 21:479–493

Larkindale J, Knight MR (2002) Protection against heat stress-induced oxidative damage in Arabidopsis involves calcium, abscisic acid, ethylene, and salicylic acid. Plant Physiol 128:682–695

Leckie CP, McAinsh MR, Allen GJ, Sanders D, Hetherington AM (1998) Abscisic acid-induced stomatal closure mediated by cyclic ADP-ribose. Proc Natl Acad Sci USA 95: 15837–15842

Leclerc E, Corti C, Schmid H, Vetter S, James P, Carafoli E (1999) Serine/threonine phosphorylation of calmodulin modulates its interaction with the binding domains of target enzymes. Biochem J 344(Pt 2):403–411

Lecourieux D, Mazars C, Pauly N, Ranjeva R, Pugin A (2002) Analysis and effects of cytosolic free calcium increases in response to elicitors in Nicotiana plumbaginifolia cells. Plant Cell 14:2627–2641

Lecourieux D, Lamotte O, Bourque S, Wendehenne D, Mazars C, Ranjeva R, Pugin A (2005) Proteinaceous and oligosaccharidic elicitors induce different calcium signatures in the nucleus of tobacco cells. Cell Calcium 38:527–538

Lee JS, Mulkey TJ, Evans ML (1983) Reversible Loss of Gravitropic Sensitivity in Maize Roots After Tip Application of Calcium Chelators. Science 220:1375–1376

Lee SH, Seo HY, Kim JC, Heo WD, Chung WS, Lee KJ, Kim MC, Cheong YH, Choi JY, Lim CO, Cho MJ (1997) Differential activation of NAD kinase by plant calmodulin isoforms. The critical role of domain I. J Biol Chem 272:9252–9259

Lee JY, Yoo BC, Harmon AC (1998) Kinetic and calcium-binding properties of three calcium-dependent protein kinase isoenzymes from soybean. Biochemistry 37:6801–6809

Lee SH, Johnson JD, Walsh MP, Van Lierop JE, Sutherland C, Xu A, Snedden WA, Kosk-Kosicka D, Fromm H, Narayanan N, Cho MJ (2000) Differential regulation of Ca^{2+}/calmodulin-dependent enzymes by plant calmodulin isoforms and free Ca^{2+} concentration. Biochem J 350(Pt 1):299–306

Lee SM, Kim HS, Han HJ, Moon BC, Kim CY, Harper JF, Chung WS (2007) Identification of a calmodulin-regulated autoinhibited Ca^{2+}-ATPase (ACA11) that is localized to vacuole membranes in Arabidopsis. FEBS Lett 581:3943–3949

Legue V, Blancaflor E, Wymer C, Perbal G, Fantin D, Gilroy S (1997) Cytoplasmic free Ca^{2+} in Arabidopsis roots changes in response to touch but not gravity. Plant Physiol 114:789–800

Lemtiri-Chlieh F, MacRobbie EA, Webb AA, Manison NF, Brownlee C, Skepper JN, Chen J, Prestwich GD, Brearley CA (2003) Inositol hexakisphosphate mobilizes an endomembrane store of calcium in guard cells. Proc Natl Acad Sci USA 100:10091–10095

Levine A, Pennell RI, Alvarez ME, Palmer R, Lamb C (1996) Calcium-mediated apoptosis in a plant hypersensitive disease resistance response. Curr Biol 6:427–437

Li L, Kim BG, Cheong YH, Pandey GK, Luan S (2006) A Ca^{2+} signaling pathway regulates a K^+ channel for low-K response in Arabidopsis. Proc Natl Acad Sci USA 103:12625–12630

Li X, Chanroj S, Wu Z, Romanowsky SM, Harper JF, Sze H (2008) A distinct endosomal Ca^{2+}/Mn^{2+} pump affects root growth through the secretory process. Plant Physiol 147:1675–1689

Liang F, Cunningham KW, Harper JF, Sze H (1997) ECA1 complements yeast mutants defective in Ca^{2+} pumps and encodes an endoplasmic reticulum-type Ca^{2+}-ATPase in Arabidopsis thaliana. Proc Natl Acad Sci USA 94:8579–8584

Logan DC, Knight MR (2003) Mitochondrial and cytosolic calcium dynamics are differentially regulated in plants. Plant Physiol 133:21–24

Long JC, Jenkins GI (1998) Involvement of plasma membrane redox activity and calcium homeostasis in the UV-B and UV-A/blue light induction of gene expression in Arabidopsis. Plant Cell 10:2077–2086

Love J, Dodd AN, Webb AA (2004) Circadian and diurnal calcium oscillations encode photoperiodic information in Arabidopsis. Plant Cell 16:956–966

Lu SX, Hrabak EM (2002) An Arabidopsis calcium-dependent protein kinase is associated with the endoplasmic reticulum. Plant Physiol 128:1008–1021

Ludwig AA, Saitoh H, Felix G, Freymark G, Miersch O, Wasternack C, Boller T, Jones JD, Romeis T (2005) Ethylene-mediated cross-talk between calcium-dependent protein kinase and MAPK signaling controls stress responses in plants. Proc Natl Acad Sci USA 102: 10736–10741

Luo GZ, Wang HW, Huang J, Tian AG, Wang YJ, Zhang JS, Chen SY (2005) A putative plasma membrane cation/proton antiporter from soybean confers salt tolerance in Arabidopsis. Plant Mol Biol 59:809–820

Lynch J, Polito VS, Lauchli A (1989) Salinity Stress Increases Cytoplasmic Ca Activity in Maize Root Protoplasts. Plant Physiol 90:1271–1274

Ma SY, Wu WH (2007) AtCPK23 functions in Arabidopsis responses to drought and salt stresses. Plant Mol Biol 65:511–518

Ma L, Xu X, Cui S, Sun D (1999) The presence of a heterotrimeric G protein and its role in signal transduction of extracellular calmodulin in pollen germination and tube growth. Plant Cell 11:1351–1364

Ma W, Smigel A, Tsai YC, Braam J, Berkowitz GA (2008) Innate immunity signaling: cytosolic Ca^{2+} elevation is linked to downstream nitric oxide generation through the action of calmodulin or a calmodulin-like protein. Plant Physiol 148:818–828

MacRobbie EA (1989) Calcium influx at the plasmalemma of isolated guard cells of Commelina communis. Planta 178:231–241

Magnan F, Ranty B, Charpenteau M, Sotta B, Galaud JP, Aldon D (2008) Mutations in AtCML9, a calmodulin-like protein from Arabidopsis thaliana, alter plant responses to abiotic stress and abscisic acid. Plant J 56:575–589

Martens S, Kozlov MM, McMahon HT (2007) How synaptotagmin promotes membrane fusion. Science 316:1205–1208

Martin ML, Busconi L (2000) Membrane localization of a rice calcium-dependent protein kinase (CDPK) is mediated by myristoylation and palmitoylation. Plant J 24:429–435

Martinec J, Feltl T, Scanlon CH, Lumsden PJ, Machackova I (2000) Subcellular localization of a high affinity binding site for D-myo-inositol 1, 4, 5-trisphosphate from Chenopodium rubrum. Plant Physiol 124:475–483

Maser P, Thomine S, Schroeder JI, Ward JM, Hirschi K, Sze H, Talke IN, Amtmann A, Maathuis FJ, Sanders D, Harper JF, Tchieu J, Gribskov M, Persans MW, Salt DE, Kim SA, Guerinot ML

(2001) Phylogenetic relationships within cation transporter families of Arabidopsis. Plant Physiol 126:1646–1667
Mazars C, Bourque S, Mithofer A, Pugin A, Ranjeva R (2009) Calcium homeostasis in plant cell nuclei. New Phytol 181:261–274
McAinsh MR, Brownlee C, Hetherington AM (1990) Abscisic acid-induced elevation of guard cell cytosolic Ca^{2+} precedes stomatal closure. Nature 343:186–188
McCormack E, Braam J (2003) Calmodulins and related potential calcium sensors of Arabidopsis. New Phyt 159:585–598
McCormack E, Tsai YC, Braam J (2005) Handling calcium signaling: Arabidopsis CaMs and CMLs. Trends Plant Sci 10:383–389
McCubbin AG, Ritchie SM, Swanson SJ, Gilroy S (2004) The calcium-dependent protein kinase HvCDPK1 mediates the gibberellic acid response of the barley aleurone through regulation of vacuolar function. Plant J 39:206–218
McNamara VP, Gounaris K (1995) Granal photosystem II complexes contain only the high redox potential form of cytochrome b-559 which is stabilized by the ligation of calcium. Biochim Biophys Acta 1231:289–296
Mei H, Zhao J, Pittman JK, Lachmansingh J, Park S, Hirschi KD (2007) In planta regulation of the Arabidopsis Ca^{2+}/H^{+} antiporter CAX1. J Exp Bot 58:3419–3427
Menteyne A, Burdakov A, Charpentier G, Petersen OH, Cancela JM (2006) Generation of specific Ca^{2+} signals from Ca^{2+} stores and endocytosis by differential coupling to messengers. Curr Biol 16:1931–1937
Miedema H, Bothwell JH, Brownlee C, Davies JM (2001) Calcium uptake by plant cells–channels and pumps acting in concert. Trends Plant Sci 6:514–519
Miedema H, Demidchik V, Very AA, Bothwell JH, Brownlee C, Davies JM (2008) Two voltage-dependent calcium channels co-exist in the apical plasma membrane of Arabidopsis thaliana root hairs. New Phytol 179:378–385
Miller AJ, Sanders D (1987) Depletion of cytosolic free calcium induced by photosynthesis. Nature 326:397–400
Miller DD, Callaham DA, Gross DJ, Hepler PK (1992) Free Ca^{2+} gradient in growing pollen tubes of Lilium. J Cell Sci 101:7–12
Mills RF, Doherty ML, Lopez-Marques RL, Weimar T, Dupree P, Palmgren MG, Pittman JK, Williams LE (2008) ECA3, a Golgi-localized P2A-type ATPase, plays a crucial role in manganese nutrition in Arabidopsis. Plant Physiol 146:116–128
Moore AL, Åkerman KEO (1984) Calcium and plant organelles. Plant Cell Environ 7:423–429
Moreno I, Norambuena L, Maturana D, Toro M, Vergara C, Orellana A, Zurita-Silva A, Ordenes VR (2008) AtHMA1 is a thapsigargin-sensitive Ca^{2+}/heavy metal pump. J Biol Chem 283:9633–9641
Mori IC, Murata Y, Yang Y, Munemasa S, Wang Y-F, Andreoli S, Tiriac H, Alonso JM, Harper JF, Ecker JR, Kwak JM, Schroeder JI (2006) CDPKs CPK6 and CPK3 function in ABA regulation of guard cell S-type anion- and Ca^{2+}-permeable channels and stomatal closure. PLOS Biology 4:1749–1762
Mortimer JC, Laohavisit A, Macpherson N, Webb A, Brownlee C, Battey NH, Davies JM (2008) Annexins: multifunctional components of growth and adaptation. J Exp Bot 59:533–544
Muir SR, Sanders D (1997) Inositol 1, 4, 5-trisphosphate-sensitive Ca2+ release across nonvacuolar membranes in cauliflower. Plant Physiol 114:1511–1521
Mulder EG (1950) Mineral nutrition of plants. Annu Rev Plant Physiol 1:1–24
Nagae M, Nozawa A, Koizumi N, Sano H, Hashimoto H, Sato M, Shimizu T (2003) The crystal structure of the novel calcium-binding protein AtCBL2 from Arabidopsis thaliana. J Biol Chem 278:42240–42246
Nakagawa Y, Katagiri T, Shinozaki K, Qi Z, Tatsumi H, Furuichi T, Kishigami A, Sokabe M, Kojima I, Sato S, Kato T, Tabata S, Iida K, Terashima A, Nakano M, Ikeda M, Yamanaka T, Iida H (2007) Arabidopsis plasma membrane protein crucial for Ca^{2+} influx and touch sensing in roots. Proc Natl Acad Sci USA 104:3639–3644

Navazio L, Bewell MA, Siddiqua A, Dickinson GD, Galione A, Sanders D (2000) Calcium release from the endoplasmic reticulum of higher plants elicited by the NADP metabolite nicotinic acid adenine dinucleotide phosphate. Proc Natl Acad Sci USA 97:8693–8698

Navazio L, Mariani P, Sanders D (2001) Mobilization of Ca^{2+} by cyclic ADP-ribose from the endoplasmic reticulum of cauliflower florets. Plant Physiol 125:2129–2138

Nemchinov LG, Shabala L, Shabala S (2008) Calcium efflux as a component of the hypersensitive response of Nicotiana benthamiana to Pseudomonas syringae. Plant Cell Physiol 49:40–46

Neuhaus G, Bowler C, Kern R, Chua NH (1993) Calcium/calmodulin-dependent and -independent phytochrome signal transduction pathways. Cell 73:937–952

Ng CK, Carr K, McAinsh MR, Powell B, Hetherington AM (2001) Drought-induced guard cell signal transduction involves sphingosine-1-phosphate. Nature 410:596–599

Nomura H, Komori T, Kobori M, Nakahira Y, Shiina T (2008) Evidence for chloroplast control of external Ca^{2+}-induced cytosolic Ca^{2+} transients and stomatal closure. Plant J 53:988–998

Ochiai E-I (1991) Why Calcium? J Chem Educ 68:10–12

Oh SH, Roberts DM (1990) Analysis of the State of Posttranslational Calmodulin Methylation in Developing Pea Plants. Plant Physiol 93:880–887

Ohta M, Guo Y, Halfter U, Zhu JK (2003) A novel domain in the protein kinase SOS2 mediates interaction with the protein phosphatase 2C ABI2. Proc Natl Acad Sci USA 100:11771–11776

Pandey GK, Cheong YH, Kim KN, Grant JJ, Li L, Hung W, D'Angelo C, Weinl S, Kudla J, Luan S (2004) The calcium sensor calcineurin B-like 9 modulates abscisic acid sensitivity and biosynthesis in Arabidopsis. Plant Cell 16:1912–1924

Pauly N, Knight MR, Thuleau P, Graziana A, Muto S, Ranjeva R, Mazars C (2001) The nucleus together with the cytosol generates patterns of specific cellular calcium signatures in tobacco suspension culture cells. Cell Calcium 30:413–421

Pei ZM, Murata Y, Benning G, Thomine S, Klusener B, Allen GJ, Grill E, Schroeder JI (2000) Calcium channels activated by hydrogen peroxide mediate abscisic acid signalling in guard cells. Nature 406:731–734

Peiter E, Maathuis FJ, Mills LN, Knight H, Pelloux J, Hetherington AM, Sanders D (2005) The vacuolar Ca^{2+}-activated channel TPC1 regulates germination and stomatal movement. Nature 434:404–408

Perez V, Wherrett T, Shabala S, Muniz J, Dobrovinskaya O, Pottosin I (2008) Homeostatic control of slow vacuolar channels by luminal cations and evaluation of the channel-mediated tonoplast Ca^{2+} fluxes in situ. J Exp Bot 59:3845–3855

Persson S, Wyatt SE, Love J, Thompson WF, Robertson D, Boss WF (2001) The Ca^{2+} status of the endoplasmic reticulum is altered by induction of calreticulin expression in transgenic plants. Plant Physiol 126:1092–1104

Pittman JK, Shigaki T, Cheng NH, Hirschi KD (2002a) Mechanism of N-terminal autoinhibition in the Arabidopsis Ca^{2+}/H^{+} antiporter CAX1. J Biol Chem 277:26452–26459

Pittman JK, Sreevidya CS, Shigaki T, Ueoka-Nakanishi H, Hirschi KD (2002b) Distinct N-terminal regulatory domains of Ca^{2+}/H^{+} antiporters. Plant Physiol 130:1054–1062

Plieth C (2005) Calcium: just another regulator in the machinery of life? Ann Bot (Lond) 96:1–8

Plieth C, Hansen UP, Knight H, Knight MR (1999) Temperature sensing by plants: the primary characteristics of signal perception and calcium response. Plant J 18:491–497

Popescu SC, Popescu GV, Bachan S, Zhang Z, Seay M, Gerstein M, Snyder M, Dinesh-Kumar SP (2007) Differential binding of calmodulin-related proteins to their targets revealed through high-density Arabidopsis protein microarrays. Proc Natl Acad Sci USA 104:4730–4735

Portis AR Jr, Heldt HW (1976) Light-dependent changes of the Mg^{2+} concentration in the stroma in relation to the Mg^{2+} dependency of CO2 fixation in intact chloroplasts. Biochim Biophys Acta 449:434–436

Pottosin II, Schonknecht G (2007) Vacuolar calcium channels. J Exp Bot 58:1559–1569

Price AH, Taylor A, Ripley SJ, Griffiths A, Trewavas AJ, Knight MR (1994) Oxidative signals in tobacco increase cytosolic calcium. Plant Cell 6:1301–1310

Putnam-Evans C, Harmon A, Palevitz BA, Fechheimer M, Cormier MJ (1989) Calcium-dependent protein kinase is localized with F-actin in plant cells. Cell Motil Cytoskel 12:12–22

Qi Z, Stephens NR, Spalding EP (2006) Calcium entry mediated by GLR3.3, an Arabidopsis glutamate receptor with a broad agonist profile. Plant Physiol 142:963–971

Qiu QS, Guo Y, Dietrich MA, Schumaker KS, Zhu JK (2002) Regulation of SOS1, a plasma membrane Na^+/H^+ exchanger in Arabidopsis thaliana, by SOS2 and SOS3. Proc Natl Acad Sci USA 99:8436–8441

Qudeimat E, Faltusz AM, Wheeler G, Lang D, Brownlee C, Reski R, Frank W (2008) A $P_{II}B$-type Ca^{2+}-ATPase is essential for stress adaptation in Physcomitrella patens. Proc Natl Acad Sci USA 105:19555–19560

Quintero FJ, Ohta M, Shi H, Zhu JK, Pardo JM (2002) Reconstitution in yeast of the Arabidopsis SOS signaling pathway for Na^+ homeostasis. Proc Natl Acad Sci USA 99:9061–9066

Ranf S, Wunnenberg P, Lee J, Becker D, Dunkel M, Hedrich R, Scheel D, Dietrich P (2008) Loss of the vacuolar cation channel, AtTPC1, does not impair Ca^{2+} signals induced by abiotic and biotic stresses. Plant J 53:287–299

Rathore KS, Cork RJ, Robinson KR (1991) A cytoplasmic gradient of Ca^{2+} is correlated with the growth of lily pollen tubes. Dev Biol 148:612–619

Raz V, Fluhr R (1992) Calcium Requirement for Ethylene-Dependent Responses. Plant Cell 4:1123–1130

Reddy VS, Ali GS, Reddy AS (2002) Genes encoding calmodulin-binding proteins in the Arabidopsis genome. J Biol Chem 277:9840–9852

Rentel MC, Knight MR (2004) Oxidative stress-induced calcium signaling in Arabidopsis. Plant Physiol 135:1471–1479

Romeis T, Piedras P, Jones JD (2000) Resistance gene-dependent activation of a calcium-dependent protein kinase in the plant defense response. Plant Cell 12:803–816

Romeis T, Ludwig AA, Martin R, Jones JD (2001) Calcium-dependent protein kinases play an essential role in a plant defence response. Embo J 20:5556–5567

Rudd JJ, Franklin-Tong VE (1999) Calcium signaling in plants. Cell Mol Life Sci 55:214–232

Sai J, Johnson CH (2002) Dark-stimulated calcium ion fluxes in the chloroplast stroma and cytosol. Plant Cell 14:1279–1291

Saijo Y, Hata S, Kyozuka J, Shimamoto K, Izui K (2000) Over-expression of a single Ca^{2+}-dependent protein kinase confers both cold and salt/drought tolerance on rice plants. Plant J 23:319–327

Sakai-Wada A, Yagi S (1993) Ultrastructural studies on the Ca^{2+} localization in the dividing cells of the maize root tip. Cell Struct Funct 18:389–397

Sanchez-Barrena MJ, Martinez-Ripoll M, Zhu JK, Albert A (2005) The structure of the Arabidopsis thaliana SOS3: molecular mechanism of sensing calcium for salt stress response. J Mol Biol 345:1253–1264

Sanchez-Barrena MJ, Fujii H, Angulo I, Martinez-Ripoll M, Zhu JK, Albert A (2007) The structure of the C-terminal domain of the protein kinase AtSOS2 bound to the calcium sensor AtSOS3. Mol Cell 26:427–435

Sanders D, Pelloux J, Brownlee C, Harper JF (2002) Calcium at the crossroads of signaling. Plant Cell 14:S401–S417

Satterlee JS, Sussman MR (1998) Unusual membrane-associated protein kinases in higher plants. J Membr Biol 164:205–213

Schaller GE, Harmon AC, Sussman MR (1992) Characterization of a calcium- and lipid-dependent protein kinase associated with the plasma membrane of oat. Biochemistry 31:1721–1727

Schapire AL, Voigt B, Jasik J, Rosado A, Lopez-Cobollo R, Menzel D, Salinas J, Mancuso S, Valpuesta V, Baluska F, Botella MA (2008) Arabidopsis synaptotagmin 1 is required for the maintenance of plasma membrane integrity and cell viability. Plant Cell 20:3374–3388

Schiott M, Romanowsky SM, Baekgaard L, Jakobsen MK, Palmgren MG, Harper JF (2004) A plant plasma membrane Ca^{2+} pump is required for normal pollen tube growth and fertilization. Proc Natl Acad Sci USA 101:9502–9507

Schroeder JI, Hagiwara S (1989) Cytosolic calcium regulates ion channels in the plasma membrane of Vicia faba guard cells. Nature 338:427–430

Schumaker KS, Sze H (1987) Inositol 1, 4, 5-trisphosphate releases Ca^{2+} from vacuolar membrane vesicles of oat roots. J Biol Chem 262:3944–3946

Seigneurin-Berny D, Gravot A, Auroy P, Mazard C, Kraut A, Finazzi G, Grunwald D, Rappaport F, Vavasseur A, Joyard J, Richaud P, Rolland N (2006) HMA1, a new Cu-ATPase of the chloroplast envelope, is essential for growth under adverse light conditions. J Biol Chem 281:2882–2892

Shabala S, Demidchik V, Shabala L, Cuin TA, Smith SJ, Miller AJ, Davies JM, Newman IA (2006) Extracellular Ca^{2+} ameliorates NaCl-induced K^+ loss from Arabidopsis root and leaf cells by controlling plasma membrane K^+ -permeable channels. Plant Physiol 141:1653–1665

Shacklock PS, Read ND, Trewavas A (1992) Cytosolic free calcium mediates red light-induced photomorphogenesis. Nature 358:753–755

Shang ZL, Ma LG, Zhang HL, He RR, Wang XC, Cui SJ, Sun DY (2005) Ca^{2+} influx into lily pollen grains through a hyperpolarization-activated Ca^{2+}-permeable channel which can be regulated by extracellular CaM. Plant Cell Physiol 46:598–608

Shi J, Kim KN, Ritz O, Albrecht V, Gupta R, Harter K, Luan S, Kudla J (1999) Novel protein kinases associated with calcineurin B-like calcium sensors in Arabidopsis. Plant Cell 11:2393–2405

Shigaki T, Rees I, Nakhleh L, Hirschi KD (2006) Identification of three distinct phylogenetic groups of CAX cation/proton antiporters. J Mol Evol 63:815–825

Shimada T, Watanabe E, Tamura K, Hayashi Y, Nishimura M, Hara-Nishimura I (2002) A vacuolar sorting receptor PV72 on the membrane of vesicles that accumulate precursors of seed storage proteins (PAC vesicles). Plant Cell Physiol 43:1086–1095

Simon EW (1978) The symptoms of calcium deficiency in plants. New Phytol 80:1–15

Staxen II, Pical C, Montgomery LT, Gray JE, Hetherington AM, McAinsh MR (1999) Abscisic acid induces oscillations in guard-cell cytosolic free calcium that involve phosphoinositide-specific phospholipase C. Proc Natl Acad Sci USA 96:1779–1784

Stoelzle S, Kagawa T, Wada M, Hedrich R, Dietrich P (2003) Blue light activates calcium-permeable channels in Arabidopsis mesophyll cells via the phototropin signaling pathway. Proc Natl Acad Sci USA 100:1456–1461

Strynadka NC, James MN (1989) Crystal structures of the helix-loop-helix calcium-binding proteins. Annu Rev Biochem 58:951–998

Subbaiah CC, Bush DS, Sachs MM (1994) Elevation of cytosolic calcium precedes anoxic gene expression in maize suspension-cultured cells. Plant Cell 6:1747–1762

Subbaiah CC, Bush DS, Sachs MM (1998) Mitochondrial contribution to the anoxic Ca^{2+} signal in maize suspension-cultured cells. Plant Physiol 118:759–771

Sun DY, Bian YQ, Zhao BH, Zhao LY (1995) The effects of CaM on cell wall regeneration and cell division of protoplasts. Plant Cell Physiol 36:133–138

Sun QP, Guo Y, Sun Y, Sun DY, Wang XJ (2006) Influx of extracellular Ca^{2+} involved in jasmonic-acid-induced elevation of $[Ca^{2+}]_{cyt}$ and JR1 expression in Arabidopsis thaliana. J Plant Res 119:343–350

Sunkar R, Kaplan B, Bouche N, Arazi T, Dolev D, Talke IN, Maathuis FJ, Sanders D, Bouchez D, Fromm H (2000) Expression of a truncated tobacco NtCBP4 channel in transgenic plants and disruption of the homologous Arabidopsis CNGC1 gene confer Pb^{2+} tolerance. Plant J 24:533–542

Szczegielniak J, Liwosz A, Jurkowski I, Loog M, Dobrowolska G, Ek P, Harmon AC, Muszynska G (2000) Calcium-dependent protein kinase from maize seedlings activated by phospholipids. Eur J Biochem 267:3818–3827

Szczegielniak J, Klimecka M, Liwosz A, Ciesielski A, Kaczanowski S, Dobrowolska G, Harmon AC, Muszynska G (2005) A wound-responsive and phospholipid-regulated maize calcium-dependent protein kinase. Plant Physiol 139:1970–1983

Sze H, Liang F, Hwang I, Curran AC, Harper JF (2000) Diversity and regulation of plant Ca^{2+} pumps: insights from expression in yeast. Annu Rev Plant Physiol Plant Mol Biol 51:433–462

Takahashi K, Isobe M, Knight MR, Trewavas AJ, Muto S (1997) Hypoosmotic Shock Induces Increases in Cytosolic Ca^{2+} in Tobacco Suspension-Culture Cells. Plant Physiol 113:587–594

Thion L, Mazars C, Nacry P, Bouchez D, Moreau M, Ranjeva R, Thuleau P (1998) Plasma membrane depolarization-activated calcium channels, stimulated by microtubule-depolymerizing drugs in wild-type Arabidopsis thaliana protoplasts, display constitutively large activities and a longer half-life in ton 2 mutant cells affected in the organization of cortical microtubules. Plant J 13:603–610

Trewavas A (1999) Le calcium, C'est la vie: calcium makes waves. Plant Physiol 120:1–6

Tripathi V, Parasuraman B, Laxmi A, Chattopadhyay D (2009) CIPK6, a CBL-interacting protein kinase is required for development and salt tolerance in plant. Plant J 58:778–790

Uno Y, Furihata T, Abe H, Yoshida R, Shinozaki K, Yamaguchi-Shinozaki K (2000) Arabidopsis basic leucine zipper transcription factors involved in an abscisic acid-dependent signal transduction pathway under drought and high-salinity conditions. Proc Natl Acad Sci USA 97:11632–11637

Urquhart W, Gunawardena AH, Moeder W, Ali R, Berkowitz GA, Yoshioka K (2007) The chimeric cyclic nucleotide-gated ion channel ATCNGC11/12 constitutively induces programmed cell death in a Ca^{2+} dependent manner. Plant Mol Biol 65:747–761

Vainonen JP, Sakuragi Y, Stael S, Tikkanen M, Allahverdiyeva Y, Paakkarinen V, Aro E, Suorsa M, Scheller HV, Vener AV, Aro EM (2008) Light regulation of CaS, a novel phosphoprotein in the thylakoid membrane of Arabidopsis thaliana. Febs J 275:1767–1777

van Der Luit AH, Olivari C, Haley A, Knight MR, Trewavas AJ (1999) Distinct calcium signaling pathways regulate calmodulin gene expression in tobacco. Plant Physiol 121:705–714

van Goor BJ, Wiersma D (1974) Redistribution of potassium, calcium, magnesium, and manganese in the plant. Physiol Plant 31:163–168

Volotovski ID, Sokolovsky SG, Molchan OV, Knight MR (1998) Second messengers mediate increases in cytosolic calcium in tobacco protoplasts. Plant Physiol 117:1023–1030

Wagner G, Rossbacher R (1980) X-ray microanalysis and chlorotetracycline staining of calcium vesicles in the green alg Mougeotia. Planta 149:298–305

Walter A, Mazars C, Maitrejean M, Hopke J, Ranjeva R, Boland W, Mithofer A (2007) Structural requirements of jasmonates and synthetic analogues as inducers of Ca^{2+} signals in the nucleus and the cytosol of plant cells. Angew Chem Int Ed Engl 46:4783–4785

Watanabe E, Shimada T, Kuroyanagi M, Nishimura M, Hara-Nishimura I (2002) Calcium-mediated association of a putative vacuolar sorting receptor PV72 with a propeptide of 2S albumin. J Biol Chem 277:8708–8715

Webb AAR, McAinsh MR, Mansfield TA, Hetherington AM (1996) Carbon dioxide induces increases in guard cell cytosolic free calcium. Plant J 9:297–304

Wegner LH, De Boer AH (1997) Properties of Two Outward-Rectifying Channels in Root Xylem Parenchyma Cells Suggest a Role in K^+ Homeostasis and Long-Distance Signaling. Plant Physiol 115:1707–1719

Weinl S, Held K, Schlucking K, Steinhorst L, Kuhlgert S, Hippler M, Kudla J (2008) A plastid protein crucial for Ca^{2+}-regulated stomatal responses. New Phytol 179:675–686

Wheeler GL, Brownlee C (2008) Ca^{2+} signalling in plants and green algae–changing channels. Trends Plant Sci 13:506–514

White PJ, Broadley MR (2003) Calcium in plants. Ann Bot (Lond) 92:487–511

White PJ, Bowen HC, Demidchik V, Nichols C, Davies JM (2002) Genes for calcium-permeable channels in the plasma membrane of plant root cells. Biochim Biophys Acta 1564:299–309

Wick SM, Hepler PK (1980) Localization of Ca^{2+}-containing antimonate precipitates during mitosis. J Cell Biol 86:500–513

Williams RJP (1970) The biochemistry of sodium, potassium, magnesium, and calcium. Quart Rev Chem Soc 24:331–365

Williams RJ (2004) Signalling: basics and evolution. Acta Biochim Pol 51:281–298

Wu Y, Kuzma J, Marechal E, Graeff R, Lee HC, Foster R, Chua NH (1997) Abscisic acid signaling through cyclic ADP-ribose in plants. Science 278:2126–2130

Wu Z, Liang F, Hong B, Young JC, Sussman MR, Harper JF, Sze H (2002) An endoplasmic reticulum-bound Ca^{2+}/Mn^{2+} pump, ECA1, supports plant growth and confers tolerance to Mn^{2+} stress. Plant Physiol 130:128–137

Xiong TC, Jauneau A, Ranjeva R, Mazars C (2004) Isolated plant nuclei as mechanical and thermal sensors involved in calcium signalling. Plant J 40:12–21

Xu H, Heath MC (1998) Role of calcium in signal transduction during the hypersensitive response caused by basidiospore-derived infection of the cowpea rust fungus. Plant Cell 10:585–598

Xu J, Li HD, Chen LQ, Wang Y, Liu LL, He L, Wu WH (2006) A protein kinase, interacting with two calcineurin B-like proteins, regulates K^+ transporter AKT1 in Arabidopsis. Cell 125:1347–1360

Yamaguchi T, Aharon GS, Sottosanto JB, Blumwald E (2005) Vacuolar Na^+/H^+ antiporter cation selectivity is regulated by calmodulin from within the vacuole in a Ca^{2+}- and pH-dependent manner. Proc Natl Acad Sci USA 102:16107–16112

Yamazaki T, Kawamura Y, Minami A, Uemura M (2008) Calcium-Dependent Freezing Tolerance in Arabidopsis Involves Membrane Resealing via Synaptotagmin SYT1. Plant Cell 20:3389–3404

Yang T, Poovaiah BW (2003) Calcium/calmodulin-mediated signal network in plants. Trends Plant Sci 8:505–512

Yoon GM, Dowd PE, Gilroy S, McCubbin AG (2006) Calcium-dependent protein kinase isoforms in Petunia have distinct functions in pollen tube growth, including regulating polarity. Plant Cell 18:867–878

Yoshioka K, Moeder W, Kang HG, Kachroo P, Masmoudi K, Berkowitz G, Klessig DF (2006) The chimeric Arabidopsis CYCLIC NUCLEOTIDE-GATED ION CHANNEL11/12 activates multiple pathogen resistance responses. Plant Cell 18:747–763

Yu IC, Parker J, Bent AF (1998) Gene-for-gene disease resistance without the hypersensitive response in Arabidopsis dnd1 mutant. Proc Natl Acad Sci USA 95:7819–7824

Zhao J, Barkla BJ, Marshall J, Pittman JK, Hirschi KD (2008) The Arabidopsis cax3 mutants display altered salt tolerance, pH sensitivity and reduced plasma membrane H^+-ATPase activity. Planta 227:659–669

Zhao J, Shigaki T, Mei H, Guo YQ, Cheng NH, Hirschi KD (2009) Interaction between Arabidopsis Ca^{2+}/H^+ Exchangers CAX1 and CAX3. J Biol Chem 284:4605–4615

Zhu SY, Yu XC, Wang XJ, Zhao R, Li Y, Fan RC, Shang Y, Du SY, Wang XF, Wu FQ, Xu YH, Zhang XY, Zhang DP (2007) Two calcium-dependent protein kinases, CPK4 and CPK11, regulate abscisic acid signal transduction in Arabidopsis. Plant Cell 19:3019–3036

Cell Biology of Copper

Christopher M. Cohu and Marinus Pilon

Abstract The transition metal, copper (Cu), is an essential micronutrient for normal plant growth and development. Copper is a cofactor of proteins involved in photosynthesis, respiration, ethylene perception, removal of superoxide radicals, and cell wall modification. The biochemical reactions catalyzed by most Cu enzymes in plants are known. However, in many cases we are not yet sure about the biological function of these Cu proteins. Copper delivery to Cu proteins has evolved with a set of evolutionarily conserved transporters and metallo-chaperones. Analysis of Cu transporter and metallo-chaperone loss of function mutants has increased our understanding of the localization and biological function of many Cu delivery mechanisms and target Cu proteins. Studies examining the regulation of Cu transporters, metallo-chaperones, and Cu proteins have revealed an elegant system to regulate Cu homeostasis. Copper in excess is toxic while Cu deficiency can lead to decreased photosynthetic activity and reproductive success. To avoid Cu deficiency or toxicity symptoms in a sub-optimal environment, plants are capable of directing Cu delivery based on their needs via regulation of Cu proteins and delivery systems. For many Cu proteins, a network of Cu microRNAs, under the control of a SPL7 transcription factor, orchestrates the prioritization of Cu delivery based on Cu availability.

1 Introduction

Copper (Cu) is an essential micronutrient for life, and it is important for many cellular processes in numerous organelles and compartments. In cells, Cu is found in two common states, Cu(I) (reduced) or Cu(II) (oxidized). Cu ions often act as

C.M. Cohu and M. Pilon (✉)
Biology Department, Colorado State University, Fort Collins, CO 80523-1878, USA
e-mail: pilon@lamar.colostate.edu

R. Hell and R.-R. Mendel (eds.), *Cell Biology of Metals and Nutrients*, Plant Cell Monographs 17,
DOI 10.1007/978-3-642-10613-2_3, © Springer-Verlag Berlin Heidelberg 2010

cofactors in Cu-proteins that are associated with electron transfer reactions and redox reactions involving oxygen (Linder and Goode 1991). However, the redox-active properties of Cu could also cause unwanted and uncontrolled reactions if it is left alone as a free ion in the cell. Free Cu ions could lead to the formation of toxic hydroxyl radicals, which are capable of damaging macromolecules (Halliwell and Gutteridge 1984). An extensive network of Cu transporters and metallo-chaperones has therefore evolved to bind and shuttle Cu in a manner that ensures proper Cu delivery to Cu proteins in various cellular organelles and compartments.

2 Functions of Cu Proteins in Plants

We know the biochemical reactions catalyzed by most Cu enzymes in plants. Remarkably, in many cases we are not yet sure about the biological function of these Cu proteins. Before describing how Cu is delivered to Cu proteins in plants we will discuss what we know about the functions of the Cu proteins.

2.1 Plastocyanin

The plastocyanin protein was first discovered in Algae (Katoh 1960). Plant (Poplar) plastocyanin was one of the first Cu proteins with a known crystal structure, which showed that the protein binds its Cu via a cysteine, a methionine and two histidine ligands (Colman et al. 1978). Plastocyanin is the most abundant protein in the thylakoid lumen where it functions as a mobile carrier of electrons from the cytochrome-b6/f complex to PSI (Kieselbach et al. 1998; Schubert et al. 2002). Therefore, it has a critical role in both linear and cyclic electron flows (Raven et al. 1999). A higher plant (*Silene pratensis*) plastocyanin cDNA sequence was first published in 1986 (Smeekens et al. 1986). *Arabidopis* has two plastocyanin genes (Vorst et al. 1993; Kieselbach et al. 2000) that encode proteins with highly similar secondary structures and seemingly redundant functions (Pesaresi et al. 2008). Unlike some cyanobacteria and green algae, plastocyanin is the only protein that can accept electrons from the cytochrome-b6/f complex in higher plants (Molina-Heredia et al. 2003), and *Arabidopsis* mutants with insertions in both plastocyanin genes are seedling-lethal on soil (Weigel et al. 2003). Arguably, plastocyanin is the most important Cu protein in photosynthetically growing plants.

2.2 Cytochrome c Oxidase

In mitochondria, Cu is required for the function of cytochrome *c* oxidase (COX), the proton-pumping terminal oxidase in the inner membrane (Carr and Winge 2003). This multi-subunit protein contains three Cu ions as cofactors in two Cu sites, in addition to heme. At least ten subunits make up the cytochrome *c* oxidase of higher plants. In yeast, three conserved mitochondrial encoded subunits (COX1, 2, and 3) form the core of the enzyme and contain the Cu binding sites (Carr and Winge 2003). Coordinated expression of nuclear and mitochondrial encoded COX genes is required for cytochrome *c* oxidase activity. Nuclear encoded COX subunit genes may help control tissue specific activity of this enzyme (Welchen et al. 2004). It is well established that a lower cytochrome *c* oxidase activity is one effect of severe Cu deficiency in plants (Marschner 1995). Along with COX, plants have the Fe-containing alternative oxidase (AOX). This protein accepts electrons from ubiquinone and reduces O_2 to water without proton pumping. AOX activity may prevent over-reduction of the quinone pool in cases where downstream electron transport is less efficient as a result of a stressful environment (Clifton et al. 2006).

2.3 *Cu/Zn Superoxide Dismutase*

Together with Zn, Cu is also a cofactor of Cu/Zn superoxide dismutase (Cu/ZnSOD) proteins that function in reactive oxygen species metabolism (Bowler et al. 1992). SOD enzymes catalyze the conversion of two superoxide ions and two protons to peroxide and molecular oxygen. Three genes encode for Cu/ZnSOD in the *Arabidopsis* genome: CSD1 is active in the cytosol, CSD2 in the stroma, and CSD3 in peroxisome (Kliebenstein et al. 1998). Homologs of these Cu/ZnSOD genes have been found in the genomes of other plant species, but they are not found in *Chlamydomonas*.

2.4 *Ethylene Receptors*

The ethylene receptors are Cu-binding proteins (Rodriguez et al. 1999) that are active in an early endomembrane system compartment, most likely the endoplasmic reticulum (Chen et al. 2002). For proper ethylene perception and responses, Cu must be delivered to the lumen of the endomembrane system (Hirayama et al. 1999).

2.5 *Phytocyanins*

The phytocyanins constitute a plant specific blue Cu protein family. Phytocyanins are structurally similar to plastocyanin and bind a single Cu per polypeptide. Phytocyanins include plantacyanin, stellacyanin, and uclacyanin (Nersissian et al. 1998). These proteins differ from plastocyanin in that their precursors harbor signal peptides that direct them into the endoplasmic reticulum and secretory pathway. The biological role of these phytocyanins is not fully clear. It has been suggested that some may mediate lignin polymerization (Nersissian et al. 1998). In Lilly, plantacyanin functions as a signaling molecule in the transmitting tract of the pistil (Kim et al. 2003), and evidence in *Arabidopsis* also suggests a role in reproduction (Dong et al. 2005).

2.6 Laccase and Ascorbate Oxidase

Multi-copper oxidases (MCO) form a super-family of proteins that bind four Cu ions (Nakamura and Go 2005) and include the ferroxidases (in yeast and *Chlamydomonas*), ceruloplasmin (in mammals), ascorbate oxidase, and laccases (in plants and fungi). In plants, apoplastic ascorbate oxidases were shown to have roles in cell expansion, plant biomass production, and salt tolerance (Pignocchi et al. 2003; Yamamoto et al. 2005). Laccase catalyzes the oxidation of a suitable substrate molecule (phenols and aromatic or aliphatic amines) to the corresponding reactive radicals with the production of water and oligomers (Gavnholt and Larsen 2002). In plants, laccases are apoplastic and encoded by a multi-gene family with 17 genes in *Arabidopsis* (McCaig et al. 2005; Cai et al. 2006). Different laccases are expressed in different organs and at different developmental stages of growth in *Arabidopsis* (McCaig et al. 2005). Insertion mutants for most laccases do not show a phenotype except for Lac2 (altered root elongation during de-hydration), Lac8 (early flowering), and Lac 15 (altered seed color) (Cai et al. 2006). The mutation in Transparent Testa-10 (TT10), an *Arabidopsis* mutant with a lack in seed color, was mapped to Lac15 which is expressed in the developing seed and functions in the formation of proanthocyanidin or tannin (Pourcel et al. 2005). Another possible role of laccase is in Fe acquisition (Hoopes and Dean 2004). All other proposed biological functions of laccases in plants can be summarized as "cell wall modeling"; this includes roles in lignin synthesis, maintenance of cell wall structure and integrity, response to stress, and wound healing (Sterjiades et al. 1992; Bao et al. 1993; Dean and Eriksson 1994; Ranocha et al. 2002; Liang et al. 2006).

2.7 Polyphenol Oxidase

Polyphenol oxidase (PPO) or tyrosinase was the first discovered Cu enzyme in plant plastids (Arnon 1949). It is found in the thylakoid lumen and contains a dinuclear Cu center. PPO catalyzes the conversion of monophenols to ortho-diphenols and ortho-dihydroxyphenols to ortho-quinones, resulting in black or brown pigment deposits (for review see Mayer 2006). In tomato, the PPOs are encoded by a gene family with seven members (Newman et al. 1993) that are differentially expressed (Thipyapong et al. 1997). Wounding, stress, pathogen, and herbivore attack have been shown to induce PPO activity in different plant species, suggesting a role for PPO in plant resistance to stress and pathogens (Mayer 2006). PPO is not ubiquitous and there is no homolog for PPO in *Arabidopsis* (Schubert et al. 2002).

2.8 Amine Oxidase

Amine oxidases contain a single Cu as well as a special topa quinone cofactor that is formed by post-translational modification of a conserved tyrosine residue (Kumar et al. 1996). These secreted enzymes catalyze the oxidative deamination of primary amines to aldehydes in a reaction that requires free radicals and also produces hydrogen-peroxide (Frebort et al. 2000). Poly-amines such as spermine are likely substrates for these enzymes. The Cu atom is bound by three histidine residues and is required not only for the post-translational formation of the topa quinone cofactor but also for the regular catalytic cycle (Kumar et al. 1996). Amine oxidases are reported to be the most abundant Cu proteins in the apoplastic space of peas. Proposed functions include roles in cell wall differentiation, which in turn could be significant for stomatal closure (An et al. 2008), wound healing (Rea et al. 2002; Angelini et al. 2008), and responses to pathogen attack (Rea et al. 2002; Marina et al. 2008). The observed amine oxidase expression pattern and timing in tobacco would be consistent with a role in either peroxide dependent protein cross-linking or lignification (Paschalidis and Roubelakis-Angelakis 2005).

2.9 Other Roles of Cu in Plants

While determining the structure of CNX1, an enzyme that functions in molybdenum cofactor synthesis, it was found that a Cu ion temporarily occupies the site for molybdenum insertion in the bound molybdopterin substrate (Kuper et al. 2004). This observation now links Cu metabolism to nitrogen assimilation and phytohormone biosynthesis, which are functions of molybdenum cofactor requiring enzymes. Cu may also play a role in thylakoid grana stacking (Bernal et al. 2006). Unlike what has been found for yeast and *Chlamydomonas*, there is no direct requirement for Cu in Fe acquisition.

3 Cu Movement in and out of Root Cells

3.1 Cu Uptake

In root cells, Cu enters the cytosol by a cell membrane COPT-family transporter (Kampfenkel et al. 1995). The family of COPT transporters belongs to a highly conserved Ctr-like Cu transporter family also found in yeast and humans

(Dancis et al. 1994). COPT transporters have three transmembrane domains, a likely *N*-terminal metal binding domain, and an essential MXXXM transmembrane domain (Puig et al. 2002). *Arabidopsis* encodes five COPT transporters (COPT1 – 5). Of these, four are expressed and these proteins likely import Cu into the cytosol, though their sub-cellular location is not determined. Information on COPT1 promoter fusion and antisense lines suggest that COPT1 is involved in Cu uptake from the surrounding growth medium at root tips (Sancenon et al. 2004). COPT1 antisense lines exhibited elongated root growth when compared to wild-type plants, and the phenotype could be partially restored upon Cu feeding in the medium (Sancenon et al. 2004). COPT2 is also likely involved in cellular uptake considering its expression in root and leaf tissues, along with up-regulation of transcripts during limited Cu growth, similarly seen for COPT1 (Sancenon et al. 2003). COPT3 and COPT5 are highly expressed in aerial tissues (Sancenon et al. 2003), and may serve to transport Cu from intracellular stores.

Ctr-like proteins transport Cu in its reduced form (Eisses and Kaplan 2005), but most extracellular Cu in soil is oxidized as Cu(II). *Arabidopsis* and dicot species utilize root surface ferric reductases, such as FRO2, for uptake of Fe in its reduced form (Robinson et al. 1999). It is also possible that ferric reductases could reduce Cu for import (Welch et al. 1993). When plants are fed an excess of Cu, Fe concentrations decrease; the opposite is also true during limited Cu growth (Welch et al. 1993; Chen et al. 2004). Interestingly, FRO3, localized in roots and vasculature, exhibits increased expression during Cu deficient growth (Mukherjee et al. 2006). However, FRO activity has not been reported to reduce Cu(II). In addition to COPT transporters, ZIP2 and ZIP4 (ZIP family transporters) have been reported to complement the yeast *ctr1* mutant, that is deficient in Cu uptake (Wintz et al. 2003). ZIP2 transcript expression is highest in root tissue while ZIP4 expression is high in both root and leaf tissue, and they respond to Cu status (Wintz et al. 2003).

3.2 Cu Export and Intercellular Reallocation

The HMA5 (Heavy Metal Associated 5) Cu transporter likely supplies Cu to apoplastic Cu oxidases and laccases (see below). It also plays an important role in removing excess Cu from the cytosol of root tissues (Andrés-Colas et al. 2006; Kobayashi et al. 2008). Root tissues in *hma5* loss of function mutants accumulate elevated levels of Cu when compared to wild-type plants, and *hma5* mutants are more sensitive to Cu feeding (Andrés-Colas et al. 2006). Delivery of Cu within the cytosol of plants to RAN1 (HMA7, see below) (Hirayama et al. 1999) and HMA5 may be accomplished by two homologs of the yeast Atx1 Cu chaperone, ATX1 (Andrés-Colas et al. 2006; Puig et al. 2007) and CCH (Himelblau et al. 1998). Both ATX1 and CCH from *Arabidopsis* are able to complement the yeast *atx1* mutant, and they interact with the *N*-terminal domain of *Arabidopsis* HMA5 and RAN1 (ATX1 only) in a yeast two hybrid system (Andrés-Colas et al. 2006; Puig et al. 2007). The yeast

Atx1 and ATX1 in *Arabidopsis* are similar, however, CCH contains an added plant specific *C*-terminal extension (Mira et al. 2001a; Puig et al. 2007). Interestingly, this *C*-terminal addition negatively affects interactions with HMA5, but a positive interaction was observed when the *C*-terminal region of CCH was removed (Andrés-Colas et al. 2006; Puig et al. 2007). CCH has been found in phloem-endonucleated cells, and it is possible that the additional *C*-terminal region allows for symplastic intercellular Cu trafficking through plasmodesmata (Mira et al. 2001b; Andrés-Colas et al. 2006). Up-regulation of ATX1 and CCH has been reported for plants undergoing Cu deficiency, senescence, mechanical and oxidative stress, along with jasmonic acid treatments in *Arabidopsis* (Himelblau et al. 1998; Mira et al. 2001b; Puig et al. 2007), and in poplar (Lee et al. 2005).

3.3 Root to Shoot Cu Translocation

Since HMA5 is involved with Cu movement from the symplast to apoplast, and is highly expressed in roots, it is possible that HMA5 is also involved in transporting Cu into the xylem. If so, it is not the only mechanism to load Cu into the xylem considering that *hma5* loss of function mutants were able to maintain much of the Cu translocation to shoot tissues (Andrés-Colas et al. 2006). No other mechanism for Cu loading into the xylem has been suggested or identified. Once in the xylem, long distance Cu translocation to aerial tissues may involve the chelator nicotianamine. As a methionine-derived compound, nicotianamine chelation of Fe in xylem sap for translocation has been suggested (for review see Briat et al. 2007). Nicotianamine has also been shown to have a high affinity for Cu binding in tomato xylem sap, and less than 0.5% of total xylem Cu was found as free Cu(II) ions (Liao et al. 2000). This suggests that Cu in xylem sap is mostly chelated. The tomato mutant *chloronerva*, which lacks nicotianamine, also supports the idea that nicotianamine is involved with long distance transport of heavy metals. *chloronerva* mutant plants exhibit increased Cu concentrations in root tissues and decreased xylem and shoot levels compared to wild-type (Pich and Scholz 1996). Upon application of nicotianamine to these mutants it was observed that root Cu concentrations decreased while xylem and shoot levels increased, especially in young leaves (Pich and Scholz 1996). In addition, tobacco plants over-expressing a nicotianamine aminotransferase (NAAT) gene, which creates a nicotianamine shortage in tobacco, led to Cu deficiencies in leaves and problems associated with reproduction (Takahashi et al. 2003).

3.4 Excess Cu

In some cases plant cells may have to deal with excessive Cu. Plants such as *Arabidopsis* do not accumulate high levels of Cu in tissues and are often sensitive to elevated Cu. During sub-toxic Cu excess, plants may be able to chelate Cu using a cysteine-rich metallothionein (MT). *Arabidopsis* contains several MT genes, some of which are up-regulated during Cu excess (Zhou and Goldsbrough 1994; Guo et al. 2003). Another possible Cu chelator is phytochelatins, which are derived from glutathione (for review see Cobbett and Goldsbrough 2002). When plants lack both MT1a/MT2b and phytochelatin they exhibit a more severe phenotype on elevated Cu than MT or phytochelatin mutants alone (Guo et al. 2008). Simply moving Cu out of the cell may also help maintain normal cellular Cu levels. Considering that *hma5* mutants are sensitive to Cu feeding, HMA5 is likely involved in detoxifying cells of excess Cu by moving the ions into extracellular spaces (Andrés-Colas et al. 2006; Kobayashi et al. 2008). HMA5, COPT1, and COPT2 transporters are regulated by Cu differently. HMA5 increases during Cu excess (Andrés-Colas et al. 2006) while the Cu importers COPT1 and COPT2 decrease (Sancenon et al. 2003, 2004), consistent with the role of HMA5 in preventing excess ions in the cell while also avoiding Cu toxic conditions for neighboring cells.

4 Intracellular Cu Delivery to Cu Protein Targets

4.1 Chloroplast: Cu Import into the Chloroplast

Import of Cu into stroma and thylakoid lumen is the most understood of any organelle in plants. The inner envelope membrane contains a metal-transporting P-type ATPase for *Arabidopsis*, PAA1 (HMA6), and is responsible for Cu import into the stroma (Tabata et al. 1997; Shikanai et al. 2003); while PAA2 (HMA8) imports Cu from the stroma into the thylakoid lumen (Abdel-Ghany et al. 2005; Bernal et al. 2007). Both are P_{1B} type pumps and members of the Heavy Metal Associated (HMA) transporter family (Axelsen and Palmgren 2001; Baxter et al. 2003). There are eight members in the HMA family. HMA1 to 4 are classified as possible Zn, Cd, Co, and Pb transporters; while HMA5 to 8 are classified as Cu and Ag transporters (Arguello 2003; Baxter et al. 2003). Both PAA1 and PAA2 Cu transporters have sub-cellular targeting information in the *N*-terminal region of the peptide, but the mechanism of protein import is not yet identified (Abdel-Ghany et al. 2005). PAA1 and PAA2 Cu transporters have eight predicted transmembrane domains with a heavy metal binding motif in the *N*-terminal region. In addition, they contain ATP binding, phosphatase, phosphorylation, and transmembrane CPC (amino acid) ion transduction domains (Mandal et al. 2004; Abdel-Ghany et al. 2005). Upon Cu metal binding and phosphorylation of P_{1B} type transporters, the Cu ion is transported across the membrane through changes in protein conformation (Arguello et al. 2007; González-Guerrero and Argüello 2008). Transport of heavy metals in most of these P_{1B} type transporters is thought to initiate in the sub-cellular compartment containing the heavy metal binding *N*-terminal region (Arguello et al. 2007; González-Guerrero and Argüello 2008). This would place the *N*-terminal domains for PAA1 and PAA2 in the chloroplast envelope intermembrane space and stroma respectively. However, the orientation and mechanism for accepting and donating Cu is not yet known for PAA1 and PAA2.

Plastocyanin import into chloroplasts and thylakoids is conducted using the Tic/Toc and SecA/SecY-mediated pathways (for review see Schnell 1998), which translocate proteins in an unfolded state. Upon import, plastocyanin acquires its Cu cofactor for final assembly and stability. Mutants with impaired Cu transport (*paa1* and *paa2* loss of function mutants) exhibited reduced plastocyanin accumulation even though transcript levels remained high (Abdel-Ghany et al. 2005); supporting that plastocyanin requires Cu for final assembly and for stability, as similarly suggested for *Chlamydomonas* (Li and Merchant 1995). Interestingly, even though both plastocyanin forms are seemingly similar in function (Pesaresi et al. 2008), new evidence suggests that plastocyanin 2 (PC2) accumulates during increased Cu feeding even though photosynthetic benefits were not observed. Plastocyanin could, therefore, have a secondary role as a Cu buffer (Abdel-Ghany 2009). The mechanism for Cu delivery between PAA2 and plastocyanin has not been determined, and a Cu chaperone in the thylakoid lumen has not been identified. It is possible that plastocyanin receives its Cu directly from PAA2 or from a Cu pool. Along with PAA1, another possible Cu transporter in the chloroplast envelope membrane is, HMA1, that may supply some Cu to CSD2 in the stroma (Seigneurin-Berny et al. 2006). Though HMA1, PAA1, and PAA2 are in the HMA family, they do contain some differences. Unlike PAA1 and PAA2, HMA1 does not have conserved MxCxxC *N*-terminal heavy metal binding domains; instead it contains a poly-histidine domain. In addition, HMA1 contains a SPC ion transduction domain instead of CPC found in PAA1 and PAA2 (Axelsen and Palmgren 2001). Chloroplast Cu concentrations and SOD activity levels decrease in *hma1* mutants, and a photo-oxidative stress phenotype was reported when plants were grown in elevated light (Seigneurin-Berny et al. 2006). However, *hma1* plants did not exhibit defects in total plastocyanin levels. Instead, it was suggested that a decrease in CSD2 activity led to the phenotype observed (Seigneurin-Berny et al. 2006). A defect in plastocyanin levels was observed in a *paa1* mutant that also exhibited an electron transport phenotype (Shikanai et al. 2003; Abdel-Ghany et al. 2005). In the case of *paa1*, the phenotype was partially restored by Cu feeding (Shikanai et al. 2003; Abdel-Ghany et al. 2005), but the *hma1* mutant phenotype was not. (Seigneurin-Berny et al. 2006). It is possible that HMA1, with lower Cu transport activity, mediates the delivery of Cu to plastocyanin in *paa1* mutants during Cu feeding, but the link, if any, between HMA1 and Cu transport to plastocyanin is still unclear.

Like plastocyanin, Cu/ZnSODs also require Cu for final assembly, activity, and stability. When Cu delivery to the chloroplast stroma is disrupted by a *paa1* loss of function mutant, CSD2 proteins do not accumulate to wild-type levels while CSD2 transcript levels increase (Abdel-Ghany et al. 2005). The Cu Chaperone for SOD (CCS) delivers Cu to Cu/ZnSODs (Culotta et al. 1997), and is also active in the cytosol and plastids of plants (Chu et al. 2005). The stromal and cytosolic versions of CCS are encoded by one gene in *Arabidopsis* with two in-frame ATG sites that span a chloroplast transit peptide (Chu et al. 2005). In a T-DNA knock-out mutant of *CCS* (CCS-KO), Cu delivery to Cu/ZnSODs was dramatically reduced and Cu/ZnSOD proteins did not accumulate, further suggesting that Cu/ZnSODs require Cu delivery for protein stability (Chu et al. 2005). It is not known if CCS in stroma acquires Cu directly from PAA1 for delivery to Cu/ZnSOD.

While CSD1 and CSD2 receive their Cu from CCS in the compartment in which they are active, it is likely that CSD3 in the peroxisome does not. CSD3 has a peroxisomal targeting sequence but it likely receives its Cu cofactor in the cytosol prior to import since the peroxisome can import proteins in a folded state. Complementation of a CCS loss of function mutant (CCS-KO) using a CCS version without the chloroplast targeting sequence rescues both CSD1 and CSD3 activities, but not CSD2 in the stroma (Chu et al. 2005).

4.2 Delivery of Cu to other Compartments

4.2.1 Mitochondria

The Cu delivery mechanisms involved in the mitochondria are known mainly for yeast and mammalian cells; however, some homologous proteins have been found in plants. Delivery of Cu to cytochrome *c* oxidase in yeast is accomplished by Cox11, Cox17, Cox19, and Sco1 (Carr and Winge 2003). The mitochondrial matrix in yeast stores Cu as a pool of soluble low molecular weight ligand complexes (Cobine et al. 2004). It was suggested that the Cu pool in the matrix may supply Cu to the intermembrane space and Cox17. Cox17 is a metallo-chaperone that delivers Cu to Cox11 and Sco1 in the intermembrane space, which in turn deliver Cu to different cytochrome *c* oxidase subunits (Horng et al. 2004). *Arabidopsis* functional homologs for AtCOX17 (Balandin and Castresana 2002) and AtCOX19 (Attallah et al. 2007) have been identified. However, homologs of Cox11 and Sco1 chaperones in plants, along with Cu transport across the mitochondria membranes, have not been characterized.

4.2.2 Endomembrane and Secretory Pathway

The RAN1 (HMA7) (responsive-to-antagonist 1) Cu transporter is a functional homolog of yeast and human P-type ATPase Cu transporters active in the endomembrane system (Hirayama et al. 1999). Homologs of RAN1 in yeast and mammals act in Cu transport from the cytosol into the secretory pathway (Lutsenko et al. 2007). A mild loss of function in a *ran1* mutant in *Arabidopsis* lowered the plant's ability to respond to an antagonistic ethylene signal. This evidence suggests that RAN1 is involved in Cu delivery to ethylene receptors (Hirayama et al. 1999). A dramatic loss of RAN1 function led to phenotypes associated with reduced cell wall elongation (Woeste and Kieber 2000), perhaps due to defects in extracellular Cu oxidases and laccases.

Another member of the HMA family of Cu transporters, homologous to RAN1, is HMA5 (Williams and Mills 2005). HMA5 is mainly localized in root and flower tissues. In *hma5* loss of function mutants, a phenotype associated with cell wall elongation was observed. However, *hma5* had no defects in ethylene reception, as observed for *ran1* (Hirayama et al. 1999; Woeste and Kieber 2000; Andrés-Colas et al. 2006). Together, RAN1 and HMA5 could be involved in supplying Cu to many or all extracellular Cu proteins, but their specificity likely involves organ and intracellular locations relative to where extracellular oxidases and laccases receive their Cu.

5 Senescence, Reallocation, and Delivery to Reproductive Tissues

Copper is not readily reallocated from older leaves to younger tissues. During Cu deficiency, young leaves, shoot meristems, and reproductive tissues are affected before older leaves show signs (Marschner 1995). Therefore, a significant proportion of the Cu allocated to reproductive tissues likely comes directly from the roots (Waters and Grusak 2008). However, there are several indications that some Cu from older leaves is reallocated to newer leaves and reproductive tissues by chelators via symplastic movement. During senescence, the transcripts of the CCH chaperone increase (Mira et al. 2001b). As a chelator with a putative *C*-terminal peptide to facilitate symplastic movement, CCH may chelate Cu in the cytosol during senescence for movement to vascular bundle tissues or movement within the phloem (Mira et al. 2001b; Andrés-Colas et al. 2006). Up-regulation of MT1 has also been reported during senescence (Mira et al. 2002), and it is possible that MT1a and MT2b are involved in phloem reallocation of Cu (Guo et al. 2003).

Another mechanism for reallocating Cu likely involves nicotianamine and the Yellow Stripe-Like (YSL) transporters. In addition to the xylem, nicotianamine is also found in phloem sap. Nicotianamine is a precursor of phytosiderophores which together with YSL transporters is involved in the strategy-II Fe uptake in monocot roots (Briat et al. 2007). Dicots, like *Arabidopsis*, use a strategy-I Fe uptake system that utilizes the root surface FRO2 ferric reductase and the IRT1 ZIP-family transporter, yet *Arabidopsis* encodes eight YSL transporters (Briat et al. 2007). These YSL transporters, likely function to import nicotianamine metal complexes (Schaaf et al. 2004; DiDonato et al. 2004), which could then act as a metal ion redistribution system between tissues via the phloem (Briat et al. 2007; Waters and Grusak 2008). In a *ysl1 ysl3* double loss of function mutant, Cu concentrations in seeds were reduced by 82% when compared to the parental line (Waters et al. 2006). In addition, this mutant line did not efficiently reallocate Cu and Fe from rosette and cauline leaves (Waters and Grusak 2008).

For proper seed set adequate Cu is required (Marschner 1995; Epstein and Bloom 2005). Cu delivery to cells involved in reproduction likely receives most of their Cu directly sent from the roots, but it appears that Cu delivery is via the xylem and the phloem (see above). Cu, along with other metals, bound to nicotianamine could be imported by YSL transporters, and moved through the symplast by CCH. Cells that do not have plasmodesmata for intercellular Cu trafficking by CCH, like pollen, would require COPT1, which is highly expressed in pollen (Sancenon et al. 2004). COPT1 mutant plants with reduced expression exhibit defects in pollen development (Sancenon et al. 2004). Extracellularly, plantancyanin has been implicated in pollen tube guidance (Kim et al. 2003; Dong et al. 2005), and HMA5 is highly expressed in flowering tissues, most likely pollen (Andrés-Colas et al. 2006). HMA5 may deliver Cu to plantacyanin in pollen, but this connection between the two has not been verified experimentally.

6 Regulation of Copper Homeostasis

Plants that are Cu deficient exhibit photosynthetic deficiencies, shoot apical meristem death, curling of leaves, and poor seed set (Marschner 1995; Yruela et al. 1996). To avoid Cu deficiency or toxicity symptoms in a sub-optimal environment, plants are capable of directing Cu delivery based on needs via regulation of Cu delivery systems. Copper toxicity leads to increased expression of some transporters and Cu chelators (see above). However, during Cu deficiency, post-transcriptional regulation of many Cu proteins is mediated by microRNA directed cleavage of Cu protein mRNA. Transcriptional activation of microRNAs, and possibly transporters, during Cu limited growth is mediated by a SPL7 transcription factor. Together, this mechanism to down-regulate Cu proteins and delivery systems may allow for prioritized delivery to the most essential Cu proteins during limited Cu availability.

6.1 Transcription Factors

Transcriptional responses to Cu require transcription factors that can sense Cu. *Chlamydomonas* are capable of switching between two functionally similar photosynthetic proteins, cytochrome *c6* (heme protein) and plastocyanin (Cu-protein), when Cu levels are limited or sufficient, respectively. This switch is mediated by the transcription factor copper response regulator (CRR1) that activates transcription of the cytochrome *c6* when Cu is limited (Kropat et al. 2005). Interestingly, *Crr1* mRNA and expression is not regulated by Cu, suggesting a post-translational change in the CRR1 protein in response to Cu availability (Kropat et al. 2005). While higher plants cannot substitute plastocyanin with cytochrome *c6*, Cu/ZnSOD and FeSOD regulation exhibit a similar reciprocal expression pattern (Abdel-Ghany et al. 2005; Cohu and Pilon 2007; Yamasaki et al. 2007).

Higher plants contain a homolog to CRR1 know as SPL7 which has recently been shown to be a key regulator of Cu homeostasis by binding to GTAC promoter core motifs (Yamasaki et al. 2009). There are 12 members in the *Arabidopsis* SPL family that contain a conserved SBP (*SQUAMOSA* promoter-binding protein) DNA binding domain and a nuclear localization signal (Cardon et al 1999). SPL transcription factors have been reported to be involved in development and nutrient homeostasis. In *Arabidopsis*, SPL7 likely mediates regulation of some Cu, Zn, and Fe transporters. Wild-type plants increase *COPT1*, *COPT2*, *ZIP2*, *FRO3*, and *YSL2* mRNA when Cu is limited, but in a *spl7* mutant the mRNA of these transporters did not increase (Yamasaki et al. 2009). The *YSL2* promoter contains 5 GTAC core motifs indicating that the transporter may be directly regulated by Cu via SPL7. The Cu-chaperone CCS, which has been shown to decrease during Cu deficiency (Wintz et al. 2003), did not decrease in the *spl7* mutant (Yamasaki et al. 2009). When the *spl7* mutant was grown on low Cu it exhibited a severe growth phenotype, supporting that SPL7 is an important regulator during Cu-limitation. On the other hand, HMA5 and FRO6, which are regulated by Cu, and ATX1 (constitutively expressed), were not identified as being regulated by SPL7 (Yamasaki et al. 2009). It is possible that yet another Cu sensitive regulatory mechanism for HMA5 and FRO6 exists. While SPL7 may regulate some Cu transporters and chaperones directly, SPL7 has also been shown to activate specific microRNA transcription during Cu limited growth, leading to the cleavage of many Cu protein mRNAs (Yamasaki et al. 2009). In the *spl7* mutant, miR397, miR398, miR408, and miR857 (the Cu microRNAs) were not detected even when Cu was limited (Fig. 1).

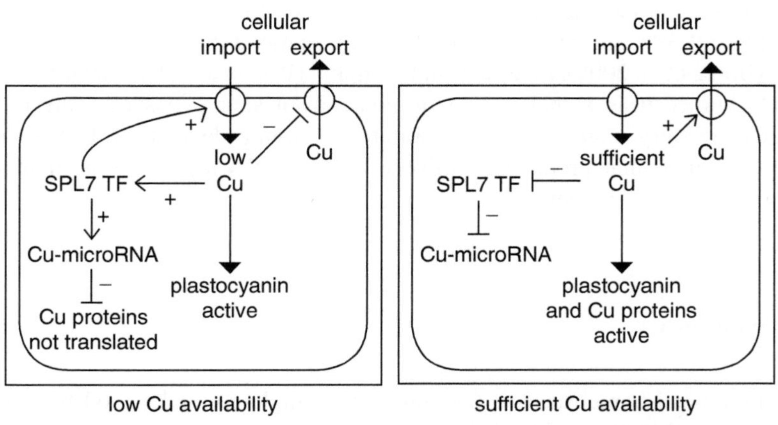

Fig. 1 Model of Cu homeostasis regulation during low and sufficient Cu availability. *SQUAMOSA* promoter-binding protein like-7 transcription factor (SPL7 TF) and open circles for plasma membrane transporters are indicated. Cu uptake, export, and delivery are represented with closed arrow heads. Up-regulation of Cu delivery proteins is represented by open arrows and (+) to indicate activation, while down-regulation of proteins is represented by a perpendicular line and (−)

6.2 The Cu microRNAs

MicroRNAs belong to a highly conserved group of small 20-21-nt RNAs that can disrupt mRNA translation by guiding the cleavage of target mRNAs (Jones-Rhoades and Bartel 2004; Jones-Rhoades et al. 2006). miR398 was the first microRNA shown to target mRNAs that encode Cu/Zn superoxide dismutases in the cytosol (CSD1) and the chloroplast (CSD2) of *Arabidopsis* (Sunkar et al. 2006). Oxidative stress was shown to reduce mature miR398 levels that led to increased CSD1 and CSD2 mRNA and enzyme activity. During non-stress growth conditions, Cu availability was also shown to regulate Cu/ZnSOD expression and activity (Abdel-Ghany et al. 2005; Cohu and Pilon 2007). Linking Cu availability with the regulation of CSD1 and CSD2 by miR398 was established when Cu-supplemented *Arabidopsis* plants, demonstrated an absence of miR398 while CSD1 and CSD2 mRNA abundance increased (Yamasaki et al. 2007). The transcripts of plantacyanin and several members of the laccase family were identified as targets of miR397, miR408, and miR857 directed cleavage during Cu-limited growth (Abdel-Ghany and Pilon 2008). Together these studies suggest that Cu microRNA mediated down-regulation of many Cu-proteins is a mechanism to allow for Cu delivery to the most essential of the Cu-proteins, such as plastocyanin. Another interesting observation of microRNA in Cu homeostasis was the observation that sucrose in tissue culture medium elevated miR398 levels regardless of Cu levels (Dugas and Bartel 2008). This suggests that there is additional regulation on Cu microRNAs from other signaling sources.

SPL7 is mainly found in the roots yet microRNAs are found throughout the plant and sometimes only in above-ground tissues (Yamasaki et al. 2009; Abdel-Ghany and Pilon 2008). High expression of SPL7 in the roots indicates a role in detecting Cu availability at the site of Cu entry, then orchestrating whole plant Cu delivery. Recently, miR398, among other microRNAs, was found in the phloem of *Brassica napus* (Buhtz et al. 2008), rapeseed, and pumpkin (Pant et al. 2008), which suggests that Cu homeostasis signals could originate from source tissues. This method of signal delivery could be very important for young developing leaves during Cu-limitation so that proper Cu delivery to essential Cu-proteins is maintained during initial development.

7 Overview

Transition metal homeostasis is perhaps more completely understood for Cu than any other metal in plants. Identification of Cu transporters and metallo-chaperones that are important in Cu delivery to ethylene receptors, for photosynthesis, and Cu movement into and out of the cell has provided a more complete understanding of Cu homeostasis mechanisms. Information on Cu delivery mechanisms has also allowed for studies that examine how Cu homeostasis is regulated during development and the changing Cu status, both at a cellular and whole plant level. While Cu delivery mechanisms and regulation of Cu homeostasis is becoming clearer, the biological function of Cu proteins remains unclear in many cases. As the regulation pattern for Cu proteins and delivery mechanisms is resolved, it may help in identifying the biological role of many Cu proteins, both essential and seemingly non-essential.

Acknowledgements Work in the author's laboratory was supported by grants from the U.S. National Science Foundation to MP (NSF-IBN-0418993; NSF-IOS-0847442). We apologize to colleagues whose work could not be cited due to space limitations.

References

Abdel-Ghany SE (2009) Contribution of plastocyanin isoforms to photosynthesis and copper homeostasis in *Arabidopsis thaliana* grown at different copper regimes. Planta 229:767–779

Abdel-Ghany SE, Pilon M (2008) MicroRNA-mediated systemic down-regulation of copper protein expression in response to low copper availability in *Arabidopsis*. J Biol Chem 283:15932–15945

Abdel-Ghany SE, Müller-Moulé P, Niyogi KK, Pilon M, Shikanai T (2005) Two P-type ATPases are required for copper delivery in *Arabidopsis thaliana* chloroplasts. Plant Cell 17:1233–1251

An Z, Jing W, Liu Y, Zhang W (2008) Hydrogen peroxide generated by copper amine oxidase is involved in abscisic acid-induced stomatal closure in *Vicia faba*. J Exp Bot 59:815–825

Andrés-Colas N, Sancenon V, Rodriguez-Navarro S, Mayo S, Thiele DJ, Ecker JR, Puig S, Penarrubia L (2006) The *Arabidopsis* heavy metal P-type ATPase HMA5 interacts with metallochaperones and functions in copper detoxification of roots. Plant J 45:225–236

Angelini R, Tisi A, Rea G, Chen MM, Botta M, Federico R, Cona A (2008) Involvement of polyamine oxidase in wound healing. Plant Physiol 146:162–177

Arguello JM (2003) Identification of ion-selectivity determinants in heavy-metal transport P_{1B}-type ATPases. J Membrane Biol 195:93–108

Arguello JM, Eren E, Gonzalez-Guerrero M (2007) The structure and function of heavy metal transport P(1B)-ATPases. Biometals 20:233–248

Arnon DI (1949) Copper enzymes in isolated chloroplasts. Polyphenoloxidase in Beta vulgaris. Plant Physiol 24:1–15

Attallah CV, Welchen E, Pujol C, Bonnard G, Gonzalez DH (2007) Characterization of *Arabidopsis thaliana* genes encoding functional homologues of the yeast metal chaperone Cox19p, involved in cytochrome *c* oxidase biogenesis. Plant Mol Biol 65:343–355

Axelsen KB, Palmgren MG (2001) Inventory of the superfamily of P-type ion pumps in *Arabidopsis*. Plant Physiol 126:696–706

Balandin T, Castresana C (2002) AtCOX17, an *Arabidopsis* homolog of the yeast copper chaperone COX17. Plant Physiol 129:1852–1857

Bao W, O'Malley DM, Whetten R, Sederoff RR (1993) A laccase associated with lignification in Loblolly Pine xylem. Science 260:672–674

Baxter I, Tchieu J, Sussman MR, Boutry M, Palmgren MG, Gribskov M, Harper JF, Axelsen KB (2003) Genomic comparison of P-type ATPase ion pumps in *Arabidopsis* and rice. Plant Physiol 132:618–628

Bernal M, Ramiro MV, Cases R, Picorel R, Yruela I (2006) Excess copper effect on growth, chloroplast ultrastructure, oxygen-evolution activity and chlorophyll fluorescence in Glycine max cell suspensions. Physiol Planta 127:312–325

Bernal M, Testillano PS, Alfonso M, del Carmen Risueño M, Picorel R, Yruela I (2007) Identification and subcellular localization of the soybean copper P1B-ATPase GmHMA8 transporter. J Struc Biol 158:46–58

Bowler C, Van Montagu M, Inze D (1992) Superoxide dismutase and stress tolerance. Ann Rev Plant Physiol Plant Mol Biol 43:83–116

Briat JF, Curie C, Gaymard F (2007) Iron utilization and metabolism in plants. Curr Opin Plant Biol 10:276–282

Buhtz A, Springer F, Chappell L, Baulcombe DC, Kehr J (2008) Identification and characterization of small RNAs from the phloem of *Brassica napus*. Plant J 53:739–749

Cai X, Davis EJ, Ballif J, Liang M, Bushman E, Haroldsen V, Torabinejad J, Wu Y (2006) Mutant identification and characterization of the laccase gene family in *Arabidopsis*. J Exp Bot 57:2563–2569

Cardon G, Hohmann S, Klein J, Nettesheim K, Saedler H, Huijser P (1999) Molecular characterisation of the *Arabidopsis* SBP-box genes. Gene 237:91–104

Carr HS, Winge DR (2003) Assembly of cytochrome *c* oxidase within the mitochondrion. Acc Chem Res 36:309–316

Chen YF, Randlett MD, Findell JL, Schaller GE (2002) Localization of the ethylene receptor ETR1 to the endoplasmic reticulum of *Arabidopsis*. J Biol Chem 277:19861–19866

Chen Y, Shi J, Tian G, Zheng S, Lina Q (2004) Fe deficiency induces Cu uptake and accumulation in Commelina communis. Plant Sci 166:1371–1377

Chu CC, Lee WC, Guo WY, Pan SM, Chen LJ, Li HM, Jinn TL (2005) A copper chaperone for superoxide dismutase that confers three types of copper/zinc superoxide dismutase activity in *Arabidopsis*. Plant Physiol 139:425–436

Clifton R, Millar AH, Whelan J (2006) Alternative oxidases in *Arabidopsis*: A comparative analysis of differential expression in the gene family provides new insights into function of non-phosphorylating bypasses. Biochim Biophys Acta 1757:730–741

Cobbett C, Goldsbrough P (2002) Phytochelatins and metallothioneins: roles in heavy metal detoxification and homeostasis. Annl Rev Plant Biol 53:159–182

Cobine PA, Ojeda LD, Rigby KM, Winge DR (2004) Yeast contain a non-proteinaceous pool of copper in the mitochondrial matrix. J Biol Chem 279:14447–14455

Cohu CM, Pilon M (2007) Regulation of superoxide dismutase expression by copper availability. Physiol Planta 129:747–755

Colman PM, Freeman HC, Guss JM, Murata M, Norris VA, Ramshaw JAM, Venkatappa MP (1978) X-ray crystal structure analysis of plastocyanin at 2.7 Å resolution. Nature 272:319–324

Culotta VC, Klomp LWJ, Strain J, Casareno RLB, Krems B, Gitlin JD (1997) The copper chaperone for superoxide dismutase. J Biol Chem 272:23469–23472

Dancis A, Yuan DS, Haile D, Askwith C, Eide D, Moehle C, Kaplan J, Klausner RD (1994) Molecular characterization of a copper transport protein in *S. cerevisiae*: an unexpected role for copper in iron transport. Cell 76:393–402

Dean JFD, Eriksson K-EL (1994) Laccase and the deposition of lignin in vascular plants. Holzforschung 48:21–33

DiDonato RJ, Roberts LA, Sanderson T, Eisley RB, Walker EL (2004) *Arabidopsis* yellow stripe-like2 (YSL2): a metal-regulated gene encoding a plasma membrane transporter of nicotianamine-metal complexes. Plant J 39:403–414

Dong J, Kim ST, Lord EM (2005) Plantacyanin plays a role in reproduction in *Arabidopsis*. Plant Physiol 138:778–789

Dugas DV, Bartel B (2008) Sucrose induction of *Arabidopsis* miR398 represses two Cu/Zn superoxide dismutases. Plant Mol Biol 67:403–417

Eisses JF, Kaplan JH (2005) The mechanism of copper uptake mediated by human CTR1, a mutational analysis. J Biol Chem 280:37159–37168

Epstein E, Bloom AJ (2005) Mineral nutrition of plants: principles and perspectives, 2nd edn. Sinauer Associates, Inc, Sunderland, MA

Frebort I, Sebela M, Svendsen I, Hirota S, Endo M, Yamauchi O, Bellelli A, Lemr K, Pec P (2000) Molecular mode of interaction of plant amine oxidase with the mechanism-based inhibitor 2-butyne-1, 4-diamine. European J Biochem 267:1423–1433

Gavnholt B, Larsen K (2002) Molecular biology of plant laccases in relation to lignin formation. Physiol Planta 116:273–280

González-Guerrero M, Argüello JM (2008) Mechanism of Cu+-transporting ATPases: soluble Cu+ chaperones directly transfer Cu+ to transmembrane transport sites. Proc Natl Acad Sci USA 105:5992–5997

Guo WJ, Bundithya W, Goldsbrough PB (2003) Characterization of the *Arabidopsis* metallothionein gene family: tissue specific expression and induction during senescence and in response to copper. New Phytol 159:369–381

Guo WJ, Meetam M, Goldsbrough PB (2008) Examining the specific contributions of individual *Arabidopsis* metallothioneins to copper distribution and metal tolerance. Plant Physiol 146:1697–1706

Halliwell B, Gutteridge JM (1984) Oxygen toxicity, oxygen radicals, transition metals and disease. Biochem J 219:1–14

Himelblau E, Mira H, Lin SJ, Culotta VC, Penarrubia L, Amasino RM (1998) Identification of a functional homolog of the yeast copper homeostasis gene ATX1 from *Arabidopsis*. Plant Physiol 117:1227–1234

Hirayama T, Kieber JJ, Hirayama N, Kogan M, Guzman P, Nourizadeh S, Alonso JM, Dailey WP, Dancis A, Ecker JR (1999) RESPONSIVE-TO-ANTAGONIST1, a Menkes/Wilson disease-related copper transporter, is required for ethylene signaling in *Arabidopsis*. Cell 97:383–393

Hoopes JT, Dean JFD (2004) Ferroxidase activity in a laccase-like multicopper oxidase from *Liriodendron tulipifera*. Plant Physiol Biochem 42:27–33

Horng Y-C, Cobine PA, Maxfield AB, Carr HS, Winge DR (2004) Specific copper transfer from the Cox17 metallochaperone to both Sco1 and Cox11 in the assembly of yeast cytochrome *c* oxidase. J Biol Chem 279:35334–35340

Jones-Rhoades MW, Bartel DP (2004) Computational identification of plant MicroRNAs and their targets, including a stress-induced miRNA. Mol Cell 14:787–799

Jones-Rhoades MW, Bartel DP, Bartel B (2006) MicroRNAs and their regulatory roles in plants. Ann Rev Plant Biol 57:19–53

Kampfenkel K, Kushnir S, Babiychuk E, Inze D, Van Montagu M (1995) Molecular characterization of a putative *Arabidopsis thaliana* copper transporter and its yeast homologue. J Biol Chem 270:28479–28486

Katoh S (1960) A new copper protein from *Chlorella ellipsoidea*. Nature 186:533–534

Kieselbach T, Hagman Å, Andersson B, Schröder WP (1998) The thylakoid lumen of the chloroplasts: isolation and characterization. J Biol Chem 273:6710–6716

Kieselbach T, Bystedt M, Hynds P, Robinson C, Schröder WP (2000) A peroxidase homologue and novel plastocyanin located by proteomics to the *Arabidopsis* chloroplast thylakoid lumen. FEBS Lett 480:271–276

Kim S, Mollet JC, Dong J, Zhang K, Park SY, Lord EM (2003) Chemocyanin, a small basic protein from the lily stigma, induces pollen tube chemotropism. Proc Natl Acad Sci USA 100:16125–16130

Kliebenstein DJ, Monde RA, Last RL (1998) Superoxide dismutase in *Arabidopsis*: an ecletic enzyme family with disparate regulation and protein localization. Plant Physiol 118:637–650

Kobayashi Y, Kuroda K, Kimura K, Southron-Francis JL, Furuzawa A, Kimura K, Iuchi S, Kobayashi M, Taylor GJ, Koyama H (2008) Amino acid polymorphisms in strictly conserved domains of a P-type ATPase HMA5 are involved in the mechanism of copper tolerance variation in *Arabidopsis*. Plant Physiol 148:969–980

Kropat J, Tottey S, Birkenbihl RP, Depege N, Huijser P, Merchant S (2005) A regulator of nutritional copper signaling in *Chlamydomonas* is an SBP domain protein that recognizes the GTAC core of copper response element. Proc Natl Acad Sci USA 102:18730–18735

Kumar V, Dooley DM, Freeman HC, Guss JM, Harvey I, McGuirl MA, Wilce MC, Zubak VM (1996) Crystal structure of a eukaryotic (pea seedling) copper-containing amine oxidase at 2.2 A resolution. Structure 4:943–955

Kuper J, Llamas A, Hecht HJ, Mendel RR, Schwarz G (2004) Structure of the molybdopterin-bound Cnx1G domain links molybdenum and copper metabolism. Nature 430:803–806

Lee H, Lee JS, Bae EK, Choi YI, Noh EW (2005) Differential expression of a poplar copper chaperone gene in response to various abiotic stresses. Tree Physiol 25:395–401

Li HH, Merchant S (1995) Degradation of plastocyanin in copper-deficient *Chlamydomonas reinhardtii*. Evidence for a protease-susceptible conformation of the apoprotein and regulated proteolysis. J Biol Chem 270:23504–23510

Liang M, Haroldsen V, Cai X, Wu Y (2006) Expression of a putative laccase gene, ZmLAC1, in maize primary roots under stress. Plant Cell Environ 29:746–753

Liao MT, Hedley MJ, Woolley DJ, Brooks RR, Nichols MA (2000) Copper uptake and translocation in chicory (*Cichorium intybus* L. cv Grasslands Puna) and tomato (*Lycopersicon esculentum* Mill. cv Rondy) plants grown in NFT system. II. The role of nicotianamine and histidine in xylem sap copper transport. Plant Soil 223:1573–5036

Linder MC, Goode CA (1991) Biochemistry of copper. Plenum Press, New York

Lutsenko S, Barnes NL, Bartee MY, Dmitriev OY (2007) Function and regulation of human copper-transporting ATPases. Physiol Reviews 87:1011–1046

Mandal AK, Yang Y, Kertesz TM, Argüello JM (2004) Identification of the transmembrane metal binding site in Cu+-transporting PIB-type ATPases. J Biol Chem 279:54802–54807

Marina M, Maiale SJ, Rossi FR, Romero MF, Rivas EI, Gárriz A, Ruiz OA, Pieckenstain FL (2008) Apoplastic polyamine oxidation plays different roles in local responses of tobacco to infection by the necrotrophic fungus *Sclerotinia sclerotiorum* and the biotrophic bacterium *Pseudomonas viridiflava*. Plant Physiol 147:2164–2178

Marschner H (1995) Mineral nutrition of higher plants. Academic Press, London

Mayer AM (2006) Polyphenol oxidases in plants and fungi: going places? A review. Phytochem 67:2318–2331

McCaig BC, Meagher RB, Dean JFD (2005) Gene structure and molecular analysis of the laccase-like multicopper oxidase (LMCO) gene family in *Arabidopsis thaliana*. Planta 221:619–636

Mira H, Vilar M, Perez-Paya E, Penarrubia L (2001a) Functional and conformational properties of the exclusive C-domain from the *Arabidopsis* copper chaperone (CCH). Biochem J 357:545–549

Mira H, Martinez-Garcia F, Penarrubia L (2001b) Evidence for the plant-specific intercellular transport of the *Arabidopsis* copper chaperone CCH. Plant J 25:521–528

Mira H, Martínez N, Peñarrubia L (2002) Expression of a vegetative-storage-protein gene from Arabidopsis is regulated by copper, senescence and ozone. Planta 214:939–946

Molina-Heredia FP, Wastl J, Navarro JA, Bendall DS, Hervás M, Howe CJ, De la Rosa MA (2003) A new function for an old cytochrome? Nature 424:33–34

Mukherjee I, Campbell NH, Ash JS, Connolly EL (2006) Expression profiling of the *Arabidopsis* ferric chelate reductase (FRO) gene family reveals differential regulation by iron and copper. Planta 223:1178–1190

Nakamura K, Go N (2005) Function and molecular evolution of multicopper blue proteins. Cell Mol Life Sci 62:2050–2066

Nersissian AM, Valentine JS, Immoos C, Hill MG, Hart PJ, Williams G, Herrmann RG (1998) Uclacyanins, stellacyanins, and plantacyanins are distinct subfamilies of phytocyanins: plant-specific mononuclear blue copper proteins. Protein Sci 7:1915–1929

Newman SM, Eannetta NT, Yu H, Prince JP, de Vicente MC, Tanksley SD, Steffens JC (1993) Organisation of the tomato polyphenol oxidase gene family. Plant Mol Biol 21:1035–1051

Pant BD, Buhtz A, Kehr J, Scheible WR (2008) MicroRNA399 is a long-distance signal for the regulation of plant phosphate homeostasis. Plant J 53:731–738

Paschalidis KA, Roubelakis-Angelakis KA (2005) Sites and regulation of polyamine catabolism in the tobacco plant. Correlations with cell division/expansion, cell cycle progression, and vascular development. Plant Physiol 138:2174–2184

Pesaresi P, Scharfenberg M, Weigel M, Granlund I, Schroder WP, Finazzi G, Rappaport F, Masiero S, Furini A, Jahns P, Leister D (2008) Mutants, overexpressors, and interactors of *Arabidopsis* plastocyanin isoforms: revised roles of plastocyanin in photosynthetic electron flow and thylakoid redox state. Mol Plant 1:1–13 in press, out online

Pich A, Scholz I (1996) Translocation of copper and other micronutrients in tomato plants (*Lycopersicon esculentum* Mill.): nicotianamine-stimulated copper transport in the xylem. J Exp Bot 47:41–47

Pignocchi C, Fletcher JM, Wilkinson JE, Barnes JD, Foyer CH (2003) The function of ascorbate oxidase in tobacco. Plant Physiol 132:1631–1641

Pourcel L, Routaboul JM, Kerhoas L, Caboche M, Lepiniec L, Debeaujon I (2005) TRANSPARENT TESTA10 encodes a laccase-like enzyme involved in oxidative polymerization of flavonoids in *Arabidopsis* seed coat. Plant Cell 17:2966–2980

Puig S, Lee J, Lau M, Thiele DJ (2002) Biochemical and genetic analyses of yeast and human high affinity copper transporters suggest a conserved mechanism for copper uptake. J Biol Chem 277:26021–26030

Puig S, Mira H, Dorcey E, Sancenon V, Andres-Colas N, Garcia-Molina A, Burkhead JL, Gogolin KA, Abdel-Ghany SE, Thiele DJ, Ecker JR, Pilon M (2007) Higher plants possess two different types of ATX1-like copper chaperones. Biochem Biophys Res Commun 354:385–390

Ranocha P, Chabannes M, Chamayou S, Danoun S, Jauneau A, Boudet AM, Goffner D (2002) Laccase down-regulation causes alterations in phenolic metabolism and cell wall structure in poplar. Plant Physiol 129:145–155

Raven JA, Evans MCW, Korb RE (1999) The role of trace metals in photosynthetic electron transport in O_2-evolving. Photosynth Res 60:111–150

Rea G, Metoui O, Infantino A, Federico R, Angelini R (2002) Copper amine oxidase expression in defense responses to wounding and *Ascochyta rabiei* invasion. Plant Physiol 128:865–875

Robinson NJ, Procter CM, Connolly EL, Guerinot ML (1999) A ferric-chelate reductase for iron uptake from soils. Nature 397:694–697

Rodriguez FI, Esch JJ, Hall AE, Binder BM, Schaller GE, Bleecker AB (1999) A copper cofactor for the ethylene receptor ETR1 from *Arabidopsis*. Science 283:996–998

Sancenon V, Puig S, Mira H, Thiele DJ, Penarrubia L (2003) Identification of a copper transporter family in *Arabidopsis thaliana*. Plant Mol Biol 51:577–587

Sancenon V, Puig S, Mateu-Andres I, Dorcey E, Thiele DJ, Penarrubia L (2004) The *Arabidopsis* copper transporter COPT1 functions in root elongation and pollen development. J Biol Chem 279:15348–15355

Schaaf G, Ludewig U, Erenoglu BE, Mori S, Kitahara T, von Wiren N (2004) ZmYS1 functions as a proton-coupled symporter for phytosiderophore- and nicotianamine-chelated metals. J Biol Chem 279:9091–9096

Schnell DJ (1998) Protein targeting to the thylakoid membrane. Ann Rev Plant Physiol Plant Mol Biol 49:97–126

Schubert M, Petersson UA, Haas BJ, Funk C, Schroder WP, Kieselbach T (2002) Proteome map of the chloroplast lumen of *Arabidopsis thaliana*. J Biol Chem 277:8354–8365

Seigneurin-Berny D, Gravot A, Auroy P, Mazard C, Kraut A, Finazzi G, Grunwald D, Rappaport F, Vavasseur A, Joyard J, Richaud P, Rolland N (2006) HMA1, a new Cu-ATPase of the chloroplast envelope, is essential for growth under adverse light conditions. J Biol Chem 281:2882–2892

Shikanai T, Müller-Moulé P, Munekage Y, Niyogi KK, Pilon M (2003) PAA1, a P-type ATPase of *Arabidopsis*, functions in copper transport in chloroplasts. Plant Cell 15:1333–1346

Smeekens S, Bauerle C, Hageman J, Keegstra K, Weisbeek P (1986) The role of the transit peptide in the routing of precursors toward different chloroplast compartments. Cell 46: 365–375

Sterjiades R, Dean JFD, Eriksson K-EL (1992) Laccase from Sycamore maple (*Acer pseudoplatanus*) polymerizes monolignols. Plant Physiol 99:1162–1168

Sunkar R, Kapoor A, Zhu J-K (2006) Posttranscriptional induction of two Cu/Zn superoxide dismutase genes in *Arabidopsis* is mediated by downregulation of miR398 and important for oxidative stress tolerance. Plant Cell 18:2051–2065

Tabata K, Kashiwagi S, Mori H, Ueguchi C, Mizuno T (1997) Cloning of a cDNA encoding a putative metal-transporting P-type ATPase from *Arabidopsis thaliana*. Biochim Biophys Acta 1326:1–6

Takahashi M, Terada Y, Nakai I, Nakanishi H, Yoshimura E, Mori S, Nishizawa NK (2003) Role of nicotianamine in the intracellular delivery of metals and plant reproductive development. Plant Cell 15:1263–1280

Thipyapong P, Joel DM, Steffens JC (1997) Differential expression and turnover of the tomato polyphenol oxidase gene family during vegetative and reproductive development. Plant Physiol 113:707–718

Vorst O, Kock P, Lever A, Weterings B, Weisbeek P, Smeekens S (1993) The promoter of the *Arabidopsis thaliana* plastocyanin gene contains a far upstream enhancer-like element involved in chloroplast-dependent expression. Plant J 4:933–945

Waters BM, Grusak MA (2008) Whole-plant mineral partitioning throughout the life cycle in *Arabidopsis thaliana* ecotypes Columbia, Landsberg erecta, Cape Verde Islands, and the mutant line ysl1ysl3. New Phytol 177:389–405

Waters BM, Chu HH, Didonato RJ, Roberts LA, Eisley RB, Lahner B, Salt DE, Walker EL (2006) Mutations in *Arabidopsis* yellow stripe-like1 and yellow stripe-like3 reveal their roles in metal ion homeostasis and loading of metal ions in seeds. Plant Physiol 141:1446–1458

Weigel M, Varotto C, Pesaresi P, Finazzi G, Rappaport F, Salamini F, Leister D (2003) Plastocyanin is indispensable for photosynthetic electron flow in *Arabidopsis thaliana*. J Biol Chem 278:31286–31289

Welch RM, Norvell WA, Schaefer SC, Shaff JE, Kochian LV (1993) Induction of iron(III) and copper(II) reduction in pea roots by Fe and Cu status: does the root-cell plasmalemma Fe(III)-chelate reductase perform a general role in regulation of cation uptake. Planta 190:555–561

Welchen E, Chan RL, Gonzalez DH (2004) The promoter of the Arabidopsis nuclear gene COX5b–1, encoding subunit 5b of the mitochondrial cytochrome c oxidase, directs tissue-specific expression by a combination of positive and negative regulatory elements. J Exp Bot 55:1997–2004

Williams LE, Mills RF (2005) P(1B)-ATPases: an ancient family of transition metal pumps with diverse functions in plants. Trends Plant Sci 10:491–502

Wintz H, Fox T, Wu YY, Feng V, Chen W, Chang HS, Zhu T, Vulpe C (2003) Expression profiles of *Arabidopsis thaliana* in mineral deficiencies reveal novel transporters involved in metal homeostasis. J Biol Chem 278:47644–47653

Woeste KE, Kieber JJ (2000) A strong loss-of-function mutation in RAN1 results in constitutive activation of the ethylene response pathway as well as a rosette-lethal phenotype. Plant Cell 12:443–455

Yamamoto A, Bhuiyan MN, Waditee R, Tanaka Y, Esaka M, Oba K, Jagendorf AT, Takabe T (2005) Suppressed expression of the apoplastic ascorbate oxidase gene increases salt tolerance in tobacco and *Arabidopsis* plants. J Exp Bot 56:1785–1796

Yamasaki H, Abdel-Ghany SE, Cohu CM, Kobayashi Y, Shikanai T, Pilon M (2007) Regulation of copper homeostasis by micro-RNA in *Arabidopsis*. J Biol Chem 282:16369–16378

Yamasaki H, Hayashi M, Fukazawa M, Kobayashi Y, Shikanai T (2009) SQUAMOSA promoter-binding protein-like 7 is a central regulator for copper homeostasis in *Arabidopsis*. Plant Cell 21:347–361

Yruela I, Pueyo JJ, Alonso PJ, Picorel R (1996) Photoinhibition of photosystem II from higher plants. Effect of copper inhibition. J Biol Chem 271:27408–27415

Zhou J, Goldsbrough PB (1994) Functional homologs of fungal metallothionein genes from *Arabidopsis*. Plant Cell 6:875–884

Iron

Mary Lou Guerinot

Abstract Fe deficiency commonly limits plant growth and crop yields. If the mechanisms of Fe uptake, distribution, and regulation were clearly understood, it might be feasible to engineer plants better able to grow in soils now considered marginal and to increase crop biomass in soils now in cultivation. Furthermore, plants that serve as better sources of this essential element would improve human nutrition because most people rely on plants as their dietary source of Fe. Here we review our current understanding of Fe homeostasis in plants, emphasizing the challenges of safely transporting and storing this essential redox-active metal.

1 Introduction

Fe is essential for plants, but this redox-active element is not readily accessible in the rhizosphere. Fe is only sparingly soluble in aerobic soils at neutral-basic pH. Indeed, the concentration of free Fe^{3+} in the soil is estimated to be 10^{-17} M, which is well below that required for plants. This low bioavailability limits growth and productivity, making Fe the third most limiting nutrient for plants, after nitrogen and phosphorus. Furthermore, in response to Fe deficiency, plants are known to accumulate other metals, some of which are toxic to both plants and animals. As plants are a major dietary source of Fe worldwide, understanding plant Fe homeostasis is pivotal not only for improving crop yields but also for improving human nutrition. According to the World Health Organization, Fe deficiency is the most prevalent nutritional disorder in the world today, affecting almost 3 billion people (http://www.who.int/nutrition/topics/ida/en/index.html).

M.L. Guerinot
Department of Biological Sciences, Dartmouth College, Hanover, NH 03755, USA
e-mail: Guerinot@Dartmouth.edu

R. Hell and R.-R. Mendel (eds.), *Cell Biology of Metals and Nutrients*,
Plant Cell Monographs 17,
DOI 10.1007/978-3-642-10613-2_4,

Plants, like most organisms, use Fe as a cofactor in vital metabolic pathways such as the electron transport chain of respiration. As photosynthetic organisms, however, plants have an additional need for Fe because Fe serves as a cofactor in the photosynthetic electron transport chain and is essential for chlorophyll biosynthesis. Indeed, chloroplasts contain up to 90% of the Fe found in leaf cells, with about half in the stroma and the rest in the thylakoid membranes (Bughio et al. 1997a; Landsberg 1984; Shikanai et al. 2003; Terry and Abadia 1986). Not surprisingly, Fe deficiency results in chlorosis, which is attributed to the selective reduction and remodeling of photosynthetic components. For example, Photosystem I, which contains 12 atoms of Fe, is the primary target for reduction in photosynthesis due to Fe deficiency (Moseley et al. 2002). However, many other components of photosynthesis, or closely associated processes, also require large amounts of Fe. These include, photosystems II, cytochromes b_{563}, f, and c_6, ferrodoxin NAD(P)H/PQ oxidoreductase, catalase, and ascorbate peroxidase (Raven et al. 1999). And while Fe is required for life-sustaining processes from respiration to photosynthesis, too much Fe can be toxic. Fe catalyzes the formation of highly reactive hydroxyl radicals that can damage cellular components such as DNA, proteins, lipids, and sugars (Halliwell and Gutteridge 1992). Thus, Fe metabolism in plants is highly regulated to prevent excess accumulation.

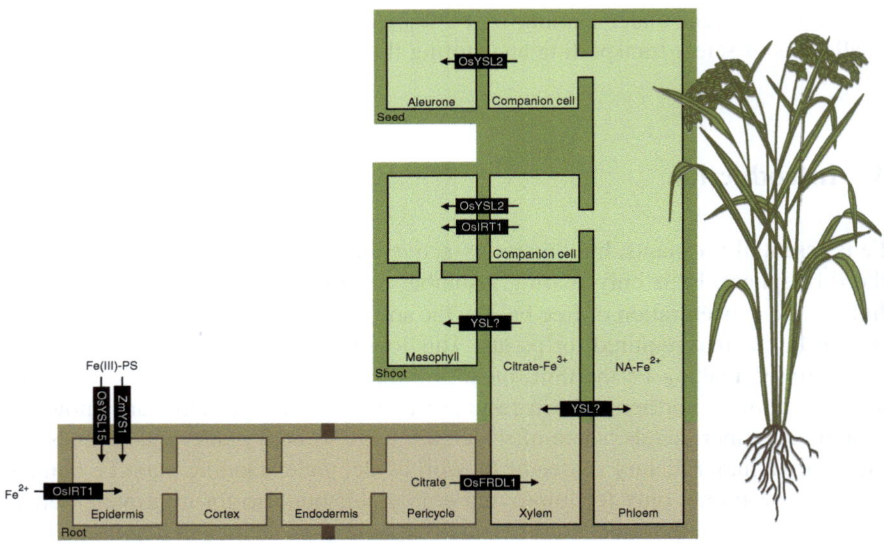

Fig. 1 Fe metal transport in monocots. Fe is taken up as Fe(III)-phytosiderophore chelates by YSL transporters in the epidermis. Fe^{2+} can also be taken up by OsIRT1. Fe moves through the symplastic space to the vasculature, bypassing the waxy Casparian strip on the endodermis. The citrate effluxer FRDL1 is important for loading of citrate into the xylem and subsequent Fe transport to the shoot through the transpiration stream. YSL transporters may also play a role in unloading the xylem into the shoot and the phloem. Fe is unloaded from the phloem by OsYSL2 and OsIRT1 into shoot and seed tissue. *Dark brown* boxes represent the Casparian strip. *MA* mugineic acid; *NA* nicotianamine. Modified from Palmer and Guerinot (2009)

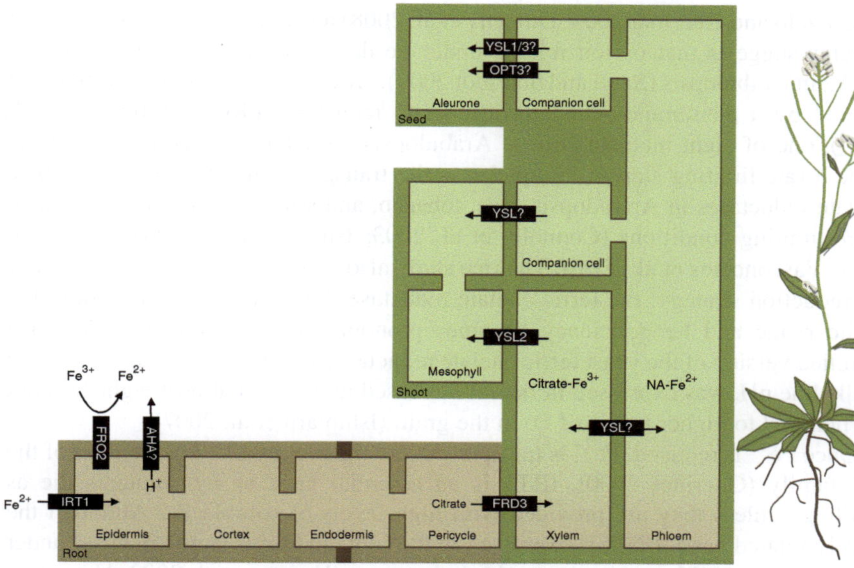

Fig. 2 Fe transport in dicots. Fe is taken up into the symplast by IRT1 transporters in the epidermis. Reduction of Fe by FRO2 and acidification of the soil by AHA2 contribute to increased Fe uptake. Fe can then travel through the symplastic space to the vasculature, bypassing the waxy Casparian strip on the endodermis. The citrate effluxer FRD3 is important for loading of citrate into the xylem and Fe-citrate is then transported to the shoot via the transpiration stream. Unloading into the shoot most likely involves a member of the YSL family. YSLs/OPT3 may translocate Fe to the phloem, where it can then be delivered to the seed. Fe is likely to be complexed to NA in the phloem. *Dark brown boxes* represent the Casparian strip. *NA* nicotianamine. Modified from Palmer and Guerinot (2009)

Upon sensing Fe limitation, plants induce a coordinated set of responses to maximize Fe mobilization and uptake from the soil, to utilize Fe stores and to prioritize the allocation of Fe to critical cellular processes. The grasses are typified by using a strategy based on chelation to retrieve Fe^{3+} from the soil (Fig. 1). This has been termed Strategy II and is best studied in rice and barley. Non-grasses, on the other hand, use a Reduction Strategy, termed Strategy I, whereby Fe^{3+} is reduced to Fe^{2+} prior to uptake (Fig. 2). Strategy I has been best studied in Arabidopsis.

2 The Reduction Strategy

When faced with Fe deficiency, plants that employ the reduction strategy release protons into the rhizosphere to increase the solubility of Fe; each unit drop in pH (above pH 4.0) results in a 1,000-fold increase in Fe availability. Several Arabidopsis H^+-ATPase (AHA) family members are induced in Fe-deficient roots

(Colangelo and Guerinot 2004; Dinneny et al. 2008) and analysis of loss of function mutants suggests that proton release under Fe deficiency is chiefly mediated by AHA2 in Arabidopsis (Santi and Schmidt 2009). After acidification, Fe^{3+} is reduced to Fe^{2+} by a membrane-bound ferric chelate reductase, FRO2 (Robinson et al. 1999), one of eight members of the Arabidopsis FRO family. Reduction appears to be a rate-limiting step in Fe uptake as the transgenic overexpression of ferric chelate reductases in Arabidopsis, rice, tobacco, and soybeans increases tolerance to Fe-limiting conditions (Connolly et al. 2003; Ishimaru et al. 2007; Oki et al. 2004; Vasconcelos et al. 2006). The rice study also addressed a major limitation of the reduction strategy: the ferric chelate reductase itself has a pH optimum in the acidic range and Fe deficiency becomes pronounced in alkaline soil. When a modified version of the yeast ferric chelate reductase, selected for enhanced activity at alkaline pH, was expressed in rice, it increased grain yield almost eight-fold but did not lead to higher levels of Fe in the grain (Ishimaru et al. 2007).

Once Fe^{3+} is reduced, Fe^{2+} is transported into the root by IRT1, a member of the ZIP family (Guerinot 2000). IRT1 is an essential gene as *irt1* mutants die as seedlings unless they are provided with high levels of soluble Fe. Although the closely related gene, *IRT2*, is expressed in the root epidermis and is induced under Fe deficiency, IRT2 cannot rescue the *irt1* mutant (Varotto et al. 2002; Vert et al. 2009). Furthermore, an *irt2* mutant shows no chlorosis under Fe deficiency (Vert et al. 2009). Transient expression in cultured cells indicated that IRT2 may be localized to intracellular vesicles, leading the authors to suggest that IRT2 may prevent toxicity from IRT1-mediated Fe influx via compartmentalization. However, this is not supported by the observation that plants overexpressing IRT2 show a clear upregulation of IRT1, arguing in favor of depletion of cytosolic Fe by IRT2. Analysis of plants overexpressing IRT1 shows that IRT1 protein is present only in Fe-deficient roots (Connolly et al. 2002). The Fe-induced turnover of IRT1 requires two lysine residues located in the intracellular loop of IRT1 between transmembrane domains III and IV (Kerkeb et al. 2008). This is consistent with the turnover of ZRT1, a yeast zinc transporter that also belongs to the ZIP family (Gitan and Eide 2000). ZRT1 has a lysine residue in its variable loop region, which is ubiquitinated to target it for protein degradation under zinc-sufficient conditions. When either IRT1 lysine residue was substituted with arginine and the variant was overexpressed, the plants accumulated higher levels of Fe than did wild-type plants; by contrast, plants over-expressing wild-type IRT1 contained Fe levels similar to those of wild-type plants (Kerkeb et al. 2008). Whether IRT2 also shows Fe-regulated turnover has not been addressed.

3 The Chelation Strategy

To acquire Fe, grasses use a mechanism based on chelation, in which phytosiderophores (PS) with a high affinity for Fe^{3+} are released into the rhizosphere, perhaps via anionic channels or vesicles (Negishi et al. 2002). The resulting Fe(III)-PS complexes are then transported into the roots via proteins belonging to the Yellow Stripe (YS) family, named for YS1, the phytosiderophore transporter of maize (Curie et al. 2001, 2009). YS1 is expressed in maize roots in response to Fe deficiency, and its loss results in decreased Fe uptake, and a constitutive Fe deficiency response; in the leaf, the decrease in Fe-containing proteins impairs chlorophyll synthesis, resulting in a yellowing between the veins, or interveinal chlorosis (Bell et al. 1958; Curie et al. 2001; von Wirén et al. 1994). In rice, OsYSL15 is the primary transporter responsible for uptake of Fe(III)-PS from the rhizosphere (Inoue et al. 2009; Lee et al. 2009). OsYSL15 is up-regulated in response to Fe deficiency and is expressed in the root epidermis, in addition to the stele, flowers, and developing seeds. Two *osysl15* insertional mutants exhibited chlorotic phenotypes under Fe deficiency and had reduced Fe concentrations in their shoots, roots, and seeds (Lee et al. 2009). Reducing OsYSL15 expression with RNAi resulted in severe germination defects, indicating an important role for this transporter in Fe homeostasis, although the defects could relate more to Fe loading of seeds than Fe(III)-PS uptake by roots (Inoue et al. 2009).

The mugineic acid (MA) family of PS is synthesized from L-methionine. First, S-adenosyl-methionine (SAM) synthetase converts methionine into SAM. Subsequently, nicotianamine (NA) synthase (NAS) condenses three molecules of SAM to form one molecule of NA. NA is then converted to a 3'-keto acid by NA aminotransferase (NAAT), and 2'-deoxymugineic acid (DMA) is synthesized by the subsequent action of a reductase. In barley, a further series of hydroxylations of DMA is catalyzed by two dioxygenases, Fe-deficiency-specific clones 2 and 3 (IDS2 and IDS3) (Nakanishi et al. 2000). In barley, the genes required for sulfur uptake, methionine synthesis, and PS synthesis are all dramatically up-regulated in the first 24 h of Fe deficiency (Nagasaka et al. 2009). In rice, expression of the *OsIRO2* transcription factor increases dramatically over the course of the first five days of Fe starvation, and is believed to activate the expression of genes related to PS synthesis and Fe uptake (Ogo et al. 2006).

There is a strong correlation between the volume of PS released and resistance to Fe limiting soils. For instance, barley, which is adapted to alkaline soils, releases a much greater volume of PS than most rice species (Nagasaka et al. 2009) that are adapted for growing in anaerobic soils where Fe is more soluble. Indeed, in *Oryza sativa* var. *japonica*, which grows poorly on calcareous soils, the overexpression of enzymes in the barley PS synthesis pathway greatly increased PS secretion (Takahashi et al. 2001). This resulted in a four-fold increase in grain yield by rice grown on Fe limited soil.

Grasses can also take up Fe^{2+} in addition to Fe(III)-PS (Cheng et al. 2007; Ishimaru et al. 2006). Rice plants that cannot synthesize PS owing to a mutation in the nicotianamine aminotransferase (*NAAT*) gene do not show growth defects if Fe^{2+} is supplied (Cheng et al. 2007). However, unlike Strategy I plants, H^+-ATPase or Fe^{3+}-chelate reductase activity is not induced under Fe deficiency. This likely reflects an adaptation to flooded rice paddies, where Fe^{2+} is more abundant than Fe^{3+} due to reduced levels of oxygen (Ishimaru et al. 2006). Transgenic rice plants over-expressing the Fe^{2+} transporter OsIRT1 showed enhanced tolerance to Fe deficiency as seedlings, demonstrating that Fe^{2+} uptake can be an important source of this essential nutrient (Lee and An 2009). Furthermore, Fe and Zn levels were elevated in the shoots, roots, and mature seeds of over-expressing plants, demonstrating that OsIRT1 can be used for enhancing micronutrient levels in rice grain.

4 Regulation of the Reduction Strategy

The *fer* mutant of tomato is severely chlorotic and is unable to induce the Fe deficiency response (Brown et al. 1971). *FER* encodes a basic helix-loop-helix (bHLH) transcription factor that is expressed in the root epidermis, the outer cortical layer of root tips, and in the vascular cylinder of the mature root-hair zone suggesting that FER plays a role in regulating both uptake from the soil as well as Fe distribution (Ling et al. 2002). Expression of *FER* is controlled by Fe availability both transcriptionally and post-transcriptionally (Brumbarova and Bauer 2005). Studies in Arabidopsis identified FIT as the functional ortholog of FER (Colangelo and Guerinot 2004; Jakoby et al. 2004; Yuan et al. 2005). *fit* is severely chlorotic, has reduced Fe content, and is unable to induce the Strategy I response (Colangelo and Guerinot 2004; Jakoby et al. 2004; Yuan et al. 2005). The *fit* mutation is seedling lethal unless plants are watered with supplemental Fe. *FRO2* transcript abundance is dramatically reduced and ferric chelate reductase activity is not induced by Fe deficiency in *fit*. However, while *IRT1* mRNA abundance is somewhat decreased in roots of Fe deficient *fit* plants (Jakoby et al. 2004; Yuan et al. 2005), IRT1 protein is not detectable in *fit* (Colangelo and Guerinot 2004). These results suggested that FIT may function to control the Fe uptake machinery at multiple levels and it was proposed that in addition to its role in induction of transcription of *FRO2* and *IRT1*, FIT may also act indirectly to prevent turnover of IRT1 protein when Fe is limiting (Colangelo and Guerinot 2004). This model is appealing because IRT1 is known to be subject to Fe-induced protein turnover (Connolly et al. 2002; Kerkeb et al. 2008). *FIT* mRNA itself is up regulated by Fe deficiency, thus upstream regulatory components in the signaling cascade also remain to be discovered. In addition, FIT regulates only a subset of Fe deficiency-inducible genes, indicating that FIT-independent regulatory mechanisms also operate in Fe deficiency responses (Colangelo and Guerinot 2004).

Constitutive high-level expression of *FIT* is not sufficient to induce high level expression of *FRO2* and *IRT1* in roots under Fe-sufficient conditions (Colangelo and Guerinot 2004; Jakoby et al. 2004; Yuan et al. 2005), implying that FIT acts with a binding partner that is expressed only in response to Fe limitation. Expression profiling experiments have implicated additional bHLH family members (BHLH38, BHLH39, BHLH100, and BHLH101) in the Fe deficiency response (Wang et al. 2007; Yuan et al. 2008; Yuan et al. 2005). These family members are not regulated by FIT. BHLH38 and BHLH39 which physically interact with FIT and transgenic plants that constitutively co-express either *bHLH38* or *bHLH39* with *FIT* show Fe-independent high-level expression of *FRO2* and *IRT1* and accumulate more Fe than wild type plants (Yuan et al. 2008). These data suggest that FIT functions together with either AtbHLH38 or AtbHLH39 to induce expression of the Strategy I Fe uptake machinery. Presumably, FIT and BHLH38/39 directly induce expression of *FRO2* and *IRT1* because co-expression of FIT with either BHLH38 or BHLH39 in yeast cells leads to activation of GUS expression driven from the *IRT1* and *FRO2* promoters (Yuan et al. 2008).

How do plants sense Fe status and formulate the Fe deficiency signal? We know from reciprocal grafting and split root experiments that there are both local and systemic signals (Grusak and Pezeshgi 1996; Schikora and Schmidt 2001; Vert et al. 2003). Once generated, the signal(s) must be transmitted via a signal transduction cascade resulting in the activation or repression of transcription factors that control downstream effector genes such as ferric chelate reductases and Fe transporters. There are several candidates for the Fe signal in plants. Nitric oxide (NO) is synthesized during Fe deficiency and promotes Fe uptake (Arnaud et al. 2006; Graziano et al. 2002; Graziano and Lamattina 2005, 2007). NO reverted the chlorotic phenotype of the maize *ys1* mutant (Graziano et al. 2002) and the rice *osysl15* mutant (Lee et al. 2009). However, NO treatment did not increase the total Fe content but rather somehow increased Fe availability. In tomato, NO was rapidly produced in roots as an early response to Fe deficiency; this helped facilitate Fe uptake, presumably by regulating root hair growth and by enhancing expression of Fe uptake-related genes, as treatment with NO enhanced *FER*, *LeFRO1*, and *LeIRT1* mRNA levels (Graziano and Lamattina 2007). As the NO-induced response was FER-dependent, it does appear that the NO is upstream of this essential transcription factor.

NA also has been proposed as an Fe sensor (Curie and Briat 2003). NA is a non-proteogenic amino acid that chelates both Fe^{2+} and Fe^{3+} and is thought to carry Fe in the phloem (Haydon and Cobbett 2007). NA has a higher affinity for Fe^{3+}, but forms a more stable complex with Fe^{2+} (von Wiren et al. 1999). NA also readily binds Cu^{2+}, Ni^{2+}, Co^{2+}, Zn^{2+}, and Mn^{2+}, in decreasing order of affinity (Curie et al. 2009). As previously mentioned, it is a precursor of MA but it is synthesized and used in all plants, regardless of their Fe uptake strategy (Curie et al. 2009; Haydon and Cobbett 2007). A characteristic phenotype of plants lacking NA is interveinal chlorosis in young growing leaves, as seen in the tomato *chloronerva* (*chln*) mutant defective in nicotianamine synthase (Bell et al. 1958; Ling et al. 1999; von Wirén et al. 1994). These plants have a constitutive Fe uptake response and accumulate high levels of Fe in the shoot.

One approach for identifying the Fe sensor in plants is to ask how other organisms sense Fe. In many bacterial species, the Fur protein transcriptionally represses promoters of Fe-regulated genes in an Fe^{2+}-dependent manner (Carpenter et al. 2009). As Fur directly binds Fe, it serves directly as the Fe sensor. In yeast, the identity of the sensor is unknown although evidence points to a role for Fe-S biogenesis in the sensing mechanism (Chen et al. 2004; Kumanovics et al. 2008; Lill 2009; Rutherford et al. 2005).

In mammals, the cytosolic form of the Fe-S cluster protein aconitase (also referred to as IRP) is the Fe sensor within cells and hepcidin, a small peptide hormone released by the liver, controls long distance coordination of Fe homeostasis (Ganz 2008; Muckenthaler et al. 2008). Arabidopsis aconitase does not appear to act as an Fe sensor as loss of function mutants in the three aconitase genes do not have Fe phenotypes (Arnaud et al. 2007). No obvious Arabidopsis hepcidin ortholog can be found by sequence analysis but, as discussed above, NA may serve in long distance signaling.

Not surprisingly, hormones also play a role in Fe deficiency signaling, with good experimental evidence for both ethylene and cytokinin being involved. Ethylene precursors increase ferric chlelate reductase activity whereas ethylene inhibitors decrease activity (Romera and Alcantara 2004). Similarly, treatment with ethylene precursors enhanced expression of *IRT1*, *FRO2,* and *FIT*, whereas expression of these genes was repressed by ethylene inhibitors (Lucena et al. 2006; Waters et al. 2007). Cytokinins had the opposite effect, negatively regulating *IRT1*, *FRO2,* and *FIT* expression at the transcript level, independent of the Fe status of the plants (Seguela et al. 2008). This repression required cytokinin receptors CRE1/Wol/AHK4 and AHK3, but not FIT, and conditions that inhibit root growth, such as osmotic stress induced by mannitol or NaCl, and hormonal treatments with auxin or abscissic acid, repressed Fe-deficiency response genes (Seguela et al. 2008). Because cytokinins inhibit root growth (Dello Ioio et al. 2007), and the results imply cytokinin treatment restricted nutrient uptake via a growth-dependent pathway, by transiently arresting root elongation to reduce nutrient demand.

Using high resolution expression profiling to report how each cell layer in the root responds to Fe deficiency, the Benfey lab determined that Fe deficiency modulates as much as 85% of the root transcriptome (Dinneny et al. 2008). Large transcriptional differences between layers were identified. The expression of genes related to metal transport and chelation was increased in the epidermis, while in the stele, genes associated with signaling and stress responses were upregulated. This suggests that sensing of Fe levels and control of the Fe deficiency response occurs in the vasculature, while regulation of Fe levels in the root is facilitated by modulating uptake in the epidermis. This dataset is consistent with the expression patterns based on GUS assays or mRNA in situ studies previously reported for important genes, including *IRT1*, *FRO2,* and *FRD3* (Connolly et al. 2003; Green and Rogers 2004; Vert et al. 2002).

5 Regulation of the Chelation Strategy

In rice, three transcription factors have been identified to date as playing a role in regulation of genes involved in the chelation strategy. Using microarray expression profiling, the bHLH protein OsIRO2 was identified (Ogo et al. 2006). Unlike the bHLH proteins, AtFIT and LeFER whose expression is largely root specific, OsIRO2 is expressed in both roots and shoots. Plants overexpressing OsIRO2 show improved growth compared to wild type under Fe deficient conditions, whereas OsIRO2 RNAi knockdown lines show the opposite phenotype of reduced biomass and chlorophyll content when grown under Fe deficiency (Ogo et al. 2007). As might be expected, expression of many genes involved in PS synthesis and transport is enhanced in the IRO2 overexpressor lines, and diminished in the corresponding RNAi lines. Interestingly, the expression of *OsIRT1* is unchanged, possibly indicating that OsIRO2 regulates Fe(III)-PS uptake, but not the uptake of Fe^{2+}. Many of the genes regulated by OsIRO2 do not have a consensus IRO2 binding sequence (CTCGTGG) in their promoters, suggesting that OsIRO2 may be acting by regulating other transcription factors. Two other transcription factors, IDEF1 (Kobayashi et al. 2007) and IDEF2 (Ogo et al. 2008), belonging to the ABI3/VP1 and NAC families of transcription factors, respectively, have been implicated in Fe deficiency. They were both identified based on their ability to bind to sequences that confer Fe regulation on the barley *IDS2* gene (Kobayashi et al. 2003). Overexpression of IDEF1 leads to enhanced expression of the OsIRO2 transcription factor and improved tolerance to Fe deficiency when plants are grown either in hydroponic culture or in calcareous soil (Kobayashi et al. 2007). There is no information as yet on overexpression of IDEF2 but reducing IDEF2 expression via RNAi did not affect the expression of genes directly involved in PS synthesis or transport (Ogo et al. 2008). The only affected gene that had previously been shown to be involved in iron homeostasis was OsYSL2. The IDEF2 RNAi lines did show altered Fe levels, presumably due to decreased transport of Fe-NA by OsYSL2. The role of the transcription factors involved in the Strategy II response has been recently reviewed (Walker and Connolly 2008).

6 Fe Transport within the Plant

For proper storage and use, Fe must be safely translocated to multiple parts of the plant, and compartmentalized into organelles such as chloroplasts and mitochondria. These organelles require Fe to carry out various metabolic processes, and serve as Fe reservoirs. This is essential to regulate Fe not only at the cellular level, but also at the organismal level. For example, defects in organellar Fe homeostasis can cause a lethal phenotype, as seen in the *Arabidopsis* frataxin mutant, which is defective in mitochondrial Fe homeostasis (Vazzola et al. 2007).

6.1 Intercellular Fe Transport

6.1.1 Citrate

Fe that is taken up from the rhizosphere must be loaded into the xylem for transport to the shoot via transpiration. Fe(III)-citrate is the major form of Fe present in xylem exudates (Grotz and Guerinot 2006). FRD3, a member of the multidrug and toxin efflux (MATE) family, is localized to the plasma membrane of cells in the pericycle and vasculature (Green and Rogers 2004) and functions in Fe translocation from roots to shoots by loading citrate into the xylem (Durrett et al. 2007). *frd3* xylem exudate contained less citrate and Fe than exudate from wild-type, and *frd3* mutant phenotypes were rescued by supplementing with citrate (Durrett et al. 2007), consistent with the role of FRD3 as a citrate transporter. Heterologous studies in Xenopus oocytes confirmed that FRD3 does indeed transport citrate. Once the citrate loaded into the xylem chelates Fe, Fe(III)-citrate complexes are either taken up at different locations via yet unidentified transporters, or alternatively, the complexes might be reduced by FROs and then transported into various cells of the plant. Rice has six orthologs of FRD3. OsFRDL1, was found to transport citrate when expressed in *Xenopus* oocytes, and the loss of *OsFRDL1* results in chlorotic plants with Fe precipitation in the xylem, similar to what has been observed with *frd3* mutants in Arabidopsis (Yokosho et al. 2009). The *osfrdl1* loss of function insertion mutant has increased *OsIRT1* expression, and accumulates more Zn and Mn in the shoot, again similar to what is seen with the Arabidopsis *frd3* mutant. The loss of OsFRDL1 reduced the concentration of Fe^{3+} in the xylem sap, but not Fe^{2+}, suggesting that there is an additional chelator besides citrate involved in moving Fe in the xylem (Yokosho et al. 2009). Several other members of the MATE family have also been shown to efflux citrate but, in these cases, the citrate plays a role in mitigating aluminum toxicity (Furukawa et al. 2007; Liu et al. 2009; Magalhaes et al. 2007). Neither FRD3 nor OsFRDL1 appear to function in Al tolerance (Liu et al. 2009; Yokosho et al. 2009). At present, it is not known which transporters efflux Fe into the xylem.

6.1.2 Nicotianamine

NA probably functions as the Fe chelator in phloem. As was mentioned earlier, NA is found in all plants and can chelate Fe^{2+} as well as Fe^{3+}. In Arabidopsis, there are four *NAS* genes. During Fe deficiency, *NAS2* and *NAS4* were upregulated in the root, (Klatte et al. 2009) suggesting a role in Fe translocation to the shoot. *NAS3* expression increased four-fold after the transition from vegetative to reproductive growth, suggesting NA also mediates Fe movement to the flowers. Despite the varied patterns and Fe regulation, all the single mutants had wild type NA levels, indicating functional redundancy, presumably because NA is mobile. In fact, interveinal chlorosis and sterility were observed only when the quadruple mutant was created (Klatte et al. 2009).

Yellow Stripe Like (YSL) family members are thought to transport metal-NA complexes (Curie et al. 2009; Haydon and Cobbett 2007). There are eight YSLs in *Arabidopsis* and their proposed functions have been recently reviewed (Curie et al. 2009). We note here that YSL1 and YSL3 are suggested to be involved in mobilizing metals, including Fe, from leaves for use in developing seeds (Waters et al. 2006). *ysl1* and *ysl3* are functionally redundant as the single mutants lack visible phenotypes, whereas *ysl1 ysl3* double mutants show severe interveinal chlorosis, lower Fe content in roots, leaves, and seeds, decreased fertility, arrested pollen and embryo development and defects in mobilizing metals from leaves during senescence (Waters et al. 2006).

The expression pattern of the rice YSLs also suggests a role in the long-distance transport of Fe complexes, including delivery to the seeds (Inoue et al. 2009). OsYSL15 and OsYSL2 are both upregulated in response to Fe deficiency, and may coordinate long distance Fe transport from root to shoot to seed, via the phloem: OsYSL15 in the root vasculature, flower, and developing seed; and OsYSL2 in the phloem companion cells of the shoot (Inoue et al. 2009; Koike et al. 2004). Interestingly, expression in oocytes showed that OsYSL2 transports Fe-NA but not Fe-PS (Koike et al. 2004), whereas OsYSL15 transports Fe-PS but not Fe-NA (Inoue et al. 2009). OsYSL18, like OsYSL15, also transports Fe-PS but does not appear to be involved in uptake from the rhizosphere. Rather, based on its expression pattern, it may be involved in DMA-mediated Fe distribution in reproductive organs, lamina joints, and phloem cells at the base of the sheath (Aoyama et al. 2009).

In Arabidopsis, the YSL proteins represent a subfamily of the oligopeptide transporter (OPT) family. One member, AtOPT3, is involved in supplying Fe for seed development (Stacey et al. 2008). *OPT3* is expressed in the vasculature, pollen and developing embryos (Stacey et al. 2002; Stacey et al. 2006). In *opt3-2* mutant plants, where *OPT3* expression is reduced, the yield and Fe content of *opt3-2* seeds both decreased (Stacey et al. 2008). The mutant roots exhibited constitutive Fe deficiency responses and the leaves were necrotic and accumulated high levels of Fe.

6.1.3 Iron Transport Protein (ITP)

The ITP is a member of the late embryogenesis abundant (LEA) protein family. It was originally identified as an Fe-binding ligand found in the phloem of castor beans and is able to bind Fe, Cu, Mn, and Zn in vitro (Kruger et al. 2002). The most similar genes in Arabidopsis have annotations related to stress, and several are highly upregulated in response to Fe deficiency in the root, although none are specifically expressed in the stele. It has been proposed that NA serves as a shuttle, facilitating Fe movement in and out of the phloem (via the YSLs), while the actual movement of Fe within the phloem occurs via ITP.

6.2 Subcellular Fe Transport

6.2.1 Vacuoles

Arabidopsis VIT1 was recently identified as an Fe^{2+} transporter that functions in vacuolar Fe storage (Kim et al. 2006). VIT1 is 62% similar to its yeast ortholog CCC1p, a transporter that can mediate Fe and Mn transport into vacuoles (Li et al.

2001). Yeast overexpressing *CCC1* accumulate more Fe in their vacuoles, and conversely, deletion mutants accumulate less Fe and are sensitive to elevated levels of Fe. *VIT1* complements the sensitivity of *ccc1* yeast mutants to Fe toxicity (Kim et al. 2006). VIT1 localizes to the vacuolar membrane, and it is expressed in the vasculature with increased expression seen during embryo and seed development. When visualized by synchrotron X-ray fluorescence microtomography, Fe is shown to localize to the provascular strands of wild-type seeds. In *vit1*, this Fe distribution is completely abolished, suggesting VIT1-mediated vacuolar Fe transport plays an important role in Fe localization in seeds. Furthermore, *vit1* plants grow poorly in Fe-limiting soils, emphasizing the critical role of vacuolar Fe storage for the growth of germinating seedlings.

Nramp3 and Nramp4 are divalent cation transporters known to function in Fe remobilization from the vacuole (Lanquar et al. 2005; Thomine et al. 2003). Expression of *Nramp3* and *Nramp4* is upregulated in response to Fe deficiency and both can mediate Fe transport when assayed in yeast (Curie et al. 2000; Thomine et al. 2003). In *Nramp3* overexpressing plants, *IRT1* and *FRO2* are down-regulated, indicating that Nramp3 remobilizes vacuolar Fe into the cytosol, thereby down-regulating Fe uptake genes. Studies of an *nramp3 nramp4* double mutant demonstrate that Fe mobilization mediated by Nramp3 and Nramp4 is crucial during early seedling development (Lanquar et al. 2005). Mutant seeds contain wild-type levels of Fe; however, the mutant displays retarded root growth and cotyledon greening during seed germination under low Fe. Electron microscopy showed the disappearance of Fe-associated globoids in wild type vacuoles during germination while globoids of the mutant remained unaltered, suggesting mutant seeds fail to retrieve Fe from the vacuole. Interestingly, like the *VIT1* gene, both *Nramp3* and *Nramp4* are expressed in the vasculature. Growth of the double *atnramp3 atnramp4* mutant and the *vit1* single mutant are arrested on Fe limiting soil. Taken together, these data suggest that vacuoles in the vascular cells are an important site of Fe storage and that Fe remobilization during germination is crucial for the seedling development when Fe supply is low. Most recently, it has been shown that alteration of vacuolar Fe transport leads to a decrease in ferritin protein abundance in seeds (Ravet et al. 2009). It appears that ferritin stability depends on the proper allocation of Fe between the vacuole and the plastid.

6.2.2 Chloroplasts

Photosynthesis, heme biosynthesis, and Fe-S cluster assembly all take place in the chloroplast and all require Fe. Indeed, as mentioned earlier, chloroplasts contain up to 90% of the Fe found in leaf cells. Despite the quantitative and qualitative significance of Fe in chloroplasts, our understanding of Fe transport and homeostasis in this organelle is limited. Fe uptake studies with isolated barley chloroplasts indicated that this process is light dependent and requires Fe(III) chelate reductase activity (Bughio et al. 1997b). Further support for uptake of Fe^{2+} by chloroplasts was provided by direct measurements of Fe^{2+} transport across vesicles prepared from the chloroplast inner envelope (Shingles et al. 2001; Shingles et al. 2002). An Fe(III) chelate reductase encoded by *FRO7* plays a role in chloroplast Fe acquisition and is required for efficient photosynthesis in young seedlings and for survival under Fe-limiting conditions (Jeong et al. 2008). Chloroplasts isolated from *fro7* mutants had 75% less ferric chelate reductase activity and contained 33% less Fe than wild-type chloroplasts. A presumptive Fe transporter, PIC1, has been identified that localizes to the chloroplast envelope (Duy et al. 2007). Although PIC1 was also reported to be part of the chloroplast inner envelope translocon (Teng et al. 2006), expression of PIC1 complements the phenotype of a yeast mutant defective in Fe uptake, and *pic1* mutants show severe chlorosis, only grow heterotrophically, and accumulate ferritin – all phenotypes consistent with a defect in Fe transport. Although it is not yet known whether PIC1 transports Fe^{2+} or Fe^{3+}, we speculate that chloroplasts might take up both Fe^{2+} and Fe^{3+} via multiple pathways as observed in modern day cyanobacteria.

Because the photosynthetic electron transport chain produces ROS, Fe should be tightly regulated in chloroplasts to avoid oxidative damage via the Fenton reaction. In plants, the Fe storage protein ferritin, which stores up to 4,500 Fe atoms, is found in plastids (Briat et al. 2009). There are four ferritin (*FER*) genes in *Arabidopsis*. FER1 is proposed to be involved in senescence (Murgia et al. 2007). Age-dependent senescence was accelerated in *fer1* loss-of-function mutants, due to Fe toxicity under excessive ROS accumulation. A recent study with mutants that lack seed (*fer2*) or leaf ferritins (*fer1 fer3 fer4*), showed that ferritins are essential for protection against oxidative damage, but are not the major Fe pool for either seedling development or proper functioning of the photosynthetic apparatus (Ravet et al. 2008).

6.2.3 Mitochondria

Virtually nothing is known about Fe import into plant mitochondria despite the fact that Fe is required for a large number of mitochondrial enzymes and Fe-S cluster biosynthesis occurs there (in addition to chloroplasts) (Balk and Lobreaux 2005; Briat et al. 2007). FRO8 was detected in a mitochondrial proteomics study (Heazlewood et al. 2004), implying that ferric chelate reductase(s) might also be involved in mitochondrial Fe transport as seen for chloroplast Fe transport (Jeong et al. 2008). Recently, mitochondrial Fe uptake proteins, named mitoferrins, were identified in zebrafish and mice (Shaw et al. 2006). Mitoferrin is a member of the mitochondrial solute carrier family and functions as a Fe importer for heme synthesis during erythropoiesis. Yeast orthologs (MRS3 and MRS4) of vertebrate mitoferrin have also been identified and the *mrs3/4* mutant shows defects in Fe homeostasis and in mitochondrial Fe-S cluster biogenesis (Foury and Roganti 2002; Li and Kaplan 2004; Muhlenhoff et al. 2003; Zhang et al. 2006). A search of the Arabidopsis genome reveals two putative mitoferrin orthologs that belong to the 45-member Arabidopsis mitochondrial carrier protein family (Millar and Heazlewood 2003). Analysis of publicly available microarray data shows that both genes are ubiquitously expressed in plants and neither is Fe-regulated. Both proteins are predicted to localize to mitochondria [Aramemnon; (Schwacke et al. 2003)].

Studies in Arabidopsis have identified putative mitochondrial Fe-S cluster effluxers *STA1 (ATM3)* and *STA2 (ATM1)* (Chen et al. 2007; Kushnir et al. 2001) and a putative Fe-chaperone called frataxin *AtFH* (Busi et al. 2006; Vazzola et al. 2007).

In animals, mitochondrial ferritins have been identified (Levi and Arosio 2004; Missirlis et al. 2006), and proteomics and electron microscopy suggest that mitochondrial ferritins are also present in *Arabidopsis* (Zancani et al. 2004). As is the case for chloroplasts, mitochondria must deal with ROS generated from the electron transport chain and maintain Fe homeostasis.

Acknowledgements Work in my laboratory is supported by grants from the National Science Foundation (IBN-0344305; IBN-0419695; DBI-0606193), the National Institutes of Health (RO1 GM 078536), the Department of Energy (DE-FG-2-06ER15809) and the National Institute of Environmental Health Sciences (5 P42 ES007373). I thank Christine Palmer for drawing the original figures on which the ones in this chapter are based.

References

Aoyama T, Kobayashi T, Takahashi M, Nagasaka S, Usuda K, Kakei Y, Ishimaru Y, Nakanishi H, Mori S, Nishizawa NK (2009) OsYSL18 is a rice iron(III)-deoxymugineic acid transporter specificially expressed in reproductive organs and phloem of lamina joints. Plant Mol Biol 70:681–692

Arnaud N, Murgia I, Boucherez J, Briat J-F, Cellier F, Gaymard F (2006) An iron-induced nitric oxide burst precedes ubiquitin-dependent protein degradation for Arabidopsis *AtFer1* ferritin gene expression. J Biol Chem 281:23579–23588

Arnaud N, Ravet K, Borlotti A, Touraine B, Boucherez J, Briat JF, Cellier F, Gaymard F (2007) The iron responsive element (IRE)/iron regulatory protein1 (IRP1)-cytosolic aconitase iron-regulatory switch does not operate in plants. Biochem J 405:523–531

Balk J, Lobreaux S (2005) Biogenesis of iron-sulfur proteins in plants. Trends Plant Sci 10:324–331

Bell WD, Bogorad L, McIlrath WJ (1958) Response of the yellow-stripe maize mutant (*ys1*) to ferrous and ferric iron. Bot Gaz 119:36–39

Briat JF, Curie C, Gaymard F (2007) Iron utilization and metabolism in plants. Curr Opin Plant Biol 10:276–282

Briat J-F, Ravet K, Arnaud N, Duc C, Boucherez J, Touraine B, Cellier F, Gaymard F (2009) New insights into ferritin synthesis and function highlight a link between iron homeostasis and oxidative stress in plants. An Bot (in press) doi:10.1093/aob/mcp128

Brown JC, Chaney RL, Ambler JE (1971) A new tomato mutant inefficient in the transport of iron. Physiol Plant 25:48–53

Brumbarova T, Bauer P (2005) Iron-mediated control of the basic helix-loop-helix protein FER, a regulator of iron uptake in tomato. Plant Physiol 137:1018–1026

Bughio N, Takahashi M, Yoshimura E, Nishizawa NK, Mori S (1997a) Characteristics of light-regulated iron transport system in barley chloroplasts. Soil Sci Plant Nutr 43:959–963

Bughio N, Takahashi M, Yoshimura E, Nishizawa NK, Mori S (1997b) Light-dependent iron transport into isolated barley chloroplasts. Plant Cell Physiol 38:101–105

Busi MV, Maliandi MV, Valdez H, Clemente M, Zabaleta EJ, Araya A, Gomez-Casati DF (2006) Deficiency of *Arabidopsis thaliana* frataxin alters activity of mitochondrial Fe-S proteins and induces oxidative stress. Plant J 48:873–882

Carpenter BM, Whitmire JM, Merrell DS (2009) This is not your mother's repressor: the complex role of Fur in pathogenesis. Infect Immun 77:2590–2601

Chen OS, Crisp RJ, Valachovic M, Bard M, Winge DR, Kaplan J (2004) Transcription of the yeast iron regulon does not respond directly to iron but rather to iron-sulfur cluster biosynthesis. J Biol Chem 279:29513–29518

Chen S, Sanchez-Fernandez R, Lyver ER, Dancis A, Rea PA (2007) Functional characterization of AtATM1, AtATM2 and AtATM3, a subfamily of Arabidopsis half-molecule ABC transporters implicated in iron homeostasis. J Biol Chem 282:21561–21571

Cheng L, Wang F, Shou H, Huang F, Zheng L, He F, Li J, Zhao FJ, Ueno D, Ma JF, Wu P (2007) Mutation in nicotianamine aminotransferase stimulated the Fe(II) acquisition system and led to iron accumulation in rice. Plant Physiol 145:1647–1657

Colangelo EP, Guerinot ML (2004) The essential bHLH protein FIT1 is required for the iron deficiency response. Plant Cell 16:3400–3412

Connolly EL, Fett JP, Guerinot ML (2002) Expression of the IRT1 metal transporter is controlled by metals at the levels of transcript and protein accumulation. Plant Cell 14:1347–1357

Connolly EL, Campbell N, Grotz N, Prichard CL, Guerinot ML (2003) Overexpression of the FRO2 iron reductase confers tolerance to growth on low iron and uncovers post-transcriptional control. Plant Physiol 133:1102–1110

Curie C, Briat J-F (2003) Iron transport and signaling in plants. Annu Rev Plant Biol 54:183–206

Curie C, Alonso JM, Le Jean M, Ecker JR, Briat J-F (2000) Involvement of NRAMP1 from *Arabidopsis thaliana* in iron transport. Biochem J 347:749–755

Curie C, Panaviene Z, Loulergue C, Dellaporta SL, Briat J-F, Walker EL (2001) Maize yellow stripe1 encodes a membrane protein directly involved in Fe(III) uptake. Nature 409:346–349

Curie C, Cassin G, Counch D, Divol F, Higuchi K, Le Jean M, Misson J, Schikora A, Czernic P, Mari S (2009) Metal movement within the plant: contribution of nictotianamine and yellow stripe 1-like transporters. Ann Bot 103:1–11

Dello Ioio R, Linhares FS, Scacchi E, Casamitjana-Martinez E, Heidstra R, Costantino P, Sabatini S (2007) Cytokinins determine Arabidopsis root-meristem size by controlling cell differentiation. Curr Biol 17:678–682

Dinneny JR, Long TA, Wang JY, Jung JW, Mace D, Pointer S, Barron C, Brady SM, Schiefelbein J, Benfey PN (2008) Cell identity mediates the response of Arabidopsis roots to abiotic stress. Science 320:942–945

Durrett TP, Gassmann W, Rogers EE (2007) The FRD3-mediated efflux of citrate into the root vasculature is necessary for efficient iron translocation. Plant Physiol 144:197–205

Duy D, Wanner G, Meda AR, von Wiren N, Soll J, Philippar K (2007) PIC1, an ancient permease in Arabidopsis chloroplasts, mediates iron transport. Plant Cell 19:986–1006

Foury F, Roganti T (2002) Deletion of the mitochondrial carrier genes *MRS3* and *MRS4* suppresses mitochondrial iron accumulation in a yeast frataxin-deficient strain. J Biol Chem 277:24475–24483

Furukawa J, Yamaji N, Wang H, Mitani N, Murata Y, Sato K, Katsuhara M, Takeda K, Ma JF (2007) An aluminum-activated citrate transporter in barley. Plant Cell Physiol 48:1081–1091

Ganz T (2008) Iron homeostasis: fitting the puzzle pieces together. Cell Metab 7:288–290

Gitan RS, Eide DJ (2000) Zinc-regulated ubiquitin conjugation signals endocytosis of the yeast ZRT1 zinc transporter. Biochem J 346:329–336

Graziano M, Lamattina L (2005) Nitric oxide and iron in plants: an emerging and converging story. Trends Plant Sci 10:4–8

Graziano M, Lamattina L (2007) Nitric oxide accumulation is required for molecular and physiological responses to iron deficiency in tomato roots. Plant J 52:949–960

Graziano M, Beligni MV, Lamattina L (2002) Nitric oxide improves internal iron availability in plants. Plant Physiol 130:1852–1859

Green LS, Rogers EE (2004) *FRD3* controls iron localization in *Arabidopsis thaliana*. Plant Physiol 136:2523–2531
Grotz N, Guerinot ML (2006) Molecular aspects of Cu, Fe and Zn homeostasis in plants. Biochim Biophys Acta 1763:595–608
Grusak MA, Pezeshgi S (1996) Shoot-to-root signal transmission regulates root Fe(III) reductase activity in the *dgl* mutant of pea. Plant Physiol 110:329–334
Guerinot ML (2000) The ZIP family of metal transporters. Biochim Biophys Acta 1465:190–198
Halliwell B, Gutteridge JMC (1992) Biologically relevant metal ion-dependent hydroxyl radical generation. FEBS Lett 307:108–112
Haydon MJ, Cobbett CS (2007) Transporters of ligands for essential metal ions in plants. New Phytol 174:499–506
Heazlewood JL, Tonti-Filippini JS, Gout AM, Day DA, Whelan J, Millar AH (2004) Experimental analysis of the Arabidopsis mitochondrial proteome highlights signaling and regulatory components, provides assessment of targeting prediction programs, and indicates plant-specific mitochondrial proteins. Plant Cell 16:241–256
Inoue H, Kobayashi T, Nozoye T, Takahashi M, Kakei Y, Suzuki K, Nakazono M, Nakanishi H, Mori S, Nishizawa NK (2009) Rice OsYSL15 is an iron-regulated iron(III)-deoxymugineic acid transporter expressed in the roots and is essential for iron uptake in early growth of the seedlings. J Biol Chem 284:3470–3479
Ishimaru Y, Suzuki M, Tsukamoto T, Suzuki K, Nakazono M, Kobayashi T, Wada Y, Watanabe S, Matsuhashi S, Takahashi M, Nakanishi H, Mori S, Nishizawa NK (2006) Rice plants take up iron as an Fe^{3+}-phytosiderophore and as Fe^{2+}. Plant J 45:335–346
Ishimaru Y, Kim SA, Tsukamoto T, Oki H, Kobayashi T, Watanabe S, Matsuhashi S, Takahashi M, Nakanishi H, Mori S, Nishizawa NK (2007) Mutational reconstructed ferric chelate reductase confers enhanced tolerance in rice to iron deficiency in calcareous soil. Proc Natl Acad Sci USA 104:7373–7378
Jakoby M, Wang H-Y, Reidt W, Weissharr B, Bauer P (2004) FRU(BHLH029) is required for induction of iron mobilization genes in *Arabidopsis thaliana*. FEBS Lett 577:528–534
Jeong J, Cohu C, Kerkeb L, Piln M, Connolly EL, Guerinot ML (2008) Chloroplast Fe(III) chelate reductase activity is essential for seedling viability under iron limiting conditions. Proc Natl Acad Sci USA 105:10619–10624
Kerkeb L, Mukherjee I, Chatterjee I, Lahner B, Salt DE, Connolly EL (2008) Iron-induced turnover of the Arabidopsis IRT1 metal transporter requires lysine residues. Plant Physiol 146:1964–1973
Kim SA, Punshon T, Lanzirotti A, Li L, Alonzo JM, Ecker JR, Kaplan J, Guerinot ML (2006) Localization of iron in Arabidopsis seed requires the vacuolar membrane transporter VIT1. Science 314:1295–1298
Klatte M, Schuler M, Wirtz M, Fink-Straube C, Hell R, Bauer P (2009) The analysis of Arabidopsis nicotianamine synthase mutants reveals functions for nicotianamine in seed iron loading and iron deficiency responses. Plant Physiol 150:257–271
Kobayashi T, Nakayama Y, Itai RN, Nakanishi H, Yoshihara T, Mori S, Nishizawa NK (2003) Identification of novel cis-acting elements, IDE1 and IDE2, of the barley IDS2 gene promoter conferring iron-deficiency-inducible, root-specific expression in heterologous tobacco plants. Plant J 36:780–793
Kobayashi T, Ogo Y, Itai RN, Nakanishi H, Takahashi M, Mori S, Nishizawa NK (2007) The transcription factor IDEF1 regulates the response to and tolerance of iron deficiency in plants. Proc Natl Acad Sci USA 104:19150–19155
Koike S, Inoue H, Mizuno D, Takahashi M, Nakanishi H, Mori S, Nishizawa N (2004) OsYSL2 is a rice metal-nicotianamine transporter that is regulated by iron and expressed in the phloem. Plant J 39:415–424
Kruger C, Berkowitz O, Stephan UW, Hell R (2002) A metal-binding member of the late embryogenesis abundant protein family transports iron in the phloem of *Ricinus communis* L. J Biol Chem 277:25062–25069

Kumanovics A, Chen OS, Li L, Bagley D, Adkins EM, Lin H, Dingra NN, Outten CE, Keller G, Winge D, Ward DM, Kaplan J (2008) Identification of *FRA1* and *FRA2* as genes involved in regulating the yeast iron regulon in response to decreased mitochondrial iron-sulfur cluster synthesis. J Biol Chem 283:10276–10286

Kushnir S, Babiychuk E, Storozhenko S, Davey MW, Papenbrock J, De Rucke R, Engler G, Stephan UW, Lange H, Kispai G, Lill R, Van Monagu M (2001) A mutation of the mitochondrial ABC transporter Sta1 leads to dwarfism and chlorosis in the Arabidopsis mutant *starik*. Plant Cell 13:89–100

Landsberg E (1984) Regulation of iron-stress-response by whole-plant activity. J Plant Nutr 7:609–621

Lanquar V, Lelievre F, Bolte S, Hames C, Alcon C, Neumann D, Vansuyt G, Curie C, Schroder A, Kramer U, Barbier-Brygoo H, Thomine S (2005) Mobilization of vacuolar iron by AtNRAMP3 and AtNRAMP4 is essential for seed germination on low iron. EMBO J 24:4041–4051

Lee S, An G (2009) Over-expression of OsIRT1 leads to increased iron and zinc accumulation in rice. Plant Cell Environ 32:408–416

Lee S, Chiecko JC, Kim SA, Walker EL, Lee Y, Guerinot ML, An G (2009) Disruption of OsYSL15 leads to iron inefficiency in rice plants. Plant Physiol 150:786–800

Levi S, Arosio P (2004) Mitochondrial ferritin. Int J Biochem Cell Biol 36:1887

Li L, Kaplan J (2004) A mitochondrial-vacuolar signaling pathway in yeast that affects iron and copper metabolism. J Biol Chem 279:33653–33661

Li L, Chen OS, Ward DM, Kaplan J (2001) CCC1 is a transporter that mediates vacuolar iron storage in yeast. J Biol Chem 276:29515–29519

Lill R (2009) Function and biogenesis of iron-sulfur proteins. Nat Chem Biol 460:831–838

Ling HQ, Koch G, Baulein H, Ganal MW (1999) Map-based cloning of *chloronerva*, a gene involved in iron uptake of higher plants encoding nicotianamine synthase. Proc Natl Acad Sci USA 96:7098–7103

Ling H-Q, Bauer P, Bereczky Z, Keller B, Ganal M (2002) The tomato *fer* gene encoding a bHLH protein controls iron-uptake responses in roots. Proc Natl Acad Sci USA 99:13938–13943

Liu J, Magalhaes JV, Shaff JE, Kochian LV (2009) Aluminum-activated citrate and malate transporters from the MATE and ALMT families function independently to confer Arabidopsis aluminum tolerance. Plant J 57:389–399

Lucena C, Waters BM, Romera FJ, Garcia MJ, Morales M, Alcantara E, Perez-Vicente R (2006) Ethylene could influence ferric reductase, iron transporter, and H^+-ATPase gene expression by affecting *FER* (or *FER-like*) gene activity. J Exp Bot 57:4145–4154

Magalhaes JV, Liu J, Guimarães CT, Lana UG, Alves VM, Wang YH, Schaffert RE, Hoekenga OA, Piñeros MA, Shaff JE, Klein PE, Carneiro NP, Coelho CM, Trick HN, Kochian LV (2007) A gene in the multidrug and toxic compound extrusion (MATE) family confers aluminum tolerance in sorghum. Nat Genet 39:1156–1161

Millar AH, Heazlewood JL (2003) Genomic and proteomic analysis of mitochondrial carrier proteins in Arabidopsis. Plant Physiol 131:443–453

Missirlis F, Holmberg S, Georgieva T, Dunkov BC, Rouault TA, Law JH (2006) Characterization of mitochondrial ferritin in Drosophila. Proc Natl Acad Sci USA 103:5893–5898

Moseley JL, Allinger T, Herzog S, Hoerth P, Wehinger E, Merchant S, Hippler M (2002) Adaptation to Fe-deficiency requires remodeling of the photosynthetic apparatus. EMBO J 21:6709–6720

Muckenthaler MU, Galy B, Hentze MW (2008) Systemic iron homeostasis and the iron-responsive element/iron-regulatory protein (IRE/IRP) regulatory network. Annu Rev Nutr 28:197–213

Muhlenhoff U, Stadler JA, Richhardt N, Seubert A, Eickhorst T, Schweyen RJ, Lill R, Wiesenberger G (2003) A specific role of the yeast mitochondrial carriers Mrs3/4p in mitochondrial iron acquisition under iron-limiting conditions. J Biol Chem 278:40612–40620

Murgia I, Vazzola V, Tarantino D, Cellier F, Ravet K, Briat JF, Soave C (2007) Knock-out of ferritin *AtFer1* causes earlier onset of age-dependent leaf senescence in Arabidopsis. Plant Physiol Biochem 45:898–907

Nagasaka S, Takahashi M, Nakanishi-Itai R, Bashir K, Nakanishi H, Mori S, Nishizawa NK (2009) Time course analysis of gene expression over 24 hours in Fe-deficient barley roots. Plant Mol Biol 69:621–631

Nakanishi H, Yamaguchi H, Sasakuma T, Nishizawa NK, Mori S (2000) Two dioxygenase genes, *Ids3* and *Ids2*, from *Hordeum vulgare* are involved in the biosynthesis of mugineic acid family phytosiderophores. Plant Mol Biol 44:199–207

Negishi T, Nakanishi H, Yazaki J, Kishimoto N, Fujii F, Shimbo K, Yamamoto K, Sakata K, Sasaki T, Kikuchi S, Mori S, Nishizawa N (2002) cDNA microarray analysis of gene expression during Fe-deficiency stress in barley suggests that polar transport of vesicles is implicated in phytosiderophore secretion in Fe-deficient barley roots. Plant J 30:83–94

Ogo Y, Itai RN, Nakanishi H, Inoue H, Kobayashi T, Suzuki M, Takahashi M, Mori S, Nishizawa NK (2006) Isolation and characterization of IRO2, a novel iron regulated bHLH transcription factor in graminaceous plants. J Exp Bot 57:2867–2878

Ogo Y, Itai RN, Nakanishi H, Kobayashi T, Takahashi M, Mori S, Nishizawa NK (2007) The rice bHLH protein OsIRO2 is an essential regulator of the genes involved in Fe uptake under Fe-deficient conditions. Plant J 51:366–377

Ogo Y, Kobayashi T, Nakanishi Itai R, Nakanishi H, Kakei Y, Takahashi M, Toki S, Mori S, Nishizawa NK (2008) A novel NAC transcription factor, IDEF2, that recognizes the iron deficiency-responsive element 2 regulates the genes involved in iron homeostasis in plants. J Biol Chem 283:13407–13417

Oki H, Kim S, Nakanishi H, Takahashi M, Yamaguchi H, Mori S, Nishizawa NK (2004) Directed evolution of yeast ferric reductase to produce plants with tolerance to iron deficiency in alkaline soils. Soil Sci Plant Nutr 50:1159–1165

Palmer C, Guerinot ML (2009) Facing the challenges of Cu, Fe and Zn homeostasis in plants. Nat Chem Biol 5:333–340

Raven JA, Evans MCW, Korb RE (1999) The role of trace metals in photosynthetic electron transport in O_2-evolving organisms. Photosyn Res 60:111–149

Ravet K, Touraine B, Boucherez J, Briat JF, Gaymard F, Cellier F (2008) Ferritins control interaction between iron homeostasis and oxidative stress in Arabidopsis. Plant J 57:400–412

Ravet K, Touraine B, Kim SA, Cellier F, Thmine S, Guerinot ML, Briat J-F, Gaymard F (2009) Post-translational regulation of AtFER2 ferritin in response to intracellular iron trafficking during fruit development in Arabidopsis. Mol Plant 2:1095–1106

Robinson NJ, Proctor CM, Connolly EL, Guerinot ML (1999) A ferric-chelate reductase for iron uptake from soils. Nature 397:694–697

Romera FJ, Alcantara E (2004) Ethylene involvement in the regulation of Fe-deficiency stress responses by Strategy I plants. Func Plant Biol 31:315–328

Rutherford JC, Ojeda L, Balk J, Mulenhoff U, Lill R, Winge DR (2005) Activation of the iron regulon by yeast Aft1/Aft2 transcription factors depends on mitochondrial but not cytosolic iron-sulfur protein biogenesis. J Biol Chem 280:10135–10140

Santi S, Schmidt W (2009) Dissecting iron deficiency-induced proton extrusion in Arabidopsis roots. New Phytol 183:1072–1084

Schikora A, Schmidt W (2001) Iron stress-induced changes in root epidermal cell fate are regulated independently from physiological responses to low iron availability. Plant Physiol 125:1679–1687

Schwacke R, Schneider A, van der Graaff E, Fischer K, Catoni E, Desimone M, Frommer WB, Flugge U-I, Kunze R (2003) ARAMEMNON, a novel database for Arabidopsis Integral membrane proteins. Plant Physiol 131:16–26

Seguela M, Briat J-F, Vert G, Curie C (2008) Cytokinins negatively regulate the root iron uptake machinery in Arabidopsis through a growth-dependent pathway. Plant J 55:289–300

Shaw GC, Cope JJ, Li L, Corson K, Hersey C, Ackermann GE, Gwynn B, Lambert AJ, Wingert RA, Traver D, Trede NS, Barut BA, Zhou Y, Minet E, Donovan A, Brownlie A, Balzan R, Weiss MJ, Peters LL, Kaplan J, Zon LI, Paw BH (2006) Mitoferrin is essential for erythroid iron assimilation. Nature 440:96–100

Shikanai T, Muller-Moule P, Munekage Y, Niyogi KK, Pilon M (2003) PAA1, a P-type ATPase of Arabidopsis, functions in copper transport in chloroplasts. Plant Cell 15:1333–1346

Shingles R, North M, McCarty RE (2001) Direct measurement of ferrous ion transport across membranes using a sensitive fluorometric assay. Anal Biochem 296:106–113

Shingles R, North M, McCarty RE (2002) Ferrous iron transport across chloroplast inner envelope membranes. Plant Physiol 128:1022–1030

Stacey MG, Koh S, Becker J, Stacey G (2002) AtOPT3, a member of the oligopeptide transporter family, is essential for embryo development in Arabidopsis. Plant Cell 14:2799–2811

Stacey MG, Osawa H, Patel A, Gassmann W, Stacey G (2006) Expression analyses of Arabidopsis oligopeptide transporters during seed germination. Plant J 223:291–305

Stacey MG, Patel A, McClain WE, Mathieu M, Remley M, Rogers EE, Gassmann W, Blevins DG, Stacey G (2008) The Arabidopsis AtOPT3 protein functions in metal homeostasis and movement of iron to developing seeds. Plant Physiol 146:589–601

Takahashi MT, Nakanishi H, Kawasaki S, Nishizawa NK, Mori S (2001) Enhanced tolerance of rice to low iron availability in alkaline soils using barley nicoianamine aminotransferase genes. Nat Biotechnol 19:466–469

Teng YS, Su YS, Chen LJ, Lee YJ, Hwang I, Li HM (2006) Tic21 is an essential translocon component for protein translocation across the chloroplast inner envelope membrane. Plant Cell 18:2247–2257

Terry N, Abadia J (1986) Function of iron in chloroplasts. J Plant Nutr 9:609–646

Thomine S, Lelievre F, Debarbieux E, Schroeder JI, Barbier-Bygoo H (2003) AtNRAMP3, a multispecific vacuolar metal transporter involved in plant responses to iron deficiency. Plant J 34:685–695

Varotto C, Maiwald D, Pesaresi P, Jahns P, Francesco S, Leister D (2002) The metal ion transporter IRT1 is necessary for iron homeostasis and efficient photosynthesis in *Arabidopsis thaliana*. Plant J 31:589–599

Vasconcelos M, Eckert H, Arahana V, Graef G, Grusak MA, Clemente T (2006) Molecular and phenotypic characterization of transgenic soybean expressing the Arabidopsis ferric chelate reductase gene, *FRO2*. Planta 224:1116–1128

Vazzola V, Losa A, Soave C, Murgia I (2007) Knockout of frataxin gene causes embryo lethality in Arabidopsis. FEBS Lett 581:667–672

Vert G, Grotz N, Dedaldechamp F, Gaymard F, Guerinot ML, Briat J-F, Curie C (2002) IRT1, an Arabidopsis transporter essential for iron uptake from the soil and plant growth. Plant Cell 14:1223–1233

Vert G, Briat J-F, Curie C (2003) Dual regulation of the Arabidopsis high affinity root iron uptake system by local and long-distance signals. Plant Physiol 132:796–804

Vert G, Barberon M, Zelazny E, Seguela M, Briat J-F, Curie C (2009) Arabidopsis IRT2 cooperates with the high affinity iron uptake system to maintain iron homeostasis in root epidermal cells. Planta 229:1171–1179

von Wiren N, Klair S, Bansal S, Briat J-F, Khodr H, Shioiri T, Leigh RA, Hider RC (1999) Nicotianamine chelates both Fe(III) and Fe(II). Implications for metal transport in plants. Plant Physiol 119:1107–1114

von Wirén N, Mori S, Marschner H, Römheld V (1994) Iron inefficiency in maize mutant ys1 (*Zea mays* L. cv yellow-stripe) is caused by a defect in uptake of iron phytosiderophores. Plant Physiol 106:71–77

Walker EL, Connolly EL (2008) Time to pump iron: iron-deficiency-signaling mechanisms of higher plants. Curr Opin Plant Biol 11:530–535

Wang H-Y, Klatte M, Jacoby M, Baumlein H, Weisshaar B, Bauer P (2007) Iron deficiency-mediated stress regulation of four subgroup Ib bHLH genes in *Arabidopsis thaliana*. Planta 226:897–908

Waters BM, Chu HH, Didonato RJ, Roberts LA, Eisley RB, Walker EL (2006) Mutations in Arabidopsis *Yellow Stripe-Like1* and *Yellow Stripe-Like3* reveal their roles in metal ion homeostasis and loading of metal ions in seeds. Plant Physiol 141:1446–1458

Waters BM, Lucena C, Romera FJ, Jester GG, Wynn AN, Rojas CL, Alcántara E, Pérez-Vicente R (2007) Ethylene involvement in the regulation of the H(+)-ATPase *CsHA1* gene and of the new isolated ferric reductase *CsFRO1* and iron transporter *CsIRT1* genes in cucumber plants. Plant Physiol Biochem 45:293–301

Yokosho K, Yamaji N, Ueno D, Mitani N, Ma JF (2009) OsFRDL1 is a citrate transporter required for efficient translocation of iron in rice. Plant Physiol 149:297–305

Yuan YX, Zhang J, Wang DW, Ling HQ (2005) AtbHLH29 of *Arabidopsis thaliana* is a functional ortholog of tomato FER involved in controlling iron acquisition in strategy I plants. Cell Res 15:613–621

Yuan Y, Wu H, Wang N, KLi J, Zhao W, Du J, Wang D, Ling H-Q (2008) FIT interacts with AtbHLH38 and AtbHLH39 in regulating iron uptake gene expression for iron homeostasis in Arabidopsis. Cell Res 18:385–397

Zancani M, Peresson C, Biroccio A, Federici G, Urbani A, Murgia I, Soave C, Micali F, Vianello A, Macr F (2004) Evidence for the presence of ferritin in plant mitochondria. Eur J Biochem 271:3657–3664

Zhang Y, Lyver ER, Knight SAB, Pain D, Lesuisse E, Dancis A (2006) Mrs3p, Mrs4p, and frataxin provide iron for Fe-S cluster synthesis in mitochondria. J Biol Chem 281:22493–22502

Dissecting Pathways Involved in Manganese Homeostasis and Stress in Higher Plant Cells

Lorraine E. Williams and Jon K. Pittman

Abstract Manganese (Mn) is an essential heavy metal micronutrient in plants having a range of cellular functions. Mn homeostasis involves the coordinated operation of transporters mediating cellular import and export and distribution between cell organelles. Significant progress has been made in identifying transport mechanisms in plants responsible for Mn uptake from the soil, distribution around the plant and compartmentalisation within the cell. These processes are required to ensure that Mn reaches its correct target proteins and also for cellular Mn detoxification. Transporters that have been implicated in Mn transport include members of the P_{2A}-type ATPase, ZIP, CAX, CDF, NRAMP and OPT/YSL families. Here we discuss their roles in the acquisition, distribution and homeostasis of Mn and aspects of their regulation.

1 Introduction

The mineral nutrition of higher plants is of fundamental importance to agriculture and human health. Awareness of the importance of micronutrients (most of which are heavy metals) to agriculture has been growing because in many soils micronutrient availability limits crop production and influences nutritional quality. Plants require certain heavy metals such as copper (Cu), zinc (Zn), iron (Fe) and manganese (Mn) in trace amounts as essential micronutrients for normal growth and development. These heavy metal micronutrients serve structural roles in proteins,

L.E. Williams (✉)
School of Biological Sciences, University of Southampton, Southampton SO16 7PX, UK
e-mail: lew@soton.ac.uk

J.K. Pittman
Faculty of Life Sciences, University of Manchester, Michael Smith Building, Oxford Road, Manchester M13 9PT, UK
e-mail: jon.pittman@manchester.ac.uk

R. Hell and R.-R. Mendel (eds.), *Cell Biology of Metals and Nutrients*, Plant Cell Monographs 17,
DOI 10.1007/978-3-642-10613-2_5,

act as enzyme cofactors and are components of cellular redox reactions (Hall and Williams 2003). If plants do not have sufficient quantities of these metals then deficiency symptoms develop. However, if present in excess in a bioavailable form, these metals, and other non-essential heavy metals such as Cd and Pb, can be poisonous and toxicity symptoms become apparent (Srivastava and Singh 2006). The micronutrient content of foods is a crucial factor in human health and nutrition because micronutrient imbalances in humans are the cause of many diseases (Ortiz-Monasterio et al. 2007). Plants are at the beginning of the food chain; therefore, improving the uptake of minerals from the soil and enhancing their movement to and bioavailability in the edible parts of the plant will provide benefits for animal and human nutrition (Palmgren et al. 2008). Conversely, large areas of agricultural soil are contaminated with heavy metals by natural and anthropogenic activities, and plants are directly exposed to these contaminants (Nicholson et al. 2003; Huang et al. 2007; Mico et al. 2008). A major source of exposure to deleterious metals in human populations is via the consumption of metal-contaminated foodstuffs and so understanding which mechanisms regulate absorption and storage of deleterious metals in plants is also very important.

Here we highlight recent work that has focussed on the important micronutrient, Mn, in particular its homeostasis at the cellular level. This metal is necessary throughout all stages of plant development and has numerous functions in plants (see below). Its physiological functions are discussed as are the transport mechanisms controlling cellular uptake and compartmentalisation.

2 Importance of Mn in Plants and Consequences of Mn Deficiency and Excess

Mn is an essential element required by all known organisms. Mn has numerous functions in plants (see Table 1), the most well known being its association with the oxygen evolving complex (OEC) of Photosystem II (PSII) where it is required for water oxidation and oxygen evolution (essential for aerobic life on this planet) (Barber 2009). It is also required by mitochondrial superoxide dismutase (MnSOD), an enzyme involved in protection against oxidative stress (Fig. 1), and oxalate oxidase (germin), although this enzyme activity is mainly found in cereals rather than dicot plants (Bowler et al. 1991; Requena and Bornemann 1999; Woo et al. 2000). Mn also activates a variety of other enzymes and proteins including phenylalanine ammonia-lyase, decarboxylases and glycosyltransferases (Durst 1976, Burnell 1988; Marschner 1995; White et al. 1993) (see Table 1). For many Mn-stimulated enzymes, Mn can be replaced by other metals such as magnesium (Mg), although in PSII, MnSOD and oxalate oxidase the Mn requirement is indispensable (Hebbern et al. 2009). Under conditions of excess, Mn can replace Mg in many enzymatic reactions, often inhibiting those enzymes. For some enzymes, such as those involved in glycolysis and the TCA cycle, Mn can be as or more effective

Table 1 Some key plant cell enzymes and proteins which are either dependent on Mn, sensitive to Mn, or contain Mn

Mn-affected enzyme/protein/process	Selected references
Photosystem II oxygen evolving complex	
Catalyses water-splitting reaction and O_2 evolution in the chloroplast thylakoid	Barber (2009)
Mn-dependent superoxide dismutase (MnSOD)	
Redox enzyme in mitochondria which acts on superoxide radical	Bowler et al. (1991)
Oxalate oxidase (germin)	
Catalyses the conversion of oxalate and dioxygen into CO_2 and H_2O_2 in apoplast of cereals; may also have SOD activity	Requena and Bornemann (1999); Woo et al. (2000)
IAA-amino acid hydrolases	
Releases free IAA by cleaving IAA-amino acid conjugates and Mn^{2+} is the preferred cofactor; isoforms include ILR1 and IAR3 possibly localised in the endoplasmic reticulum	LeClere et al. (2002)
Glycosyl transferases	
Golgi enzymes that catalyse the transfer of a sugar molecule to a specific acceptor such as for synthesising cell wall polysaccharides; including: xyloglucan glycosyltransferase; UDP-xylose glucuronyltransferase; arabinoxylan arabinosyltransferase	White et al. (1993); Zeng et al. (2008) Porchia et al. (2002)
Phenylalanine ammonia-lyase (PAL)	
Key enzyme in metabolism of plant phenolics, catalyses the deamination of L-phenylalanine to yield trans-cinnamic acid and ammonia; in the cytosol	Durst (1976)
NAD-malic enzyme	
Catalyses decarboxylation of malate to pyruvate releasing CO_2 in mitochondria	Anderson and Evans (1956)
Phosphoenolpyruvate (PEP) carboxykinase	
Catalyses decarboxylation of oxaloacetate to PEP releasing CO_2 in mitochondria	Chen et al. (2002)
Isocitrate dehydrogenase	
Catalyses oxidation of isocitrate to α-ketoglutarate to form NADH in mitochondria	Anderson and Evans (1956)
RNA polymerase	Guilfoyle and Hanson (1973)
Xylose accumulaton	
Increased under Mn deficiency	Bar-Akiva and Lavon (1967)
Auxin deficiency	
Caused by activation of indole-acetic acid oxidase under excess Mn	Srivastava and Singh (2006)
Mn-peroxidase	
Apoplastic bidirectional enzyme can catalyse oxidation of Mn(II) leading to formation of cytotoxic oxidant Mn(III); also catalyses oxidation of phenols which may mediate lignin formation, which increases under Mn deficiency; the reverse reaction catalyses H_2O_2 production, and is activated by excess Mn^{2+}	Fecht-Christoffers et al. (2003); Bar-Akiva and Lavon (1967); Srivastava and Singh (2006)

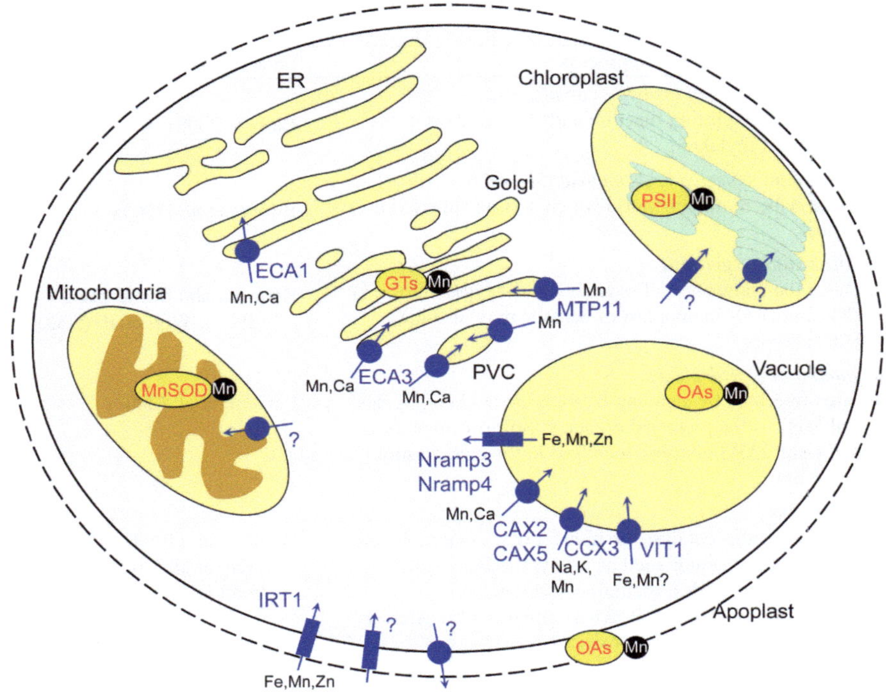

Fig. 1 Summary of the known Mn transport pathways in a typical *Arabidopsis* plant cell and the location of key Mn binding proteins and conjugates. Pathways for Mn uptake into some cell types include the Fe-deficiency-regulated ZIP-type transition metal transporter IRT1. Other Mn cell uptake pathways are unclear as are pathways for Mn efflux from the cell. Mn can be released into the cytosol via the vacuolar NRAMP3 and NRAMP4 transporters. Mn can be transported into the vacuole via the Ca^{2+} and Mn^{2+}/H^{+} antiporters CAX2 and CAX5, and possibly via the Fe transporter VIT1. A cation/H^{+} antiporter CCX3 may also transport Mn^{2+} into the vacuole. ECA1 and ECA3 are P_{2A}-type ATPases that transport Ca^{2+} and Mn^{2+} into the endoplasmic reticulum (ER) and Golgi, respectively, although ECA3 may also, or alternatively, localise at a pre-vacuolar compartment (PVC). MTP11 is a CDF-type transporter that transports Mn^{2+} from the cytosol into the Golgi and/or a PVC. Pathways for mitochondrial and chloroplastic Mn transport are currently unknown. Key Mn binding proteins and Mn conjugates include Photosystem II (PSII) in the chloroplast, Mn-dependent superoxide dismutase (MnSOD) in the mitochondria, and glycosyltransferases (GTs) in the Golgi. Mn accumulates in the vacuole and apoplast during Mn excess and can be conjugated with organic acids (OAs)

than Mg. Mn is necessary throughout all stages of plant development and is accumulated as Mn(II). Mn deficiency (most common in alkaline soils where there is reduced availability) is often seen in cereals such as wheat and barley where it is one of the most common trace element deficiencies leading to reduced growth and yield (Jiang 2006). Interveinal chlorosis is commonly reported in plants as a sign of Mn deficiency. Not only is there variability between plant species in their ability to grow on low Mn, there are also considerable differences among genotypes of the same species. Varieties displaying tolerance to Mn deficiency are termed

Mn-efficient and the mechanisms underlying this are starting to be elucidated. In barley, differential capacity for high-affinity Mn influx contributes to Mn efficiency (Pedas et al. 2005); in wheat it may be due to improved internal utilisation (Jiang 2006). Latent Mn deficiency in barley substantially increases transpiration and decreases water use efficiency and these changes are associated with a marked decrease in the epicuticular wax layer (Hebbern et al. 2009). This led to the conclusion that drought would put an additional stress on Mn-deficient plants that are already suffering from disturbances in key metabolic processes (Hebbern et al. 2009).

Mn deficiency can be overcome by Mn fertiliser application (although Mn^{2+} is rapidly oxidised when supplemented artificially), but growing genotypes that are Mn-efficient is a more cost-effective and environmentally-friendly approach (Jiang 2006). Mn availability increases at low pH and thus in acidic soils toxicity is a significant problem causing chlorosis and even necrosis (Marschner 1995). Roots sometimes show browning and crack (Foy et al. 1995). Another characteristic Mn toxicity symptom is the presence of brown spots on mature leaves which leads to chlorosis (Fig. 2). These are caused by accumulation of oxidised Mn and oxidised phenolics in the cell wall (Fecht-Christoffers et al. 2003). Oxidised Mn can be extremely toxic to the cell; thus antioxidant capacity in the apoplast is an important Mn detoxification mechanism. Both deficiency and excess Mn can result in significant losses in yield, therefore the molecular and biochemical consequences of Mn deficiency and toxicity require further investigation if we are to develop approaches to overcome these problems.

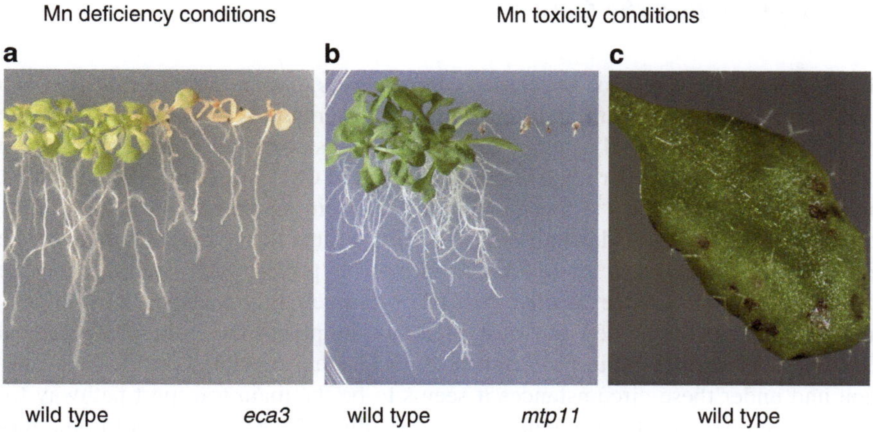

Fig. 2 Phenotypes of *Arabidopsis* plants under Mn stress conditions. (**a**) Comparison of wild type and *eca3* mutant under Mn deficiency conditions. Plants were grown on 0.5 MS low Ca medium, 1% sucrose, 0.8% agarose with no added Mn (Mills et al. 2008). (**b**) Comparison of wild type and *mtp11* mutant under conditions of Mn excess (0.5 MS medium, 1% sucrose, 0.8% agarose with 150 μM Mn (Lomax, Pittman and Williams, unpublished). (**c**) A common Mn toxicity symptom of plants. Wild type *Arabidopsis* transferred onto 0.5 MS medium, 1% sucrose, 0.8% agarose containing 1 mM Mn display brown spots on mature leaves due to excess Mn accumulation in the apoplast and oxidation of apoplastic Mn and phenols by peroxidise activity

3 Uptake, Distribution and Detoxification

Mechanisms are required to ensure sufficient uptake of Mn into the root and its correct partitioning to plant organs, cells and subcellular organelles. In particular, it is important to ensure that Mn is delivered to correct target proteins with the appropriate timing. Conversely Mn must not be allowed to accumulate where it might have damaging effects. Detoxification mechanisms may include regulation of uptake and transport, sequestration and compartmentalisation. Plants employ various mechanisms to ensure sufficient uptake and distribution of Mn from the soil into the root, and for Mn detoxification. The situation is likely to be complex but it is clear that membrane transport proteins play a crucial role in ensuring Mn homeostasis. In yeast and many bacterial systems, several pathways for Mn transport have been identified (Kehres and Maguire 2003; Culotta et al. 2005) and we are now beginning to elucidate the transport mechanisms for Mn in plants. Establishing exactly which mechanisms are involved in Mn homeostasis under deficiency, replete and excess conditions and clarifying their relative importance is essential if we wish to manipulate plants to improve Mn efficiency and tolerance. A variety of different transporters belonging to a range of transporter families have now been identified. These and some of the related genes found in other species are discussed below.

3.1 Uptake into the Cell

There are a variety of broad-specificity transporters that are able to transport Mn into some cell types, in particular root cells. Ca^{2+}-permeable channels may be a site of entry for Mn^{2+} as well as other cations. This was observed in a rye root cation channel which had similar relative permeabilities of Ca^{2+} and Mn^{2+} (White 1993), although evidence of Mn^{2+} transport by this route in maize roots was based only on competition assays (Marshall et al. 1994). It seems likely that members of the Zn-regulated transporter, Fe-regulated transporter protein (ZIP) family may be candidates for Mn uptake. *Arabidopsis* IRT1 can certainly transport Mn^{2+}, Zn^{2+} and Cd^{2+} when expressed in yeast although in plants its main physiological function is thought to be in Fe^{2+} transport. IRT1 is up-regulated during Fe limitation and under these circumstances it seems to be the main transport pathway for Mn (Fig. 1). *irt1* knockout mutants have markedly reduced Mn content in the root under Fe limitation (Vert et al. 2002). Interestingly, Yang et al. (2008) showed that although *IRT1* transcripts were decreased under Mn deficiency conditions there was an increase in Fe concentration in the root; they proposed that Fe was transported more efficiently in the absence of Mn. Under normal conditions one of the 15 ZIP genes in *Arabidopsis* may function in Mn transport. ZIPs may also function in Mn uptake in other species; barley HvIRT1 seems to play an important role in controlling Mn efficiency as its expression in roots is induced mainly by Mn or Fe deficiency and it is more highly expressed in an Mn-efficient genotype compared to an Mn-inefficient genotype (Pedas et al. 2008). Transient expression in onion epidermal cells localised HvIRT1 to the plasma membrane, and heterologous expression in yeast mutants indicated that in addition to Mn^{2+}, HvIRT1 could transport Fe^{2+}/Fe^{3+}, Zn^{2+} and Cd^{2+}. The latter was supported by a novel yeast uptake assay based on inductively coupled plasma-mass spectrometry (ICP-MS) analysis (Pedas et al. 2008). There is also evidence that certain ZIPs from other plants may contribute to Mn transport such as from the legume *Medicago truncatula*. *MtZIP3* and *MtZIP4* are down-regulated under Mn deficiency conditions whereas *MtZIP5* shows a differential response in leaves and roots. MtZIP4 and MtZIP7 can restore yeast growth on low Mn media (López-Millán et al. 2004).

3.2 *Subcellular Compartmentalisation*

Techniques such as X-ray microanalysis and synchrotron X-ray mapping have been used to localise Mn in particular organs and tissues (Mazzolini et al. 1985; Kim et al. 2006). A clear knowledge of the subcellular partitioning of Mn requires the ability to image localised Mn at the cellular and subcellular level. Unlike metals such as calcium (Ca), no obvious reporters appear to be available for imaging Mn. A number of studies have utilised differential fractionation and Mn measurement methods to indicate the localisation of Mn within cellular compartments; however, this method can only clearly resolve certain compartments accurately, such as the cell wall and vacuole (Gonzalez and Lynch 1999). Various techniques coupled to transmission electron microscopy such as micro-particle-induced X-ray emission (PIXE), and electron energy loss spectrometry (EELS), or synchrotron radiation X-ray fluorescence (SXRF) microscopy, potentially have the required high resolution and sensitivity to image and quantify metals at the subcellular level (Lobinski et al. 2006; Punshon et al. 2009). For example, Mn has been shown to be localised in rat brain mitochondria affected by Mn toxicity by EELS (Morello et al. 2008) while SXRF has been used to map Mn distribution within unicellular diatoms (Twining et al. 2003).

Various transporters are thought to have a role in moving Mn into and out of particular organelles and these are discussed further below (Fig. 1).

3.2.1 The Role of CAX (Cation Exchanger) Transporters and the MTP (Metal Tolerance Protein) Family of Transporters

The sequestration of Mn into the vacuole is one of the key mechanisms for providing Mn tolerance in many plant tissues (Gonzalez and Lynch 1999; Fernando et al. 2006). Furthermore, Mn storage in the vacuole is also likely to be important for retrieval and trafficking to other locations in the cell, as and when it is required (see below). At least two protein classes have been demonstrated to mediate vacuolar Mn influx in plants: CAX (cation exchanger) transporters and some members of the MTP (metal tolerance protein) family of transporters (Fig. 1). The vacuolar Fe transporter VIT1 (Vacuolar Iron Transporter 1), which is orthologous to the yeast Fe/Mn transporter CCC1 (Ca^{2+}-sensitive cross complementer 1), can also accumulate Mn into vacuoles when expressed in yeast (Kim et al. 2006), suggesting that it may be another pathway for vacuolar Mn sequestration in planta. Mn^{2+} can be transported into the vacuole by H^+-coupled antiport (Gonzalez et al. 1999) and this could be mediated by one of the CAX transporters. The plant CAX transporters can transport a range of divalent cations (Shigaki and Hirschi 2006). Expression of CAX2 in yeast can provide tolerance to Mn stress (Shigaki et al. 2003) and expression of *Arabidopsis* transporters CAX2 and CAX4 in tobacco shows that they can effectively mediate the vacuolar sequestration of Mn^{2+} and provide tolerance to Mn stress (Hirschi et al. 2000; Korenkov et al. 2007). The physiological relevance is indicated by the significant reduction of vacuolar Mn^{2+}/H^+ antiport activity in *Arabidopsis cax2* mutants. The fact that the *cax2* plants do not have a significant Mn-sensitive phenotype (Pittman et al. 2004) suggests that there may be functional redundancy by other Mn^{2+}/H^+ antiport pathways. One candidate is CAX5 which can also function as a tonoplast-localised Mn^{2+} transporter (Edmond et al. 2009). Testing the *cax2cax5* double mutant for vacuolar Mn^{2+}/H^+ antiport activity will be necessary to support this suggestion. Unlike *CAX2* and *CAX5* which are expressed relatively highly throughout the plant (Edmond et al. 2009), *CAX4* is specifically expressed in roots and appears to be involved in root growth and development under metal stress conditions (Mei et al. 2009). *CAX4* is highly up-regulated in the root in response to Mn and both primary and lateral root growth is reduced in *cax4* mutant lines in response to elevated Mn. Mn transport by CAX-type transporters is not unique to *Arabidopsis*. CAX transporters from rice (OsCAX1a, OsCAX3), barley (HvCAX2) and tomato (LeCAX2) are able to transport Mn^{2+} (Kamiya et al. 2005; Edmond et al. 2009), probably into the vacuole. Interestingly, Mn^{2+}/H^+ antiport activity has also been identified at the plasma membrane of cucumber roots (Migocka and Klobus 2007). It is as yet unclear whether a CAX protein is responsible for this Mn^{2+} efflux activity.

CCX3 has recently been shown to be an *Arabidopsis* endomembrane (vacuolar and unknown vesicle) H^+-dependent K^+ transporter that may also have Na^+ and Mn^{2+} transport properties (Morris et al. 2008). The five *Arabidopsis* CCX genes were previously named *CAX7-CAX11* but were renamed after phylogenetic analysis found them to be distinct from the *CAX1-CAX6* genes. Rather they show high similarity to the mammalian K^+-dependent Na^+/Ca^{2+} antiporter *NCKX6*. Expression of CCX3 rescued the Mn growth sensitivity phenotype of the yeast *smf1smf2* mutant. Mn also inhibited K^+ ($^{86}Rb^+$) uptake into yeast or *Arabidopsis* tonoplast vesicles expressing CCX3, suggesting that CCX3 can also transport Mn^{2+}. Mn also induced expression of *CCX3* in both roots and flowers. When ectopically expressed in tobacco, mature leaves accumulated Mn (as well as K and Na). This transporter does not seem to have affinity for divalent cations other than Mn^{2+}. It will be interesting in the future to determine the affinity of the other CCXs and to elucidate their potential contribution to Mn homeostasis.

MTP transporters (plant members of the cation diffusion facilitator, CDF family), can also provide H^+-coupled Mn^{2+} transport (Delhaize et al. 2007). Based on phylogenetic analysis and some functional data, it has been suggested that CDF transporters across all kingdoms can be characterised into three main groups; Zn-CDFs, Fe/Zn-CDFs and Mn-CDFs (Montanini et al. 2007). A number of plant and algal genes cluster within the Mn-CDF group, including *ShMTP1* from the Mn-tolerant tropical legume *Stylosanthes hamata.* ShMTP1 (renamed ShMTP8 to maintain consistency with *Arabidopsis* MTP gene nomenclature) provides Mn tolerance by vacuolar Mn^{2+} sequestration (Delhaize et al. 2003). ShMTP8 orthologues from *Arabidopsis* and poplar are also involved in providing significant Mn tolerance (Delhaize et al. 2007; Peiter et al. 2007). The *Arabidopsis MTP11* gene is particularly interesting because of its membrane localisation. While MTP11 appears critical for providing tolerance to excess concentrations of Mn, as demonstrated by heterologous expression in yeast and the Mn sensitivity of the *mtp11* knockout (Fig. 2), surprisingly the transporter was not observed at the tonoplast but at an endomembrane location (Delhaize et al. 2007). The exact membrane localisation of MTP11 is open to interpretation: either at the trans Golgi network, as determined by sialyl transferase marker colocalisation in *Arabidopsis* protoplasts (Peiter et al. 2007) or the prevacuolar compartment (PVC), as determined by VSR2 marker colocalisation in tobacco protoplasts (Delhaize et al. 2007). It is possible that the localisation of MTP11 is dynamic and may be present at either location depending on the cellular conditions or indeed the Mn status. What is also unclear is the exact mechanism of Mn tolerance by MTP11. Peiter et al. (2007) suggested that loading of Mn^{2+} into a secretory pathway vesicle may then allow removal of Mn^{2+} from the cell via exocytosis (Peiter et al. 2007), a mechanism that is analogous to one observed in yeast (Lapinskas et al. 1995). In contrast it is also possible that Mn^{2+} may be shuttled to the vacuole via an endosomal route. A different role in Mn homeostasis has been implicated for the *Chlamydomonas CrMTP4* gene which also clusters with the Mn-CDF group. *CrMTP4* expression is significantly enhanced following severe Mn deficiency, leading to the suggestion that this pathway may be involved in Mn distribution to an organelle where Mn delivery is essential under severe Mn deprivation to the cell (Allen et al. 2007). A key characteristic of the Mn-CDFs analysed to date appears to be that they are specific to Mn^{2+}, a characteristic so far unique to plant Mn transporters.

3.2.2 The Role of Natural Resistance Associated Macrophage Protein Transporters

Metal transporters of the natural resistance associated macrophage protein transporters (NRAMP) family play a critical role in Mn acquisition and homeostasis in many microbial organisms, including pathogenic bacteria such as *Salmonella* and yeast such as *Saccharomyces* (Nevo and Nelson 2006). For example, the yeast NRAMP Smf1 is up-regulated by Mn starvation and mediates Mn^{2+} accumulation across the plasma membrane (Culotta et al. 2005). While plants possess a number of

NRAMP isoforms, most of which have the ability to transport Mn^{2+} (reviewed in Pittman 2005), they appear principally to have a major role in regulating Fe homeostasis and are often tightly controlled by Fe deficiency (Thomine et al. 2003; Lanquar et al. 2005). Thus to date, a plant NRAMP which is analogous to Smf1 and required for Mn influx into the cell has not been identified and may indeed be lacking in higher plants. Unlike any of the plant isoforms characterised to date, *CrNRAMP1* from *Chlamydomonas* is significantly up-regulated by Mn starvation (Allen et al. 2007) although its membrane localisation is unknown. *Arabidopsis* NRAMP3 and NRAMP4 are present at the tonoplast and are required for the mobilisation of Fe from the vacuole in cotyledons during seed germination and very early seedling growth (Lanquar et al. 2005). In addition to Fe^{2+}, both NRAMPs can transport Mn^{2+} and Zn^{2+}; for example, over-expression or knockout of *NRAMP3* alters the content of Fe, Zn and Mn in seedlings (Thomine et al. 2003), although whether NRAMP3 or NRAMP4 are important for Mn release from the vacuole remains to be seen.

3.2.3 Role of P-type ATPases

Plants possess multi-gene families of transporters classified as belonging to the P_{2A} and P_{2B} type-ATPases. Generally pumps in these families are referred to as Ca^{2+}-ATPases although for many of them their full substrate-specificity has not been assessed. Type 2A ATPases show homology to the sarcoplasmic/endoplasmic reticulum (ER) Ca^{2+}-ATPases (SERCAs) found at the ER in animal cells, and Type 2B Ca^{2+}-ATPases which are stimulated by calmodulin and show homology to the calmodulin-binding plasma membrane Ca^{2+}-ATPases (PMCAs) found at the plasma membrane of animal cells (Axelsen and Palmgren 1998; Evans and Williams 1998). *Arabidopsis* contains four P_{2A} pumps (ECA1-4, Endomembrane Ca^{2+}-ATPase) and ten P_{2B} pumps (ACAs for Auto-inhibited Ca^{2+}-ATPase) and it is certain members of the ECA family that have been implicated in Mn^{2+} transport. It remains to be tested whether the ACAs can transport Mn^{2+} although studies in yeast suggest that ACA2 cannot (Wu et al. 2002).

ECA3

ECA3, one of the four P_{2A}-type ATPases in *Arabidopsis* is of particular importance in Mn homeostatic mechanisms (Mills et al. 2008). *eca3* mutants display severe deficiency symptoms when grown on media lacking Mn (chlorosis and inhibition of root and shoot growth) (Fig. 2). These striking symptoms are suppressed by low levels of Mn (1–2 μM). *eca1* and *eca2* mutants did not show this response suggesting that ECA1 and ECA2 do not play a role under deficiency conditions. Consistent with a role in Mn transport, ECA3 partially restored the growth defect on high Mn of the *S. cerevisiae pmr1* mutant, which is defective in the Golgi Ca^{2+}/Mn^{2+} pump, PMR1 (Mills et al. 2008).

ECA3 may also have a role under toxic Mn conditions as *eca3-4* was shown to be more susceptible to Mn concentrations of 50 μM (Li et al. 2008). This needs further investigation as only one mutant was shown to be susceptible. Further mutants should be tested but if they also show susceptibility then this could indicate a dual biological role for ECA3 under conditions of Mn deficiency and toxicity. When transiently expressed in tobacco, the cellular localisation of a YFP-ECA3 fusion protein overlapped with the Golgi protein, GONST1 (Mills et al. 2008). In *Arabidopsis* protoplasts it overlaps with ARA7, an endosome/PVC marker (Li et al. 2008). As with the discrepancy with MTP11 localisation studies (see above), this could reflect the different systems used for localisation studies or could indicate that ECA3 distribution may alter depending on the physiological state of the cell. The localisation of ECA3 to the Golgi/post Golgi compartments provides a clue to its possible dual function. We hypothesised that ECA3 is required to supply Mn to particular Mn-requiring enzymes and proteins in Golgi-related compartments for important biochemical processes; when it is absent, insufficient Mn is supplied and these processes are inhibited thus resulting in poor Mn nutritional problems (Mills et al. 2008). At elevated Mn levels it is possible that ECA3 loads excess Mn into these compartments and contributes to a Golgi-based detoxification mechanism. ECA3 does not appear to be transcriptionally regulated by metals (Mills et al. 2008) but there may be other forms of regulation.

Previously we proposed that the P_{2A}-type ATPase ECA3 could serve a similar function to Golgi-associated secretory pathway Ca^{2+}-ATPases (SPCAs) found in other organisms (Mills et al. 2008). Whereas in mammalian cells there are three distinct classes of P_2-type Ca-ATPases: SERCA type, PMCA type and SPCAs (also referred to as PMR1 type), which can clearly be distinguished in phylogenetic analysis, there appear to be no SPCAs in plants (Pittman et al. 1999; Mills et al. 2008). SPCAs function in the secretory pathway and transport not only Ca^{2+} but also Mn^{2+}, with high affinity, and they function both in Mn detoxification via exocytosis and also in providing Mn for Golgi-localised enzymes (Yadav et al. 2007). In yeast, the SPCA, PMR1, functions as a Ca^{2+} and Mn^{2+} transporter at the Golgi, transporting these ions into the secretory pathway where they can exert distinct functions: Mn supplied by PMR1 is required to activate various Mn-requiring enzymes involved in the addition of complex carbohydrates onto N- and O-linked glycosylated proteins (Lapinskas et al. 1995; Durr et al. 1998) while Ca is required to sustain vacuolar protein sorting. The complementation on low Ca medium by ECA3 of the yeast mutant strain K616, which lacks its endogenous Golgi (PMR1) and vacuolar (PMC1) Ca^{2+} pumps, is consistent with ECA3 functioning as a Ca^{2+} pump in yeast (Mills et al. 2008). In addition, expression of ECA3 reduced the Mn-sensitive phenotype of both the K616 yeast mutant and also the single *pmr1* mutant, consistent with it serving a similar Mn^{2+}-transporting function to PMR1 (Mills et al. 2008). Thus plants may compensate for not having SPCAs by possessing other pumps that can fulfil their divalent cation transporting function in the secretory pathway.

There are also Ca-dependent phenotypes in *Arabidopsis eca3* mutants, with evidence that ECA3 contributes to Ca-stimulated root growth (Li et al. 2008;

Mills et al. 2008). It is not clear whether ECA3 has a predominant role as a Ca^{2+} pump in some cells and an Mn^{2+} pump in others and it may depend on the relative cellular concentrations of these divalent cations in the cell. Enhanced secretion of peroxidases by *eca3* mutants has been taken to suggest that this pump promotes root growth possibly through the activities of endosomes involved in sorting, membrane trafficking and secretion (Li et al. 2008).

ECA1 and LCA1

Arabidopsis ECA1 may also play a role under conditions of Mn toxicity (Wu et al. 2002). ECA1 is localised at the ER and studies in yeast suggest it can transport Ca^{2+}, Mn^{2+} and Zn^{2+} (Liang et al. 1997; Wu et al. 2002). Although only investigated in a single T-DNA insertional mutant (*eca1-1*), growth of *eca1-1* was inhibited at elevated Mn (0.5 mM). In particular it appeared that tip growth and root hair elongation were inhibited in this mutant at elevated Mn (Wu et al. 2002). ECA2 has not been investigated as thoroughly as ECA1 and ECA3 and as yet there is no direct evidence to suggest that it functions in Mn^{2+} transport *in planta*. *eca2* mutants did not show the deficiency phenotype at low Mn displayed by *eca3* mutants nor did they show any marked difference to wild type at 50 μM Mn (Mills et al. 2008). It will be interesting to see whether deficiency or toxicity symptoms are exacerbated in combinations of *eca* mutants. For example, if the pathways by which ECA3 and ECA1 confer resistance to Mn toxicity are independent, one may see additive effects in a double mutant.

Tomato LCA1 shares higher similarity with ECA2 (77%) than with ECA1 (67%) or ECA3 (50%), and sequence and phylogenetic analyses suggest that LCA1 and ECA2 may be orthologous proteins (Pittman et al. 1999). LCA1 appears to function as both a Ca^{2+} pump and a Mn^{2+} pump when expressed in yeast (Johnson et al. 2009). Antibodies raised against LCA1 interact with two different proteins, one localised at the tonoplast and the other at the plasma membrane and it was postulated that post-transcriptional or post-translational processing may be taking place (Ferrol and Bennett 1996). Further studies are clearly required into the ECA family to determine their coordinated response to Mn.

3.3 Long-Distance Transport and Seed Loading

In recent years there have been some significant advances in our understanding of the mechanisms of vascular loading/unloading and long-distance translocation of Zn and Fe, and the genes required at many steps of these pathways (Briat et al. 2007; Palmgren et al. 2008). In contrast, we still know relatively little about the control mechanisms for long-distance Mn flux at the molecular level. Mn will be translocated to shoot tissues via the xylem, although the transporters responsible for Mn xylem loading are unknown. Whether Mn is loaded as Mn^{2+} or as a complex such as with an organic acid is speculative. Earlier studies estimated that in soybean and tomato approximately 60% of the xylem Mn is in the form of Mn^{2+} while the remainder is present as a complex, possibly with citrate or malate (White et al. 1981), but not with nicotianamine (NA) (Pich and Scholz 1996). However, studies in tobacco show that NA is important for the long distance translocation of metals including Mn to the leaves and flowers (Takahashi et al. 2003). Following transport in the xylem, Mn is unloaded in various aerial parts of the plant including leaves and flowers. Metals may also be remobilised following transfer to shoot tissues via the phloem, and for Mn such recirculation in the plant may be species dependent. For example, no phloem mobilisation was observed in young wheat (Pearson and Rengel 1994; Riesen and Feller 2005), but was detected in Douglas Fir (Ducic et al. 2006). Likewise, high concentrations of phloem Mn chelated to NA were measured in castor bean (Schmidke and Stephan 1995).

Seed loading of minerals such as Mn is of particular interest; however, our understanding of the processes are currently better at the whole organ level rather than at the cellular and molecular level. The physiological mechanisms of Mn loading into seeds have been well studied in cereal grains such as wheat. Many cereals have significant grain Mn content, particularly in the aleurone and the embryo where it is concentrated in globoid crystals in complex with phytate (Loneragan 1988). For example, X-ray microanalysis studies of wheat grains observed Mn concentrated in the radicle of the embryo plus the seed coat and scutellum (Lott and Spitzer 1980; Mazzolini et al. 1985). Mn is transferred to the developing wheat grain via the xylem where it accumulates in the spiklet structures of the wheat ear (Pearson and Rengel 1994; Pearson et al. 1995). Along with most micronutrients, Mn is then loaded into the grain itself via the grain phloem (Pearson et al. 1996). Mn then enters the crease and pericarp tissues of the grain (Pearson et al. 1998). Fe is loaded into seeds in the form of Fe–NA via yellow stripe-like (YSL) transporters and it is possible that Mn may be similarly loaded as Mn–NA. The rice OsYSL2 transporter, which is expressed in the vasculature of the developing seed, is a transporter of both Fe–NA and Mn–NA (Koike et al. 2004). *OsYSL2* expression increases during seed development with strong expression in the embryo, indicating a role of this transporter in Fe and Mn loading into the phloem and into the rice grain embryo cells. An efflux transport step may also be required at the seed coat plasma membrane prior to loading into the grain filial tissues, but the identity of such an Mn efflux transporter is unknown. As the wheat grain develops, Mn is remobilised within the grain. During development some Mn is slowly retranslocated to the embryo (Pearson et al. 1996, 1998). Mn is later remobilised again into the developing root and shoot during imbibition and germination, although this is mostly from the seed coat and the proportion of Mn remobilised from this tissue is small (Moussavi-Nik et al. 1997, 1998).

In *Arabidopsis* seeds, oligopeptide transporters (OPT) may play a role in seed mineral loading by facilitating the transport of metal chelate substrates. Knockout analysis indicates that OPT3 is critical for the mobilisation of Fe, Mn, Zn and possibly Cu into developing *Arabidopsis* seeds (Stacey et al. 2008). In dicot plants such as *Arabidopsis*, early developing seed Mn (as Mn-phytate) is present in the chalazal endosperm rather than in developing embryo globoid crystals (Otegui et al. 2002). Specifically, it is in the ER of the endosperm that the Mn is localised, leading to the speculation that the Mn^{2+}-ATPase ECA1 may be important in early Mn seed partitioning (Otegui et al. 2002). Later in *Arabidopsis* seed development, Mn is translocated from the endosperm to the embryo at a stage of development which coincides with increasing abundance of MnSOD and PSII proteins (Otegui et al. 2002). In the mature *Arabidopsis* seed, Mn is localised specifically in the hypocotyl and radicle, and lower epidermal cells of the cotyledons, as observed by synchrotron X-ray mapping (Kim et al. 2006).

3.4 Are There Transporters Yet to be Identified?

It is still unknown whether plants possess ATP-binding cassette transporters responsible for Mn transport similar to those existing in bacteria. MntABC is a high-affinity Mn transporter identified in the cyanobacterium *Synechocystis* sp PCC 6803 (Bartsevich and Pakrasi 1996) that is important during Mn starvation. Homologues of MntABC have not been found in plant chloroplasts; however, Mn-transporting members of the NRAMP and/or CDF families would be good candidates to mediate transport across the chloroplast envelope and thylakoid membranes.

The *S. cerevisiae* high affinity phosphate transporter, Pho84 acts as a low-affinity Mn transporter, possibly transporting $MnHPO_4$ when exposed to high Mn levels (Jensen et al. 2003). Likewise metal phosphate complexes including $MnHPO_4$ are transported by the Pit phosphate transporter of *E. coli* (van Veen et al. 1994). There is currently no evidence that any of the plant Pho84 homologues, such as PHT1;1 can transport metal substrates or play a major role in Mn homeostasis.

Metals including Mn are required for some aspects of auxin metabolism. Amidohydrolases that cleave amide-linked conjugates such as IAA–Leu require metal cofactors, with Mn the strongly preferred cofactor (LeClere et al. 2002). Genetic screening of IAA conjugate metabolism has identified *Arabidopsis IAA–Leu resistant* mutants with altered tolerance to Mn stress including *ilr2* and *ilr3*. The *ilr2* mutant has increased tolerance to excess Mn and microsomal membranes isolated from the plants have increased ATP-dependent Mn^{2+} transport compared to wild type (Magidin et al. 2003). *ILR2* encodes a putative soluble protein of unknown function which may serve as a regulator of Mn transport. The *ilr3* mutant has increased tolerance to Mn and has also altered metal homeostasis regulation (Rampey et al. 2006). ILR3 is a basic helix–loop–helix type transcription factor which appears to regulate metal homeostasis in part through the action of putative Fe/Mn CCC1-like (VIT1-like) transporters. At least three CCC1-like transporters are downregulated in *ilr3*, but the substrate range and membrane location of these putative transporters has still to be determined.

4 Chaperones for Mn?

Metallochaperones occur in many organisms and are metal-binding proteins that participate in intracellular trafficking of essential metals to particular compartments (Banci et al. 2006; Lin et al. 2006). This is required so that the accumulation of free metal ions in the cell is carefully controlled and the metal ions can find the correct site within the cell and interact with the appropriate target such as a metal-requiring metallo-protein. In plants the focus to date has been on Cu chaperones (Puig et al. 2007) but Mn chaperones could also exist in plants. Mtm1 from yeast was identified as a protein which is required for the proper insertion of Mn into mitochondrial MnSOD2 (Luk et al. 2003). It is a member of the mitochondrial carrier family (MCF) of transport proteins but it is unclear whether it functions as a specific Mn chaperone for SOD2 or has other functions. Mtm1 is not a general mitochondrial Mn transporter as *mtml* mutant mitochondria are not deficient in Mn, suggesting that the protein may deliver Mn directly to SOD2 or that it transports an unknown substrate that is required for Mn delivery to SOD2 (Luk et al. 2003). The *Arabidopsis* homologue *MTM1* has been identified which could be functionally equivalent to yeast Mtm1 (Su et al. 2007). MTM1 rescued the SOD2 defect in the *mtml* yeast mutant and *MTM1* expression in *Arabidopsis* seedlings was induced by paraquat, which promotes superoxide formation. An MTM1::GFP fusion construct localised to the mitochondria when transiently expressed in protoplasts, further suggesting a role of MTM1 in mitochondrial MnSOD activation (Su et al. 2007). Further work is required to confirm whether MTM1 functions as an Mn chaperone or a transporter in *Arabidopsis* and to determine whether other Mn chaperones exist in plants.

5 Homeostasis and Aspects of Sensing, Signalling and Regulation

Plants constantly monitor and respond to the availability of essential mineral nutrients in the soil. Metal stress can be caused when soil mineral levels are limiting, which inhibits plant growth and productivity, or by elevated mineral levels which can cause severe or even lethal toxicity symptoms. Therefore strict control is necessary so that plants are able to absorb sufficient micronutrients for normal growth and development but at the same time limit their toxicity when present in excess. A key objective is to determine whether mechanisms exist in plants for perceiving and responding to Mn stress both during conditions of deficiency and toxicity and to establish the roles and coordinated regulation of key genes involved.

Transcriptional regulation is a key process for the homeostatic control of a number of micronutrients such as Fe and Cu. Following exposure to micronutrient deficiency, mechanisms that enhance micronutrient acquisition and conserve utilisation are activated (Kim and Guerinot 2007; Yamasaki et al. 2007). For example, during Cu deficiency, microRNA398 down-regulates non-essential Cu proteins such as Cu/ZnSOD, possibly to maintain the functioning of plastocyanin (required for photosynthesis) and other essential Cu proteins (Yamasaki et al. 2007). Such mechanisms may be induced early in the transition from sufficiency to deficiency. There is much less information available for homeostatic control of Mn in plants. Transcriptional regulation of Mn stress responses in bacteria is controlled by the transcriptional regulator MntR, which represses the transcription of *MntABC* and *MntH* (a bacterial NRAMP) when Mn levels rise (Kehres and Maguire 2003). In the photosynthetic cyanobacteria *Synechocystis* 6803 the high affinity uptake transporter, MntABC, is negatively regulated upon Mn starvation by a two-component system comprising ManS, a histidine kinase that senses external Mn, and ManR a response regulator, that represses expression of the transporter (Ogawa et al. 2002; Yamaguchi et al. 2002). In yeast the NRAMP transporters Smf1 and Smf2 are up-regulated under Mn starvation by post-translational mechanisms (Culotta et al. 2005).

A transcriptome approach could provide an unbiased, broader picture of the molecular basis for Mn homeostatic networks revealing potential mechanisms and possible targets for further investigation. An important question is whether there is prioritisation over Mn supply. Using this approach, it may be possible to monitor whether there is any evidence that non-essential Mn-requiring proteins are down-regulated in order to maintain an Mn pool for more important Mn-dependent reactions. The OEC of PSII and MnSOD are two major Mn-requiring enzymes expected to be prime targets of Mn deficiency (Allen et al. 2007). In the alga *Chlamydomonas,* under Mn deficiency the loss of activity of the major MnSOD precedes the loss of PSII activity, indicative of regulated Mn supply to intracellular or intraorganellar compartments (Allen et al. 2007). Whether similar prioritisation of Mn supply occurs in *Arabidopsis* is unknown. The availability of antibodies to the proteins in the OEC and to MnSOD will allow an investigation of the Mn effects on protein levels and stability in the membrane which can be combined with effects on transcript levels to provide a more complete picture.

Recently Yang et al. (2008) have suggested that when external Mn falls below a threshold level, a Mn-specific pathway is induced which triggers post-embryonic developmental processes and results in a unique root hair phenotype (increased frequency of root hairs and increased numbers of bifurcated root hairs). This phenotype would help increase the absorptive surface of the root and thus may help combat low levels of Mn. A re-differentiation of atrichoblasts into root hair--forming cells was observed indicating that Mn deficiency altered the developmental programme of rhizodermal cells (Yang et al. 2008). Their results suggested that the normal patterning mechanisms are bypassed under Mn-deficient conditions. The mechanisms sensing these conditions and also orchestrating the changes in root epidermal patterning remain unknown (Yang et al. 2008). Using microarray analysis a total of 71 genes were classified as being differentially expressed in *Arabidopsis* roots in response to Mn deficiency (Yang et al. 2008). These fell into several functional groups: cell-wall related, signalling, transcriptional regulators and nucleic acid interacting, protein modification, transport and metabolism. Interestingly none of the genes encoding the candidate Mn transporters (described above) were induced in Mn-deficient roots suggesting that another as yet unidentified transporter may be responsible for uptake under these conditions or that post-transcriptional activation mechanisms may be operating (Yang et al. 2008).

6 Conclusions and Future Directions

The need to increase world agricultural output will progressively force crop production onto sub-optimal soils including those with micronutrient characteristics which can severely limit crop production, such as soils deficient in Mn or acidic soils where Mn toxicity is a problem. This has increased our awareness of the significance of micronutrients including Mn to agriculture. Plants are the basis of all foodstuffs, therefore understanding the processes of plant micronutrient nutrition is highly relevant to human health. Improving plant nutrition and developing superior crops is an important and potentially valuable biotechnological goal. Research that identifies genes important in Mn homeostasis has the potential to be exploited to improve plant Mn accumulation/distribution. For example, approaches aimed at generating plants which can extract Mn more efficiently from the soil may overcome problems of deficiency. Likewise, the generation of plants with superior Mn detoxification properties will allow them to be grown on a wider range of soils.

As described in this review, major advances have so far been made in the identification of Mn transport and homeostasis processes. We now need to understand more specifically how these processes operate and how they are coordinated to ensure that key physiological and metabolic processes operate. *Arabidopsis* is an excellent system for studying Mn transport and homeostasis, and for identifying genes and pathways that can then be targeted in crop systems. For example, *Arabidopsis* knockout mutants have proved invaluable for revealing important physiological roles related to Mn homeostasis (Peiter et al. 2007; Delhaize et al. 2007; Mills et al. 2008). When using mutants it is essential to test a wide range of metal conditions in order to reveal particular roles of transporters, chelators and chaperones (Tennstedt et al. 2009). A key question in relation to Mn homeostasis is how the plant ensures that Mn is delivered to the appropriate metal trafficking pathway. Metals such as Cu appear to have chaperones that sequester them in a non-reactive form and ensure delivery to appropriate sites. The situation for Mn is less clear.

A combination of approaches can be taken to dissect the pathways involved in Mn homeostasis. It will be vital to elucidate the mechanisms that plants use in perceiving and responding to Mn stress both during conditions of deficiency and toxicity and to establish the roles and coordinated regulation of key genes involved. The homeostasis of a number of micronutrients is controlled by coordinated transcriptional pathways, such as the regulation of Fe homeostasis via the transcription factor FIT. Following exposure to micronutrient deficiency, mechanisms that enhance micronutrient acquisition and conserve utilisation are activated (Kim and Guerinot 2007). Such mechanisms may be induced early in the transition from sufficiency to deficiency. However, there may be key players in Mn homeostasis that are not regulated transcriptionally, such as ECA3 and MTP11. Therefore other ways of identifying genes that are important in Mn homeostasis but which may or may not show Mn transcriptional regulation are important to pursue. Identifying mutants by ionomic profiles is one approach. The Purdue *Arabidopsis* Ionomics project has profiles of thousands of mutants which are altered in their ionomic profile, as determined by ICP-MS (Salt et al. 2008). Identifying mutants that are altered in their Mn levels could be particularly promising. Also laser ablation has been coupled with ICP-MS to investigate metal levels in particular tissues and this has revealed interesting data showing that the *ferric reductase defective3* (*frd3*) mutant of *Arabidopsis* accumulates a range of ions in its trichomes including elevated levels of Mn (Salt et al. 2008). This has great potential for the future. Forward genetic screens could also be employed to identify genes involved in Mn efficiency. There are multiple genetic mechanisms that may influence Mn efficiency by controlling root uptake, whole plant partitioning and subcellular targetting. Reduced growth and chlorosis is a symptom of Mn deficiency and could be used in a forward screen. Similarly, at a cellular level, essential Mn-dependent processes, and regulation of these processes, may directly control Mn accumulation and subcellular Mn partitioning. Thus screening for suppressors of a mutated Mn-dependent process may be fruitful. Such an approach was used in cyanobacteria to identify a novel regulator of Mn uptake by screening for a suppressor mutation which restores PSII activity (Chandler et al. 2003). A similar screen could be performed in plants.

Further attention should be given to the post-transcriptional and post-translational control mechanisms involved in Mn homeostasis. In particular, the potential for altered sub-cellular location of key transporters in response to deficiency and excess may help elucidate the processes that are involved in Mn homeostasis. The sensing mechanisms that are involved in the root growth responses to Mn deficiency conditions are of particular interest. Clearly much remains to be discovered and the tools are now available to make good progress in identifying components involved in Mn homeostasis and understanding their coordination and regulation.

Acknowledgements Work by L.E.W. is supported by the EU through its Sixth Framework Programme for RTD (contract no. FOOD-CT-2006-016253). It reflects only the author's views. The Community is not liable for any use that may be made of the information contained therein. J.K.P. is a recipient of a BBSRC David Phillips Fellowship (grant no. BB/B502152/1).

References

Allen MD, Kropat J, Tottey S, Del Campo JA, Merchant SS (2007) Manganese deficiency in Chlamydomonas results in loss of photosystem II and MnSOD function, sensitivity to peroxides, and secondary phosphorus and iron deficiency. Plant Physiol 143:263–277

Anderson I, Evans HJ (1956) Effect of manganese and certain metal cations on isocitric dehydrogenase and malic enzyme activities in *Phaseeolus vulgaris*. Plant Physiol 31:22–28

Axelsen KB, Palmgren MG (1998) Evolution of substrate specificities in the P-type ATPase superfamily. J Mol Evol 46:84–101

Bar-Akiva A, Lavon R (1967) Visible symptoms and metabolic patterns in micronutrient deficient Eureka lemon leaves. Israeli J Agric Res 17:7–16

Bartsevich VV, Pakrasi HB (1996) Manganese transport in the cyanobacterium Synechocystis sp PCC 6803. J Biol Chem 271:26057–26061

Barber J (2009) Photosynthetic energy conversion: natural and artificial. Chem Soc Rev 38:185–196

Bowler C, Slooten L, Vandenbranden S, De Rycke R, Botterman J, Sybesma C, van Montagu M, Inze D (1991) Manganese superoxide dismutase can reduce cellular damage mediated by oxygen radicals in transgenic plants. EMBO J 10:723–1732

Banci L, Bertini I, Cantini F et al (2006) The Atx1-Ccc2 complex is a metal-mediated protein-protein interaction. Nat Chem Bio 2:367–368

Briat JF, Curie C, Gaymard F (2007) Iron utilization and metabolism in plants. Curr Opin Plant Biol 10:276–282

Burnell JN (1988) The biochemistry of manganese in plants. In: Graham RD, Hannam RJ, Uren NC (eds) Manganese in soils and plants. Kluwer Academic, Dordrecht, pp 125–137

Chandler LE, Bartsevich VV, Pakrasi HB (2003) Regulation of manganese uptake in *Synechocystis* 6803 by RfrA, a member of a novel family of proteins containing a repeated five-residues domain. Biochemistry 42:5508–5514

Chen ZH, Walker RP, Acheson RM, Leegood RC (2002) Phosphoenolpyruvate carboxykinase assayed at physiological concentrations of metal ions has a high affinity for CO_2. Plant Physiol 128:160–164

Culotta VC, Yang M, Hall MD (2005) Manganese transport and trafficking: lessons learned from *Saccharomyces cerevisiae*. Eukaryot Cell 4:1159–1165

Delhaize E, Kataoka T, Hebb DM, White RG, Ryan PR (2003) Genes encoding proteins of the cation diffusion facilitator family that confer manganese tolerance. Plant Cell 15:1131–1142

Delhaize E, Gruber BD, Pittman JK, White RG, Leung H, Miao YS, Jiang LW, Ryan PR, Richardson AE (2007) A role for the AtMTP11 gene of *Arabidopsis* in manganese transport and tolerance. Plant J 51:198–210

Ducic T, Leinemann L, Finkeldey R, Polle A (2006) Uptake and translocation of manganese in seedlings of two varieties of Douglas fir (*Pseudotsuga menziesii* var. *viridis* and *glauca*). New Phytol 170:11–20

Durr G, Strayle J, Plemper R, Elbs S, Klee SK, Catty P, Wolf DH, Rudolph HK (1998) The medial-Golgi ion pump Pmr1 supplies the yeast secretory pathway with Ca^{2+} and Mn^{2+} required for glycosylation, sorting, and endoplasmic reticulum associated protein degradation. Mol Biol Cell 9:1149–1162

Durst F (1976) Correlation of phenylalanine ammonia-lyase and cinnamic acid hydroxylase activity changes in Jerusalem Artichoke tuber tissues. Planta 132:221–227

Edmond C, Shigaki T, Ewert S, Nelson M, Connorton J, Chalova V, Noordally Z, Pittman JK (2009) Comparative analysis of CAX2-like cation transporters indicates functional and regulatory diversity. Biochem J 418:145–154

Evans DE, Williams LE (1998) P-type calcium ATPases in higher plants – biochemical, molecular and functional properties. Biochim Biophys Acta 1376:1–25

Fecht-Christoffers MM, Maier P, Horst WJ (2003) Apoplastic peroxidases and ascorbate are involved in manganese toxicity and tolerance of *Vigna unguiculata*. Physiol Plant 117:237–244

Fernando DR, Batianoff GN, Baker AJ, Woodrow IE (2006) In vivo localization of manganese in the hyperaccumulator *Gossia bidwillii* (Benth.) N. Snow & Guymer (Myrtaceae) by cryo-SEM/EDAX. Plant Cell Environ 29:1012–1020

Ferrol N, Bennett AB (1996) A single gene may encode differentially localized Ca^{2+}-ATPases in tomato. Plant Cell 8:1159–1169

Foy CD, Weil RR, Coradetti CA (1995) Differential manganese tolerances of cotton genotypes in nutrient solution. J Plant Nutr 18:685–706

Gonzalez A, Lynch JP (1999) Subcellular and tissue Mn compartmentation in bean leaves under Mn toxicity stress. Aust J Plant Physiol 26:811–822

Gonzalez A, Koren'kov V, Wagner GJ (1999) A comparison of Zn, Mn, Cd, and Ca transport mechanisms in oat root tonoplast vesicles. Physiol Plant 106:203–209

Guilfoyle TJ, Hanson JB (1973) Increased activity of chromatin-bound ribonucleic acid polymerase from soybean hypocotyl with spermidine and high ionic strength. Plant Physiol 51: 1022–1025

Hall JL, Williams LE (2003) Transition metal transporters in plants. J Exp Bot 54:2601–2613

Hebbern CA, Laursen KH, Ladegaard AH, Schimidt SB, Pedas P, Bruhn D, Schjoerring JK, Wulfsohn D, Husted S (2009) Latent manganese deficiency increases transpiration in barley (*Hordeum vulgare*). Physiol Plant 135:307–316

Hirschi KD, Korenkov VD, Wilganowski NL, Wagner GJ (2000) Expression of Arabidopsis CAX2 in tobacco. Altered metal accumulation and increased manganese tolerance. Plant Physiol 124:125–133

Huang SS, Liao QL, Hua M, Wu XM, Bi KS, Yan CY, Chen B, Zhang XY (2007) Survey of heavy metal pollution and assessment of agricultural soil in Yangzhong district, Jiangsu Province, China. Chemosphere 67:2148–2155

Jensen LT, Ajua-Alemanji M, Culotta VC (2003) The *Saccharomyces cerevisiae* high affinity phosphate transporter encoded by *PHO84* also functions in manganese homeostasis. J Biol Chem 278:42036–42040

Jiang WZ (2006) Mn use efficiency in different wheat cultivars. Environ Exp Bot 57:41–50

Johnson NA, Liu F, Weeks PD, Hentzen AE, Kruse HP, Parker JJ, Laursen M, Nissen P, Costa CJ, Gatto C (2009) A tomato ER-type Ca^{2+}-ATPase, LCA1, has a low thapsigargin-sensitivity and can transport manganese. Arch Biochem Biophys 481:157–168

Kamiya T, Akahori T, Maeshima M (2005) Expression profile of the genes for rice cation/H^+ exchanger family and functional analysis in yeast. Plant Cell Physiol 46:1735–1740

Kehres DG, Maguire ME (2003) Emerging themes in manganese transport, biochemistry and pathogenesis in bacteria. FEMS Microbiol Rev 27:263–290

Kim SA, Guerinot ML (2007) Mining iron: iron uptake and transport in plants. FEBS Lett 81:2273–2280

Kim SA, Punshon T, Lanzirotti A, Li LT, Alonso JM, Ecker JR, Kaplan J, Guerinot ML (2006) Localization of iron in *Arabidopsis* seed requires the vacuolar membrane transporter VIT1. Science 314:1295–1298

Koike S, Inoue H, Mizuno D, Takahashi M, Nakanishi H, Mori S, Nishizawa NK (2004) OsYSL2 is a rice metal-nicotianamine transporter that is regulated by iron and expressed in the phloem. Plant J 39:415–424

Korenkov V, Hirschi K, Crutchfield JD, Wagner GJ (2007) Enhancing tonoplast Cd/H antiport activity increases Cd, Zn, and Mn tolerance, and impacts root/shoot Cd partitioning in *Nicotiana tabacum* L. Planta 226:1379–1387

Lanquar V, Lelievre F, Bolte S, Hames C, Alcon C, Neumann D, Vansuyt G, Curie C, Schroder A, Kramer U, Barbier-Brygoo H, Thomine S (2005) Mobilization of vacuolar iron by AtNRAMP3 and AtNRAMP4 is essential for seed germination on low iron. EMBO J 24:4041–4051

Lapinskas PJ, Cunningham KW, Liu XF, Fink GR, Culotta C (1995) Mutations in *PMR1* suppress oxidative damage in yeast-cells lacking superoxide-dismutase. Mol Cell Biol 15:1382–1388

LeClere S, Tellez R, Rampey RA, Matsuda SPT, Bartel B (2002) Characterization of a family of IAA-amino acid conjugate hydrolases from *Arabidopsis*. J Biol Chem 277:20446–20452

Li X, Chanroj S, Wu Z, Romanowsky SM, Harper JF, Sze H (2008) A distinct endosomal Ca^{2+}/Mn^{2+} pump affects root growth through the secretory process. Plant Physiol 147:1675–1689

Liang F, Cunningham KW, Harper JF, Sze H (1997) ECA1 complements yeast mutants defective in Ca^{2+} pumps and encodes an endoplasmic reticulum-type Ca^{2+}-ATPase in *Arabidopsis thaliana*. Proc Natl Acad Sci USA 94:8579–8584

Lin Y-F, Walmsley AR, Rosen BP (2006) An arsenic metallochaperone for an arsenic detoxification pump. Proc Natl Acad Sci USA. 103:15617–15622

Lobinski R, Moulin C, Ortega R (2006) Imaging and speciation of trace elements in biological environment. Biochimie 88:1591–1604

López-Millán AF, Ellis DR, Grusak MA (2004) Identification and characterization of several new members of the ZIP family of metal ion transporters in *Medicago truncatula*. Plant Mol Biol 54:583–596

Loneragan JF (1988) Distribution and movement of manganese in plants. In: Graham RD, Hannam RJ, Uren NC (eds) Manganese in soils and plants. Kluwer Academic, Dordrecht, pp 113–124

Lott JNA, Spitzer E (1980) X-ray studies of elements stored in protein body globoid crystals of *Triticum* grains. Plant Physiol 66:494–499

Luk E, Carroll M, Baker M, Culotta VC (2003) Manganese activation of superoxide dismutase 2 in *Saccharomyces cerevisiae* requires MTM1, a member of the mitochondrial carrier family. Proc Natl Acad Sci USA 100:10353–10357

Magidin M, Pittman JK, Hirschi KD, Bartel B (2003) *ILR2*, a novel gene regulating IAA conjugate sensitivity and metal transport in *Arabidopsis thaliana*. Plant J 35:523–534

Marschner H (1995) Mineral nutrition of higher plants, 2nd edn. Academic, London

Marshall J, Corzo A, Leigh RA, Sanders D (1994) Membrane potential-dependent calcium transport in right-side-out plasma membrane vesicles from *Zea mays* L. roots. Plant J 5:683–694

Mazzolini AP, Pallaghy CK, Legge GLF (1985) Quantitative microanalysis of Mn, Zn and other elements in mature wheat seed. New Phytol 100:483–509

Mei H, Cheng NH, Zhao J, Park S, Escareno RA, Pittman JK, Hirschi KD (2009) Root development under metal stress in *Arabidopsis thaliana* requires the H^+/cation antiporter CAX4. New Phytol 183:95–105

Mico C, Recatala L, Peris M, Sanchez J (2008) Discrimination of lithogenic and anthropogenic metals in calcareous agricultural soils: a case study of the lower Vinalopó region (SE Spain). Soil Sediment Contam 17:467–485

Mills RF, Doherty ML, López-Marqués RL, Weimar T, Dupree P, Palmgren MG, Pittman JK, Williams LE (2008) ECA3, a Golgi-localized P2A-type ATPase, plays a crucial role in manganese nutrition in Arabidopsis. Plant Physiol 146:116–128

Migocka M, Klobus G (2007) The properties of the Mn, Ni and Pb transport operating at plasma membranes of cucumber roots. Physiol Plant 129:578–587

Montanini B, Blaudez D, Jeandroz S, Sanders D, Chalot M (2007) Phylogenetic and functional analysis of the Cation Diffusion Facilitator (CDF) family: improved signature and prediction of substrate specificity. BMC Genomics 8:107

Morello M, Canini A, Mattioli P, Sorge RP, Alimonti A, Bocca B, Forte G, Martorana A, Bemardi G, Sancesario G (2008) Sub-cellular localization of manganese in the basal ganglia of normal and manganese-treated rats – an electron spectroscopy imaging and electron energy-loss spectroscopy study. Neurotoxicology 29:60–72

Morris J, Tian H, Park S, Sreevidya CS, Ward JM, Hirschi KD (2008) AtCCX3 is an *Arabidopsis* endomembrane H^+-dependent K^+ transporter. Plant Physiol 148:1471–1486

Moussavi-Nik M, Pearson JN, Hollamby GJ, Graham RD (1998) Dynamics of nutrient remobilization during germination and early seedling development in wheat. J Plant Nutr 21:421–434

Moussavi-Nik M, Rengel Z, Pearson JN, Hollamby G (1997) Dynamics of nutrient remobilisation from seed of wheat genotypes during imbibition, germination and early seedling growth. Plant Soil 197:271–280

Nevo Y, Nelson N (2006) The NRAMP family of metal-ion transporters. Biochim Biophys Acta 1763:609–620

Nicholson FA, Smith SR, Alloway BJ, Carlton-Smith C, Chambers BJ (2003) An inventory of heavy metal inputs to agricultural soils in England and Wales. Sci Total Environ 311:205–219

Ogawa T, Bao DH, Katoh H, Shibata M, Pakrasi HB, Bhattacharyya-Pakrasi M (2002) A two-component signal transduction pathway regulates manganese homeostasis in Synechocystis 6803, a photosynthetic organism. J Biol Chem 277:28981–28986

Ortiz-Monasterio JI, Palacios-Rojas N, Meng E, Pixley K, Trethowan R, Pena RJ (2007) Enhancing the mineral and vitamin content of wheat and maize through plant breeding. J Cereal Sci 46:293–307

Otegui MS, Capp R, Staehelin LA (2002) Developing seeds of *Arabidopsis* store different minerals in two types of vacuoles and in the endoplasmic reticulum. Plant Cell 14:1311–1327

Palmgren MG, Clemens S, Williams LE, Kraemer U, Borg S, Schjorring JK, Sanders D (2008) Zinc biofortification of cereals: problems and solutions. Trends Plant Sci 13:464–473

Pearson JN, Jenner CF, Rengel Z, Graham RD (1996) Differential transport of Zn, Mn and sucrose along the longitudinal axis of developing wheat grains. Physiol Plant 97:332–338

Pearson JN, Rengel Z (1994) Distribution and remobilization of Zn and Mn drug grain development in wheat. J Exp Bot 45:1829–1835

Pearson JN, Rengel Z, Jenner CF, Graham RD (1995) Transport of zinc and manganese to developing wheat grains. Physiol Plant 95:449–455

Pearson JN, Rengel Z, Jenner CF, Graham RD (1998) Dynamics of zinc and manganese movement in developing wheat grains. Aust J Plant Physiol 25:139–144

Pedas P, Hebbern CA, Schjoerring JK, Holm PE, Husted S (2005) Differential capacity for high-affinity manganese uptake contributes to differences between barley genotypes in tolerance to low manganese availability. Plant Physiol 139:1411–1420

Pedas P, Ytting CK, Fuglsang AT, Jahn TP, Schjoerring JK, Husted S (2008) Manganese efficiency in Barley: identification and characterization of the metal Ion transporter HvIRT1. Plant Physiol 148:455–466

Peiter E, Montanini B, Gobert A, Pedas P, Husted S, Maathuis FJM, Blaudez D, Chalot M, Sanders D (2007) A secretory pathway-localized cation diffusion facilitator confers plant manganese tolerance. Proc Natl Acad Sci USA 104:8532–8537

Pich A, Scholz G (1996) Translocation of copper and other micronutrients in tomato plants (*Lycopersicon esculentum* Mill): Nicotianamine-stimulated copper transport in the xylem. J Exp Bot 47:41–47

Pittman JK (2005) Managing the manganese: molecular mechanisms of manganese transport and homeostasis. New Phytol 167:733–742

Pittman JK, Mills RF, O'Connor CD, Williams LE (1999) Two additional type IIA Ca^{2+} ATPases are expressed in *Arabidopsis thaliana*: evidence that type IIA sub-groups exist. Gene 236:137–147

Pittman JK, Shigaki T, Marshall JL, Morris JL, Cheng NH, Hirschi KD (2004) Functional and regulatory analysis of the *Arabidopsis thaliana* CAX2 cation transporter. Plant Mol Biol 56:959–971

Porchia AC, Sorensen SO, Scheller HV (2002) Arabinoxylan biosynthesis in wheat. Characterization of arabinosyltransferase activity in Golgi membranes. Plant Physiol 130:432–441

Puig S, Mira H, Dorcey E, Sancenón V, Andrés-Colás N, Garcia-Molina A, Burkhead JL, Gogolin KA, Abdel-Ghany SE, Thiele DJ, Ecker JR, Pilon M, Peñarrubia L (2007) Higher plants possess two different types of ATX1-like copper chaperone. Biochem Biophys Res Commun 354:385–390

Punshon T, Guerinot ML, Lanzirotti A (2009) Using synchrotron X-ray fluorescence microprobes in the study of metal homeostasis in plants. Ann Bot 103:665–672

Rampey RA, Woodward AW, Hobbs BN, Tierney MP, Lahner B, Salt DE, Bartel B (2006) An *Arabidopsis* basic helix-loop-helix leucine zipper protein modulates metal homeostasis and auxin conjugate responsiveness. Genetics 174:1841–1857

Requena L, Bornemann S (1999) Barley (*Hordeum vulgare*) oxalate oxidase is a manganese-containing enzyme. Biochem J 343:185–190

Salt DE, Baxter I, Lahner B (2008) Ionomics and the study of the plant ionome. Annu Rev Plant Biol 59:709–733

Riesen O, Feller U (2005) Redistribution of nickel, cobalt, manganese, zinc, and cadmium via the phloem in young and maturing wheat. J Plant Nutr 28:421–430

Schmidke I, Stephan UW (1995) Transport of metal micronutrients in the phloem of castor bean (*Ricinus communis*) seedlings. Physiol Plant 95:147–153

Shigaki T, Hirschi KD (2006) Diverse functions and molecular properties emerging for CAX cation/H^+ exchangers in plants. Plant Biol 8:419–429

Shigaki T, Pittman JK, Hirschi KD (2003) Manganese specificity determinants in the *Arabidopsis* metal/H^+ antiporter CAX2. J Biol Chem 278:6610–6617

Stacey MG, Patel A, McClain WE, Mathieu M, Remley M, Rogers EE, Gassmann W, Blevins DG, Stacey G (2008) The *Arabidopsis* AtOPT3 protein functions in metal homeostasis and movement of iron to developing seeds. Plant Physiol 146:589–601

Srivastava AK, Singh S (2006) Biochemical markers and nutrient constraints diagnosis in citrus: a perspective. J Plant Nutr 29:827–855

Su Z, Chai MF, Lu PL, An R, Chen J, Wang XC (2007) AtMTM1, a novel mitochondrial protein, may be involved in activation of the manganese-containing superoxide dismutase in *Arabidopsis*. Planta 226:1031–1139

Takahashi M, Terada Y, Nakai I, Nakanishi H, Yoshimura E, Mori S, Nishizawa NK (2003) Role of nicotianamine in the intracellular delivery of metals and plant reproductive development. Plant Cell 15:1263–1280

Tennstedt P, Peisker D, Bottcher C, Trampczynska A, Clemens S (2009) Phytochelatin synthesis is essential for the detoxification of excel zinc and contributes significantly to the accumulation of zinc. Plant Physiol 149:938–948

Thomine S, Lelievre F, Debarbieux E, Schroeder JI, Barbier-Brygoo H (2003) AtNRAMP3, a multispecific vacuolar metal transporter involved in plant responses to iron deficiency. Plant J 34:685–695

Twining BS, Baines SB, Fisher NS, Maser J, Vogt S, Jacobsen C, Tovar-Sanchez A, Sanudo-Wilhelmy SA (2003) Quantifying trace elements in individual aquatic protist cells with a synchrotron X-ray fluorescence microprobe. Anal Chem 75:3806–3816

Van Veen HW, Abbe T, Kortstee GJJ, Konings WN, Zehnder AJB (1994) Translocation of metal phosphate via the phosphate inorganic transport system of *Escherichia coli*. Biochemistry 33:1766–1770

Vert G, Grotz N, Dedaldechamp F, Gaymard F, Guerinot ML, Briat JF, Curie C (2002) IRT1, an Arabidopsis transporter essential for iron uptake from the soil and for plant growth. Plant Cell 14:1223–1233

White MC, Baker FD, Chaney RL, Decker AM (1981) Metal complexation in xylem fluid. II. Theoretical equilibrium model and computational computer program. Plant Physiol 67: 301–310

White AR, Xin Y, Pezeshk V (1993) Xyloglucan glucosyltransferase in Golgi membranes from *Pisum sativum* (pea). Biochem J 15:231–238

White PJ (1993) Characterisation of a high-conductance, voltage-dependent cation channel from the plasma membrane of rye roots in planar lipid bilayers. Planta 191:541–551

Woo EJ, Dunwell JM, Goodenough PW, Marvier AC, Pickersgill RW (2000) Germin is a manganese containing homohexamer with oxalate oxidase and superoxide dismutase activities. Nat Struct Biol 7:1036–1040

Wu ZY, Liang F, Hong BM, Young JC, Sussman MR, Harper JF, Sze H (2002) An endoplasmic reticulum-bound Ca^{2+}/Mn^{2+} pump, ECA1, supports plant growth and confers tolerance to Mn^{2+} stress. Plant Physiol 130:128–137

Yadav J, Muend S, Zhang Y, Rao R (2007) A phenomics approach in yeast links proton and calcium pump function in the Golgi. Mol Biol Cell 18:1480–1489

Yamaguchi K, Suzuki L, Yamamoto H, Lyukevich A, Bodrova I, Los DA, Piven I, Zinchenko V, Kanchisa M, Murata N (2002) A two-component Mn^{2+}-sensing system negatively regulates expression of the *mntCAB* operon in *Synechocystis*. Plant Cell 14:2901–2913

Yamasaki H, Abdel-Ghany SE, Cohu CM, Kobayashi Y, Shikanai T, Pilon M (2007) Regulation of copper homeostasis by micro-RNA in *Arabidopsis*. J Biol Chem 282:16369–16378

Yang TJW, Perry PJ, Ciani S, Pandian S, Schmidt W (2008) Manganese deficiency alters the patterning and development of root hairs in Arabidopsis. J Exp Bot 59:3453–3464

Zeng W, Chatterjee M, Faik A (2008) UDP-xylose-stimulated glucuronyltransferase activity in wheat microsomal membranes: characterization and role in glucurono(arabino)xylan biosynthesis. Plant Physiol 147:78–91

Cell Biology of Molybdenum

Florian Bittner and Ralf-Rainer Mendel

Abstract The transition element molybdenum (Mo) is essential for (nearly) all biological systems as it is required by enzymes catalyzing diverse key reactions in the global carbon, sulfur,and nitrogen metabolism. The metal itself is biologically inactive unless it is complexed by a special cofactor. With the exception of bacterial nitrogenase, where Mo is a constituent of the FeMo-cofactor, Mo is bound to a pterin, thus forming the molybdenum cofactor (Moco), which is the active compound at the catalytic site of all other Mo-enzymes. In eukaryotes, the most prominent Mo-enzymes are nitrate reductase, sulfite oxidase, xanthine dehydrogenase, aldehyde oxidase, and the mitochondrial amidoxime reductase. The biosynthesis of Moco involves the complex interaction of six proteins and is a process of four steps, which also includes iron and copper in an indispensable way. Moco, as released after synthesis, is likely to be distributed to the apoproteins of Mo-enzymes by putative Moco-binding proteins. Xanthine dehydrogenase and aldehyde oxidase, but not sulfite oxidase and nitrate reductase, require the postranslational sulfuration of their Mo-site to become active.

Abbreviations

ABA	Abscisic acid
AO	Aldehyde oxidase
Cnx1-E	N-terminal domain of Cnx1
Cnx1-G	C-terminal domain of Cnx1
cPMP	Cyclic pyranopterin monophosphate
Cu	Copper

F. Bittner and R.-R. Mendel (✉)
Department of Plant Biology, Technical University of Braunschweig, Humboldtstrasse 1, 38106 Braunschweig, Germany
e-mail: r.mendel@tu-bs.de

R. Hell and R.-R. Mendel (eds.), *Cell Biology of Metals and Nutrients*, Plant Cell Monographs 17,
DOI 10.1007/978-3-642-10613-2_6,

IAA	Indole-3-acetic acid
Fe	Iron
FAD	Flavin adenine dinucleotide
mARC	Mitochondrial amidoxime reducing component
Mo	Molybdenum
MoBP	Molybdenum cofactor binding protein
Moco	Molybdenum cofactor
MPT	Molybdopterin
NR	Nitrate reductase
NO	Nitric oxide
ROS	Reactive oxygen species
SO	Sulfite oxidase
XDH	Xanthine dehydrogenase
XO	Xanthine oxidase

1 Introduction

Molybdenum (Mo) occurs in a wide range of metalloenzymes in bacteria, fungi, algae, plants, and animals, where it forms part of the active sites of these enzymes. Mo enters the cell as oxyanion and is subsequently incorporated by complex biosynthetic machinery into a pterin compound, thereby forming the molybdenum cofactor (Moco). All Mo enzymes catalyze depend on Mo catalyze redox reactions by taking advantage of the versatile redox-chemistry of the metal, which is controlled by the cofactor itself and the enzyme environment. In this article, we will review the path that Mo takes from uptake into the cell, via formation and modification of the Moco, to the distribution and final insertion of Moco into apo-metalloenzymes. In this review we will focus on plants because our understanding of the cell biology of Mo is most advanced among eukaryotes. The plant knowledge about Mo metabolism also forms the basis for deciphering Moco biosynthesis in humans (Mendel, in press; Reiss and Johnson 2003) where Moco deficiency is a severe genetic disease with fatal consequences for the affected individuals (Johnson and Duran 2001).

Yeast as model organism plays no role in Mo research as *Saccharomyces cerevisiae* does not contain Mo-enzymes. Genome-wide database analyses revealed a significant number of unicellular organisms that do not need Mo, with *S. cerevisiae* and *S. pombe* being prominent representatives. Loss of Mo-utilization is obviously connected to a host-associated lifestyle that makes Mo-enzymes unnecessary, while all multi-cellular eukaryotes are dependent on Mo (Zhang and Gladyshev 2008).

It has been long known that the transition element Mo (Bortels 1930) is an essential nutrient for plants, animals, and microorganisms. Mo is highly abundant in ocean water in the form of the MoO_4^{2-} anion. In soils, the molybdate anion is the only form of Mo that is available for plants and bacteria. Mo-containing enzymes hold key positions both in the biogeochemical redox cycles of carbon, nitrogen, and sulfur on Earth (Stiefel 2002) and in the metabolism of every organism. To this end, more than 50 enzymes are known to contain Mo and most of them occur in bacteria while in eukaryotes only six were found (Sigel and Sigel 2002). Biologically, Mo belongs to the group of trace elements, i.e. the organism needs it only in minute amounts. If, however, an organism takes up very high amounts of Mo, toxicity symptoms are observed (Turnlund 2002). On the other hand, unavailability of Mo is lethal for the organism. Further, even if Mo is available for the cell, it seems to be biologically inactive until it becomes complexed to form Moco, thus gaining biological activity.

2 Molybdenum Uptake into Cells

Organisms take up Mo in the form of its molybdate anion. It requires specific uptake systems to scavange molybdate in the presence of competing anions. In bacteria, high-affinity ABC-type transporters are described consisting of three protein components and requiring ATP-hydrolysis for operation. In some bacteria, specific molybdate-binding proteins with a capacity of up to eight anions (Pau and Lawson 2002) that store molybdate until further use by the cell are known. In contrast to bacterial molybdate homeostasis, eukaryotic molybdate transport is less well understood. Algae and higher plants are the only eukaryotes of which the molybdate uptake mechanisms have been unveiled recently. Two proteins (MOT1 and MOT2) belonging to the large sulfate carrier family were shown to transport molybdate with ultra-high affinity (nanomolar k_M value) across cellular membranes (Baxter et al. 2008; Tejada-Jimenez et al. 2007; Tomatsu et al. 2007). Surprisingly, none of them was found to reside in the plasmamembrane. Contradictory reports localized them to the endomembrane system (Tomatsu et al. 2007) or to the mitochondrial envelope (Baxter et al. 2008). Both suggested subcellular locations are questionable as the insertion of Mo into the Moco-backbone takes place in the cytosol. Still the cellular importer for Mo is missing, but it is likely that additional transporters, not only in autotrophs but also in animals, will be uncovered soon. It may be that in addition to a possible high-affinity system, molybdate could also nonspecifically enter the cell through the sulfate and/or phosphate uptake system.

3 The Molybdenum Cofactor

After uptake into the cell, molybdate has to be complexed by a unique compound in order to gain biological activity. This compound is a unique tricyclic pterin named Moco (as shown in the center of Fig. 1a). There is another type of Mo-containing

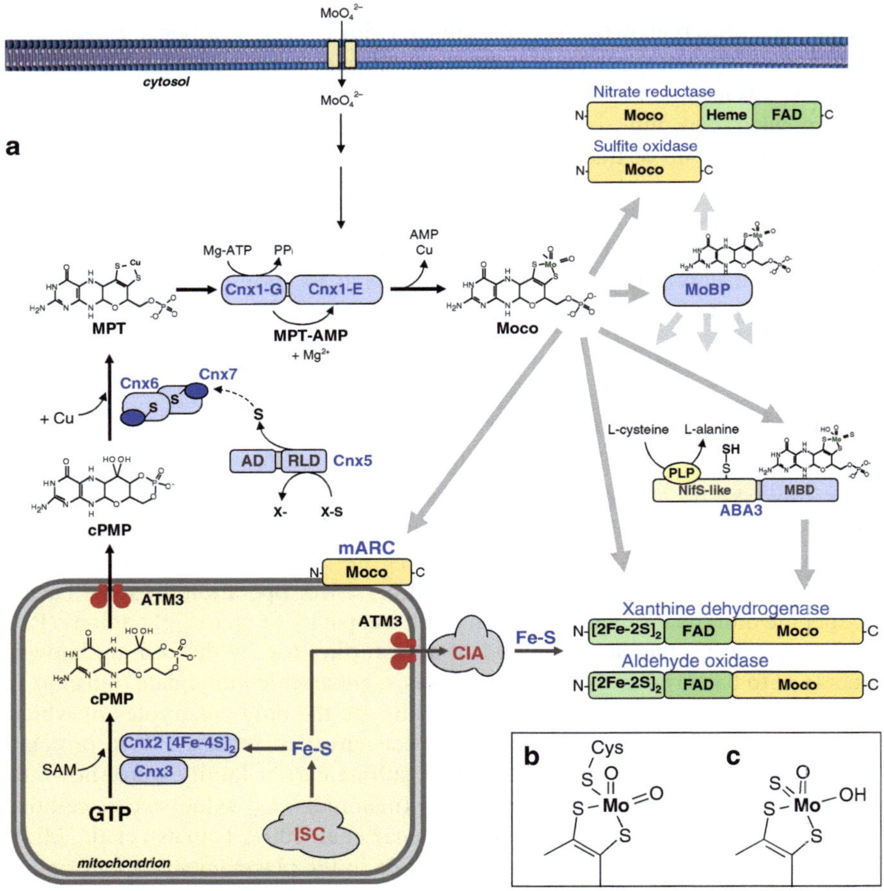

Fig. 1 Molybdenum metabolism in plant cells. (**a**) Organization of biosynthesis, distribution, and maturation of Moco in plant cells. The basic steps of Moco biosynthesis are shown starting from the conversion of GTP to cPMP in the mitochondria, to the cytosolic steps via MPT and Moco. Moco biosynthesis enzymes and Moco-binding proteins, which are not typically classified as Mo-enzymes, are shown in blue. The dependence of Cnx2 on [Fe–S] and *S*-adenosyl methionine (SAM) is indicated. MPT-synthase, consisting of Cnx6 and Cnx7, is sulfurated by Cnx5, with the primary sulfur donor (X–S) mobilized by the rhodanese-like domain of Cnx5 (RLD) being unknown. The adenylation domain of Cnx5 (AD) is required for adenylation and activation of the small MPT synthase subunit Cnx7. It is assumed that copper (Cu) is inserted directly after dithiolene formation. The individual reactions of Cnx1 and its products (Moco, pyrophosphate PP_i, AMP, copper) are indicated. Mature Moco can be either bound to a Moco-binding protein (MoBP), to the Mo-enyzmes mARC, AO, XDH, NR, and SO, or to the Moco-binding C-terminal domain of the ABA3 protein (MBD). ABA3 generates a protein-bound persulfide, which is the source of the terminal sulfur ligand of Moco in AO and XDH. Like Cnx2, AO and XDH depend on [Fe–S]-clusters, which are synthesized in mitochondria by the iron–sulfur cluster biosynthesis machinery (ISC) prior to export by ATM3 and assembly by the cytosolic iron–sulfur cluster assembly machinery (CIA). (**b**) Structure of the Mo-center in enzymes of the SO family (SO and NR). (**c**) Structure of the Mo-center in enzymes of the XO family (AO and XDH)

cofactor which is found only once in nature namely in bacterial nitrogenase, forming the so called FeMo-cofactor that consists of two partial cubanes ($MoFe_3S_3$ and Fe_4S_3) which are joined by three bridging sulfurs. Nitrogenase reduces atmospheric dinitrogen to ammonia with concomitant hydrolysis of ATP. Nitrogenase is required for biological nitrogen fixation, which is an essential step in the nitrogen cycle in the biosphere, and a major contributor to the nitrogen available to many plants species like legumes. In contrast to nitrogenase all other Mo-containing enzymes characterized to this end contain the pterin-type cofactor (Hille 1996). For this reason, and as very recently, a number of reviews about nitrogenase have been published (e.g. Dos Santos et al. 2004; Rees et al. 2005), we will focus on the ubiquitously occurring Moco in this review.

Early work with mutants of the filamentous fungus *Aspergillus nidulans* (Pateman et al. 1964) and of the higher plant *Nicotiana tabacum* (Mendel and Müller 1976) revealed a novel mutant phenotype, namely the simultaneous loss of the two Mo-enzymes nitrate reductase and xanthine dehydrogenase. Since Mo was the only common link between these two – otherwise very different – enzymes, it was suggested that both enzymes should share a common Mo-related cofactor, named Moco. The elucidation of the chemical nature of Moco is based on the work of J. Johnson and K.V. Rajoagopalan. Their final description of Moco (Johnson et al. 1980) was confirmed by crystal structures of Mo–enzymes with the only exception that a third ring, a novel pyrano ring, is formed (Fig. 1a). Due to the labile nature of Moco and its high sensitivity to oxidation most of the work was done using degradation or oxidation products of the cofactor thereby revealing the pterin nature of Moco and its C6 substitution with a unique four-carbon side chain (Johnson et al. 1984) that coordinates the metal via a dithiolene group (Fig. 1a). Due to the formation of a third pyrano ring between the C3′ hydroxy group and the pterin C7 atom, a fully reduced hydrogenated pterin (tetrahydro state) is formed (Kisker et al. 1997a). Because of the unique nature of the pterin in Moco, the metal-free form of the cofactor is called molybdopterin or metal-containing pterin (MPT). The pterin structure of Moco is unique in nature and has probably been evolved in order to control and maintain the special redox properties of Mo. The task of the cofactor is to position the catalytic metal Mo correctly within the active center, to control its redox behavior, and to participate with its pterin ring system in the electron transfer to or from the Mo atom. The pterin with its several possible reduction states as well as different structural conformations could also be important for channeling electrons to other prosthetic groups (Kisker et al. 1997b). X-ray crystallographic analyses of Mo-enzymes revealed that the cofactor is not located on the surface of the protein, but it is buried deeply within the interior of the enzyme and a tunnel-like structure makes it accessible to the appropriate substrates (Fischer et al. 2005; Kisker et al. 1997a). Once Moco is liberated from the holoenzyme, it loses the Mo atom and undergoes rapid and irreversible loss of function due to oxidation (Rajagopalan 1996). The demolybdo-forms of Mo-enzymes are catalytically inactive.

3.1 Molybdenum Cofactor Biosynthesis

A mutational block of Moco biosynthesis leads to the loss of essential metabolic functions because all enzymes depending on Mo lose their activity, which ultimately causes death of the organism. The identification of several genetic complementation groups among Moco-deficient mutants in a given organism and the conserved structure of Moco provided a basis to propose an evolutionary old multi-step biosynthetic pathway (Mendel 1992). Already in pre-genomic times, a detailed mutant characterization contributed substantially to our understanding of the genetics and biochemistry of Moco biosynthesis in bacteria, plants, fungi, and humans. Moco mutants have been described in numerous plants, e.g., in tobacco (Mendel and Müller 1976), *Nicotiana plumbaginifolia* (Müller and Mendel 1989), *Arabidopsis* (Braaksma and Feenstra 1982; Crawford 1992), barley (Kleinhofs et al. 1989), and the green alga *Chlamydomonas reinhardtii* (Fernandez and Matagne 1986). The phenotype of plant Moco mutants was best studied in *N. plumbaginifolia*, where all six Moco-specific genetic loci (*cnx*A–*cnx*F) showed a similar morphology strongly deviating from that of the wild type: stunted growth, chlorosis of leaves, and small, narrow, and crinkled leaves (Gabard et al. 1988). In contrast to animals, Moco mutants of plants can be kept alive on media containing reduced nitrogen as a N-source.

In all organisms studied so far, Moco is synthesized by a conserved biosynthetic pathway that can be divided into four steps, according to the biosynthetic intermediates cyclic pyranopterin monophosphate (cPMP, previously also known as precursor Z), MPT, adenlyated MPT, and Moco (Fig. 1a). In eukaryotes, always six gene products catalyzing Moco biosynthesis have been identified in plants (Mendel and Schwarz 2002), fungi (Millar et al. 2001), and humans (Reiss et al. 1998; Stallmeyer et al. 1999a; Stallmeyer et al. 1999b). These genes are homologous to their counterparts in bacteria and some but not all of the eukaryotic Moco biosynthesis genes are able to functionally complement the matching bacterial mutants. Genes and gene products were named in plants according to the *cnx* nomenclature (cofactor for nitrate reductase and xanthine dehydrogenase) introduced for the fungal mutants with the mutants labeled with letters (cnxA-F) and the cDNAs labeled with numbers (*cnx*1-3, *cnx*5-7).

3.1.1 Step 1: Conversion of GTP into cPMP

During the first stage, a guanosine derivative (probably GTP) is transformed into a sulfur–free pterin compound, the cPMP, already possessing the Moco-typical four-carbon side chain (Fig. 1a). In comparison to Moco and MPT, cPMP is the most stable intermediate with an estimated half life of several hours at a low pH (Wuebbens and Rajagopalan 1993). Mass spectrometry and ^{1}H NMR recently revealed that cPMP already possesses a fully reduced tetrahydropyranopterin structure and is predominantly hydrated at the C1′ position resulting in a geminal diol

(Santamaria-Araujo et al. 2004). GTP labeling studies and NMR demonstrated that each carbon atom of the ribose and of the guanine ring is incorporated into cPMP (Wuebbens and Rajagopalan 1995; Rieder et al. 1998). The detailed mechanism of this reaction step remains unclear, yet hypothetical multistep-reactions have been suggested (Wuebbens and Rajagopalan 1995; Rieder et al. 1998). In *A. thaliana* cPMP synthesis is catalyzed by the proteins Cnx2 and Cnx3 (Fig. 1a). Both, *cnx2* and *cnx3* were identified using the approach of functional complementation of *E. coli* Moco mutants *moaA* and *moaC*, thus demonstrating the functional conservation between bacterial and plant Moco synthesis (Hoff et al. 1995). Cnx2 belongs to the superfamily of *S*-adenosylmethionine-dependent radical enzymes (Hanzelmann et al. 2004). Members of this large family catalyze the formation of protein and/or substrate radicals by reductive cleavage of SAM by a [4Fe–4S] cluster (Sofia et al. 2001). Cnx2 and Cnx3 are larger than their bacterial counterparts which is attributed to N-terminal extensions carrying targeting motifs for mitochondrial or chloroplast transport (Hoff et al. 1995). Indeed it has been very recently shown that both proteins are targeted to the mitochondria (Teschner et al. 2009). The fact that Cnx2 is likely to have [4Fe–4S] clusters similar to its human homolog MOCS1A (Hanzelmann et al. 2004) might explain the mitochondrial localization as this organelle is a major site of plant Fe–S-cluster synthesis (Lill and Muhlenhoff 2006).

3.1.2 Step 2: Synthesis of Molybdopterin

In the second stage sulfur is transferred to cPMP in order to generate MPT. This reaction is catalyzed by the enzyme MPT synthase, a heterotetrameric complex of two small and two large subunits that stoichiometrically converts cPMP into MPT. Plant MPT synthase is encoded by cnx6 (large subunit) and cnx7 (small subunit). The sulfur is bound to the C-terminus of the small subunit as thiocarboxylate. Due to the fact that each small subunit of MPT synthase carries a single sulfur atom, a two-step mechanism for the formation of the MPT dithiolate has been proposed, which involves the formation of a mono-sulfurated intermediate (Gutzke et al. 2001; Wuebbens and Rajagopalan 2003). After MPT synthase has transferred the two sulfurs to cPMP, it has to be re-sulfurated by the MPT-synthase sulfurase in order to reactivate the enzyme for the next reaction cycle of cPMP conversion. This resulfuration is catalyzed by Cnx5 involving very likely an adenylation of MPT synthase followed by sulfur transfer similar to its human homolog MOCS3 (Matthies et al. 2004; Matthies et al. 2005). Cnx5 and MOCS3 are two-domain proteins consisting of a N-terminal domain responsible for adenylating MPT synthase and a C-terminal rhodanese-like domain where the sulfur is bound to a conserved cysteine in the form of persulfide (Matthies et al. 2005). The identity of the donor for the reactive mobile sulfur is as yet unknown, but the contribution of a persulfide-generating cysteine desulfurase is possible (Leimkühler et al. 2001).

3.1.3 Step 3: Adenlyation of Molybdopterin

After synthesis of the MPT moiety, the chemical backbone is built for binding and coordination of the Mo atom. In the third step, therefore, Mo has to be transferred to MPT in order to form Moco, thus linking the molybdate uptake system to the MPT pathway. Mutants defective in this step produce MPT and can be partially repaired by growing them on high-molybdate medium. In bacteria, this step is catalyzed by two proteins which during evolution to higher organisms were fused to a two-domain protein. Earlier it was assumed that one domain should be essential for generating an activated form of Mo that is incorporated by the other domain into bound MPT (Schwarz et al. 1997b; Schwarz et al. 2000). But only recently the exact mechanism was uncovered in plants where the protein Cnx1 is catalyzing this step (Llamas et al. 2004). The C-terminal Cnx1 domain (=Cnx1-G) was known to tightly bind MPT (Schwarz et al. 1997a). Yet, its crystal structure (Kuper et al. 2004) revealed an unexpected finding: a novel reaction intermediate, adenylated MPT (MPT-AMP) (Fig. 1a). Subsequently it was demonstrated that Cnx1-G adenylates MPT in an Mg^{2+}- and ATP-dependent way and forms MPT-AMP that remains bound to Cnx1-G (Llamas et al. 2004).

3.1.4 Step 4: Molybdenum Insertion into Molybdopterin and Crosstalk to Copper Metabolism

The crystal structure of the Cnx1-G revealed another unexpected finding, namely a copper bound to the MPT dithiolate sulfurs, whose nature was confirmed by anomalous scattering of the metal. In both structures the copper atom shows tetragonal coordination with two waters as additional ligands in the MPT-bound state, while one of these waters is replaced by a histidine in the MPT-AMP-bound structure. Up to now the function of this novel MPT ligand is unknown but copper might play a role in sulfur transfer to cPMP, in protecting the MPT dithiolate from oxidation, and/or presenting a suitable leaving group for Mo insertion. The origin of this copper is still unclear but it is reasonable to assume that it binds to the enedithiolate group just after the latter has been formed, i.e. at the end of step 2 of Moco biosynthesis.

In the final step of Moco biosynthesis MPT-AMP has to be converted into mature Moco: MPT-AMP is transferred to the N-terminal domain of Cnx1 (=Cnx1-E) thereby building a product–substrate channel. Cnx1-E cleaves the adenylate, releases copper, and inserts Mo, thus yielding active Moco. We found that MPT adenylate was hydrolyzed in a molybdate-dependent way (Llamas et al. 2006). This reaction was coupled to the metal exchange reaction where bound copper was released and Mo was transferred to MPT thus yielding mature Moco. As copper is always found in a protein-bound state it might be that Cnx1 interacts with a copper chaperone when the metal is released during Mo insertion. Using a split-ubiquitin based two-hybrid approach with Cnx1 as bait, a copper chaperone homolog has been recently identified (J. Winking, R. Mendel, G. Schwarz, unpublished data).

In vitro studies with Cnx1-G-bound MPT-AMP revealed an inhibition of Moco synthesis in the presence of 1 μM $CuCl_2$, providing a link between Mo and copper metabolism (Kuper et al. 2004). Copper inhibition of Moco synthesis can be explained by inhibition of the Mg-dependent Mo insertion reaction. The latter is supported by the suppression of copper inhibition with equimolar amounts of Cnx1-E and is in line with the known copper inhibition of pyrophosphatases (Moorhead et al. 2003). Our finding implies that Moco deficiency might occur when cellular copper concentrations are increased, as seen in human patients affected with Wilson's disease (Mercer 2001), where copper accumulates in liver and brain, resulting in severe damage to both organs. In plants, one should consider that the toxicity of elevated copper levels could also go back – at least in part – to a distorted Moco biosynthesis. However, it is to be noted that copper shortage also should be detrimental for Moco biosynthesis. Therefore, when studying copper homeostasis the accompanying analysis of Mo metabolism could shed further light onto the link between Mo and copper homeostases.

4 Allocation of the Molybdenum Cofactor

4.1 Storage and Transfer of the Molybdenum Cofactor

After synthesis, Moco has to be allocated to the appropriate apo-enzymes. In prokaryotes, a complex of proteins synthesizing the last steps of Moco biosynthesis donates the mature cofactor to apo-enzymes assisted by enzyme-specific chaperones (Magalon and Mendel 2008). In eukaryotes, however, no Mo-enzyme specific chaperone has been found. As free Moco is extremely sensitive to oxidation (Rajagopalan and Johnson 1992) it is also assumed that Moco occurs permanently protein-bound in the cell. Therefore, a cellular Moco distribution system should meet two demands: (1) It should bind Moco subsequent to its synthesis, and (2) it should maintain a directed flow of Moco from the Moco donor Cnx1E to the Mo-dependent enzymes. The availability of sufficient amounts of Moco is essential for the cell to meet its changing demand of Mo enzymes. Therefore the existence of a Moco carrier protein would provide a way to buffer supply and demand of Moco. Among eukaryotes, first in the green alga *C. rheinhardtii* a Moco carrier protein was described (Aguilar et al. 1992). Later it was purified and a 16 kDa protein identified which was able to bind and protect Moco against oxidation (Witte et al. 1998; Ataya et al. 2003), and the atomic structure showed that it formed a homotetramer holding four molecules of Moco (Fischer et al. 2006). Without any denaturing procedure, subsequent transfer of Moco from the carrier protein to apo-nitrate reductase from the Moco-free *Neurospora crassa* mutant *nit-1* was possible, thus indicating that carrier-bound

Moco was readily transferable. These properties of the *Chlamydomonas* protein make it a promising candidate for being part of a cellular Moco delivery system. It is however unknown whether this carrier protein is also able to donate Moco to Mo enzymes other than NR.

Utilizing the atomic structure of the *Chlamydomonas* Moco carrier protein two structural homologs could be identified in *A. thaliana* (Kruse et al. 2009). It turned out that these two structural homologs are part of a novel protein family consisting of nine members in *A. thaliana*. The recombinantly expressed proteins are able to bind Moco with k_D values in the low micromolar range and are therefore named Moco binding proteins (MoBP). They are located in the cytosol, enhance Moco transfer to apo-nitrate reductase in the Moco-free *N. crassa* mutant *nit-1*, and undergo protein-protein interaction with the cellular Moco donor protein Cnx1E, thus indicating for the MoBPs a function in cellular Moco distribution (Kruse et al. 2009). Database searches revealed that higher plants possess 8–16 MoBP-homologous sequences, and even the moss *Physcomitrella* has eight homologs. Obviously, land plants with differentiated organs need more MoBPs than a unicellular and motile alga. Indeed, silico-analyses of *Arabidopsis* expression data (Kruse, Mendel unpublished) first revealed differential and organ-specific expression of the MoBP genes.

4.2 Insertion of the Molybdenum Cofactor into Molybdenum Enyzmes

Insertion of Moco into Mo-enzymes is not understood. Using a defined in vitro-system it was shown that human apo-sulfite oxidase can directly incorporate Moco (Leimkühler et al. 2001). However, for insertion of Moco into the target apo-enzymes as it occurs in the living cell, either (still unknown) chaperone proteins would be needed or the Moco carrier and/or binding proteins could become involved at this stage. For some bacterial Mo-enzymes, system-specific chaperones are required for Moco insertion and protein folding, e.g. NarJ for *E. coli* nitrate reductase (Magalon and Mendel 2008) and XDHC for xanthine dehydrogenase from *Rhodobacter capsulatus* (Leimkühler and Klipp 1999) and *Comamonas acidovorans* (Ivanov et al. 2003).

4.3 Micro-Compartmentalization and Cytoskeleton Binding

In plants, Cnx1 catalyzes the activation of MPT followed by insertion of Mo. The primary structure of Cnx1 shows striking homologies to the mammalian protein Gephyrin that was first described as a neuroreceptor anchor protein linking glycine receptors in the postsynaptic membrane to the subcellular cytoskeleton.

It is obvious that Gephyrin combines two different functions: (1) a biosynthetic activity in Moco formation, and (2) a structural role in receptor clustering (Stallmeyer et al. 1999b). Not only Gephyrin, but also its plant homolog Cnx1 exhibits functional properties that are distinct from Moco biosynthesis. Based on the observed cytoskeleton binding of Gephyrin, a binding of Cnx1 to actin filaments, exclusively mediated by the E domain, could be demontrated (Schwarz et al. 2000). Finally, Cnx1 is essential for stabilizing the newly formed Moco (Kuper et al. 2000). What could be the functional significance of cytoskeleton binding of Cnx1 in terms of Moco biosynthesis? We assume that during evolution it became important to facilitate product–substrate flow, which could result in micro-compartmentalization of a hypothetical Moco biosynthetic multi-enzyme complex ensuring the fast and protected transfer of the labile intermediates within the reaction sequence from GTP to Moco. Therefore anchoring to cellular structures like the cytoskeleton might help organizing and stabilizing such a biosynthetic machinery.

5 Molybdenum Enzymes

Mo-enzymes are required for diverse key reactions in the global carbon, sulfur, and nitrogen metabolism, and although up to now more than 50 Mo-enzymes have been found in nature, their number is still increasing. While most of the Mo-enzymes are of bacterial origin, only a limited number is present in eukaryotes, where Mo-enzymes can be subdivided into two families: the xanthine oxidase (XO) family (Fig. 1a, c), represented by xanthine dehydrogenase (XDH), aldehyde oxidase (AO), pyridoxal oxidase, and nicotinate hydroxylase, and the sulfite oxidase (SO) family (Fig. 1a, b), represented by sulfite oxidase (SO) and nitrate reductase (NR). The mitochondrial amidoxime reducing component (mARC), which has been identified recently in pig liver mitochondria (Havemeyer et al. 2006), has not yet been integrated into one of the aforementioned families and at present its formation of a new family of Mo-enzymes cannot be excluded.

While pyridoxal oxidase and nicotinate hydroxylase were exclusively found in *Drosophila melanogaster* (Warner and Finnerty 1981) and *A. nidulans* (Lewis et al. 1978), respectively, XDH, AO, SO, and mARC are typical for all higher eukaryotes analyzed so far. As NR is required for nitrate assimilation, this enzyme is only present in autotrophic organisms like plants, algae, and fungi. In general, all these Mo-enzymes catalyze the transfer of an oxygen atom, ultimately derived from or incorporated into water, to or from a substrate (Kisker et al. 1997a; Hille 2002). Each reaction, either reduction or oxidation, is also characterized by transfer of two electrons, which enforce the Mo atom to vary its oxidation state between IV and VI. In the following sections, we will focus on plant-typic Mo-enzymes with their specific functions, physiolocial roles, and distribution within the cell.

5.1 Xanthine Dehydrogenase

Plant XDH (EC 1.17.1.4., formerly EC 1.1.1.204) has been purified from several sources such as nodules of bean (Boland and Schubert 1982), green algae (Perez-Vicente et al. 1992), wheat leaves (Montalbini 1998), Arabidopsis (Hesberg et al. 2004), leaves of legumes (Montalbini 2000), and pea seedlings (Sauer et al. 2002). The enzyme basically catalyzes the oxidative hydroxylation of a wide range of aldehydes and aromatic heterocycles, but is best known as a key enzyme of the purine degradation pathway where it oxidizes hypoxanthine to xanthine and xanthine to uric acid. The enzyme is active as a homodimer composed of two identical subunits of approximately 145 kDa, each being subdivided into three distinct domains (Fig. 1a): an N-terminal domain of 20 kDa for binding of two iron–sulfur clusters of the [2Fe–2S]-type, a 40 kDa domain harboring a FAD-binding site, and a C-terminal domain of 80 kDa required for Moco-binding and dimerization. Electrons derived from substrate hydroxylation at the Mo-center go via the iron-sulfur clusters to the flavin cofactor. At the FAD site, the electrons are transferred preferably to NAD^+ to form NADH, but alternatively can be transferred to molecular oxygen whereby superoxide anions are generated (Hesberg et al. 2004; Yesbergenova et al. 2005). In contrast to plant XDH, mammalian XDH can also form hydrogen peroxide when molecular oxygen is the final electron acceptor (Hille and Nishino 1995). Furthermore, the mammalian enzyme can be converted into the oxidase form (XO), either reversibly by oxidation of the sulfhydryl groups of two conserved cysteine residues (Nishino and Nishino 1997) or irreversibly by limited proteolysis (Amaya et al. 1990). In case of conversion into the XO form, the enzyme is no longer able to transfer its electrons to NAD^+ but instead exclusively uses molecular oxygen as electron acceptor. However, plant XDH neither possesses the conserved cysteine residues for reversible conversion (Hesberg et al. 2004), nor has limited proteolysis been observed to yield the XO form.

In addition to the hypoxanthine/xanthine-dependent production of reactive oxygen species (ROS), NADH oxidase activity with simultaneous production of superoxide has also been shown for plant XDH (Yesbergenova et al. 2005). Therefore, XDH is discussed not only to decompose purines but also to have additional physiological functions in ROS metabolism. In support of this, XDH activities and simultaneous ROS production were observed upon plant–pathogen interactions (Montalbini 1992), hypersensitive response (Montalbini and Della Torre 1996), drought stress (Yesbergenova et al. 2005), and senescence (Del Rio et al. 1998). Yet, whether this coincidence can directly be related solely to XDH or is an indirect consequence of different enzymatic pathways including XDH, remains to be shown.

The subcellular localization of plant XDH is still under debate. While a cytosolic localization was reported by (Datta et al. 1991), a peroxisomal localization was described by (Sandalio et al. 1988). However, a recent article provides an amicable agreement by claiming XDH localizes to both the cytosol and peroxisomes (Corpas et al. 2008).

5.2 *Aldehyde Oxidase*

Like XDH, AO (EC 1.2.3.1) is a molybdo-iron-flavoenzyme that catalyzes the oxidation of a variety of aromatic and nonaromatic aldehydes to their corresponding carboxylic acid. As AO proteins share many substrates as well as a significant degree of homology with XDH, it is presumed that during evolution AO is derived from XDH by gene duplication and neo-functionalization (Terao et al. 2001; Rodriguez-Trelles et al. 2003). The most prominent characteristics which distinguish AO from XDH enzymes have been found to concern the substrate binding at the Mo-center and binding of the physiological electron acceptor (Hille 2005). AO enzymes are strict oxidases that are unable to bind NAD^+ and exclusively use molecular oxygen as electron acceptor. Upon transfer of substrate-derived electrons to molecular oxygen, plant AO generates hydrogen peroxide (Yesbergenova et al. 2005), whereas animal AO, in addition, generates superoxide anions (Badwey et al. 1981) . However, the activity of the newly identified AO protein AAO4 from Arabidopsis siliques was shown to be stimulated by addition of NAD^+ (Ibdah et al. 2009), suggesting that this AO isoform, which by oxidition of benzaldehyde contributes to the synthesis of benzoic acid for incorporation into several glucosinolate compounds, represents an intermediate type between "true" AO proteins and XDH proteins.

In contrast to animal AO, whose physiological role is yet unravelled (Garattini et al. 2003), the physiological importance of AO proteins in plants is much better understood. The Arabidopsis genome harbors four AO genes, *AAO1* – *AAO4*, whose products form homodimers as well as heterodimers, thereby leading to altered substrate specificities of the respective isoenzymes. In 6-day-old seedlings, the gene products of *AAO1* and *AAO2* form three homo- and heterodimeric AO isoenzymes capable of producing indole-3-acetic acid (IAA) (Akaba et al. 1999). The fact that IAA belongs to the auxin family of plant hormones suggests a possible physiological role of these AO enzymes in auxin biosynthesis during early stages of plant development. In rosette leaves, the isoenzyme composition is altered in a way that AAO1 proteins are replaced by AAO3 proteins, thereby generating new homo- and heterodimeric isoforms (Seo et al. 2000a; Koiwai et al. 2004). The AAO3 homodimer, also referred to as AOδ, is characterized by high preference for abscisic aldehyde as substrate (Seo et al. 2000b), which is the ultimate precursor of abscisic acid (ABA) that is involved in many aspects of plant growth and development, including seed maturation, dormancy, leaf senescence, as well as adaptation to a variety of environmental stresses (Seo and Koshiba 2002; Verslues and Zhu 2005; Mauch-Mani and Mauch 2005). Arabidopsis mutants with a deficiency in AAO3 are therefore characterized by reduced ABA levels accompanied by excessive water loss and a wilty phenotype, and also by retarded vegetative growth and reduced stress tolerance (Seo et al. 2000a; González-Guzmán et al. 2004). Recently, AAO3 protein levels have been shown to be regulated by ubiquitin-dependent degradation via the 26S-proteasome to prevent premature senescence by accumulation of ABA (Raab et al. 2009). This suggests that ABA-synthesizing

AO proteins also play a critical role during the onset of senescence, which requires a tight control of AO and ABA levels.

5.3 *Sulfite Oxidase*

SO (EC 1.8.3.1), the name-giving enzyme for all members of the SO-family of Mo-enzymes, catalyzes the oxidation from sulfite to sulfate, the final step in the degradation of sulfur-containing amino acids. While the animal enzyme basically consists of an N-terminal iron/heme-containing cytochrome b_5 domain and a C-terminal domain responsible for Moco binding and dimerization (Kisker et al. 1997a) (Fig. 1a), plant SO lacks the cytochrome b_5 domain (Eilers et al. 2001; Schrader et al. 2004). Thus, besides the recently indentified mARC plant SO is the simplest Mo-enzyme found in eukaryotes with its intramolecular redox center only consisting of Moco. Like XDH and AO, plant SO liberates electrons during substrate oxidation and subsequently transfers them to molecular oxygen with simultaneous formation of hydrogen peroxide (Hänsch et al. 2006), which is in contrast to the animal enzyme, where cytochrome *c* serves as final electron acceptor.

While animal SO is located in the intermembrane space of mitochondria (Kisker et al. 1997a), plant SO has been demonstrated to reside in the peroxisomal matrix (Nowak et al. 2004), which from a physiological point of view appears reasonable as excess hydrogen peroxide generated during sulfite oxidation can easily be eliminated by catalase. The physiological role of plant SO has been clarified only recently. As sulfite is a strong nucleophile that can react with a wide variety of cellular components, it was assumed that SO is required for removing excess sulfite from the cell. In support of this, Brychkova et al. (2007) and Lang et al. (2007) independently found that in comparison to wildtype plants, SO-deficient plants are more susceptible to high concentrations of sulfite while SO-overexpressing plants are more tolerant to excess sulfite. Under normal conditions however, the lack of SO activity in plants is not related to an obvious phenotype, suggesting that SO represents a salvage enzyme rather than a metabolic housekeeping enzyme.

5.4 *Nitrate Reductase*

NR (EC 1.7.1.1., formerly EC 1.6.6.1) is a key enzyme of nitrate assimilation that catalyzes the reduction of nitrate to nitrite in the cytosol. As NR thereby provides essential nitrogen metabolites to the plant, it is obvious that plants with a deficiency in NR are no longer nitrogen autotroph and depend on alternative nitrogen sources such as ammonium. Like AO and XDH it also consists of three distinct domains, however, of different order and composition (Fig. 1a). The N-terminal domain of an NR monomer is responsible for binding of Moco and is followed by a heme-binding cytochrome b_5 domain and a C-terminal FAD-binding domain. As is true for XDH, AO, and SO, the active NR enzyme is formed by dimerization of two NR monomers. The domains are separated by solvent-exposed linker regions, called hinge I and hinge II. In plants hinge I, the linker between the cytochrome b_5 domain and the Moco/dimerization domain, contains a conserved serine residue which mediates contact with a 14-3-3 protein when phosphorylated, subsequently leading to inhibition of enzyme activity (Kaiser and Huber 2001). An N-terminal extension preceding the Moco domain was shown to be important for the posttranscriptional regulation of NR by light (Nussaume et al. 1995), but is not required for enzyme activity.

In contrast to the reactions catalyzed by XDH, AO, and SO, the process of nitrate reduction consumes rather than produces electrons deriving either from NADH or NADPH. While NADH-specific forms of NR are most abundant in higher plants and algae, NADPH-specific forms were exclusively found in fungi. However, bispecific forms, capable of using NADH as well as NADPH, have been found in all these organisms, mostly in fungi (Campbell and Kinghorn 1990). The intramolecular electron transfer starts with the reduction of FAD by NAD(P)H. The electrons are then transferred via the heme-cytochrome b_5 domain to the Mo-center, where they are required to reduce nitrate to nitrite by concomitant release of hydroxide (Skipper et al. 2001). In most plants studied thus far, NR is highly regulated on the expression level as well as on the posttranslational level, e.g. by light/dark transition and phosphorylation (Kaiser and Huber 2001) and also by nitrogen and carbon metabolites (Campbell 2001). While in terms of nitrogen assimilation, nitrite is further reduced to ammonium by nitrite reductase in the chloroplasts, it can also be reduced to nitric oxide (NO) by NR itself (Yamasaki and Sakihama 2000). It was shown that NR produces NO at saturating NADH and nitrite concentrations at about 1% of its nitrate reduction capacity in vitro (Rockel et al. 2002). However, as posttranslational modification of NR also modulated the NO-production rates, it was concluded that NR is indeed a producer of active nitrogen species also in vivo. Thus, the signaling molecule NO generated by NR might well contribute to plant growth development (Beligni et al., 2000), protection against cytotoxicity of ROS (Beligni and Lamattina 2000), and accumulation of phytoalexin (Noritake et al. 1996), and also to plant pathogen resistance (Durner and Klessig 1999) by increasing cGMP and salicylic acid levels.

5.5 *Mitochondrial Amidoxime Reducting Component*

The mitochondrial amidoxime reducing component mARC was shown to catalyze the reduction of a variety of *N*-hydroxylated compounds, some of which are widely used as prodrugs and are thus of pharmaceutical relevance (Havemeyer et al. 2006; Gruenewald et al. 2008). All eukaryotic genomes known to encode proteins for Moco biosynthesis and Mo-enzymes, likewise encode two mARC proteins, suggesting that mARC proteins form their own small protein family.

While the enzyme from pig liver has been identified in the outer membrane of mitochondria (Havemeyer et al. 2006), earlier work has shown that mARC proteins represent an abundant protein fraction in the inner membrane of mouse mitochondria (referred to as MOSC proteins Da Cruz et al. 2003). In accordance with these findings, all eukaryotic mARC proteins including the plant counterparts are characterized by the presence of N-terminal extensions which predict a mitochondrial localization of these proteins. With an average molecular mass of ~35 kDa and due to the fact that they bind Moco as the only prosthetic group, mARC proteins are the smallest Mo-enyzmes identified as yet. Although the native substrates of mARC enzymes are not known yet, the bacterial orthologs of mARC, YcbX, and YiiM, were shown to prevent the cell from DNA damage by detoxifying mutagenic *N*-hydroxylated base analogs such as 6-*N*-hydroxylaminopurine (Kozmin et al. 2008), which basically can serve as substrates for the eukaryotic enzymes as well (Wahl, Havemeyer, Clement, Mendel, Bittner, unpublished results).

In contrast to all other Mo-enzymes, eukaryotic mARC proteins do not exhibit enzymatic activity on their own but require other proteins like cytochrome b_5 and NADH/cytochrome b_5 reductase as electron transmitters and electron donors, respectively. By this composition, the mARC enzyme complex resembles the domain structure of plant nitrate reductase and in fact, the reduction mechanism of both enzymes/enzyme complexes appears to be very similar.

5.6 Posttranslational Sulfuration of Xanthine Oxidase Family-Enzymes

Plant Mo-enzymes can be subdivided into two families. While the enzymes of the SO family, SO and NR, are activated upon insertion of Moco, the members of the XO family, XDH and AO, require a final step of maturation during or after insertion of Moco (Fig. 1a,c). In addition to the dithiolene sulfurs of the pterin moiety and two oxo-groups, the Mo-atom of the Moco needs the addition of a terminal inorganic sulfur in order to provide enzymatic activity to the respective enzyme (Bordas et al. 1980; Wahl and Rajagopalan 1982). In vitro, this sulfur can be removed from active XDH/AO proteins by cyanide treatment, thereby generating an inactive desulfo enzyme (Massey and Edmondson 1970). The reaction, however, is reversible and the enzyme can be reactivated by sulfide-treatment under reducing conditions. In vivo, this terminal sulfur is added to the Moco of XDH and AO by a specific enzymatic reaction, catalyzed by the Moco sulfurase protein. The enzyme has been cloned from Arabidopsis (Bittner et al. 2001; Xiong et al. 2001) and tomato (Sagi et al. 2002). According to its corresponding mutant, which is characterized by the simultaneous loss of XDH and AO activities and by ABA deficiency, the enzyme from Arabidopsis is referred to as ABA3 (Leon-Kloosterziel et al. 1996). ABA3 is a homodimeric two-domain protein with its N-terminal domain sharing structural and functional homologies with bacterial cysteine desulfurases (Bittner et al. 2001; Heidenreich et al. 2005). In a pyridoxal phosphate-dependent manner, the N-terminal domain of ABA3 decomposes L-cysteine to yield alanine and elemental sulfur, the latter being bound as a persulfide to a highly conserved cysteine residue of ABA3 (Fig. 1a) (Heidenreich et al. 2005). The C-terminal domain of ABA3 shares a significant degree of similarity to mARC proteins and was shown to bind sulfurated Moco (Wollers et al. 2008), which receives the terminal sulfur via intramolecular persulfide-transfer from the N-terminal domain. It is still unclear whether ABA3 finally activates its target enzymes by transfering only the sulfur, which is bound to the Moco of the C-terminal domain, or whether it transfers the entire sulfurated Moco of the C-terminus in order to activate XDH and AO. In the first case the C-terminus would thus act as a scaffold for the assembly of a Mo–S center of which only the sulfur is subsequently passed to the Moco of XDH and AO. In the second case the reaction catalyzed by ABA3 would be an exchange reaction where nonsulfurated Moco of the inactive target enzymes is replaced by sulfurated Moco from the C-terminal domain of ABA3.

Under physiological aspects the terminal sulfuration step provides an efficient way of regulating the amount of active XDH and AO enzymes in the cell. The concentration of physiologically active compounds like hormones, as produced by plant AO, and ROS, as produced by the action of both AO and XDH, can rapidly be increased by changing the ratio of inactive and active XDH and AO molecules. In fact, a rapid induction of the *aba3* gene was found upon drought and salt stress in Arabidopsis as well as upon ABA treatment (Bittner et al. 2001; Xiong et al. 2001), thereby being consistent with the conditions required for induction of AO and XDH (Seo et al. 2000a; Hesberg et al. 2004; Yesbergenova et al. 2005).

5.7 Crosstalk between Molybdenum and Iron Metabolism

Iron plays an important role in the synthesis of Moco as Cnx2, which is involved in step 1 of Moco biosynthesis, and requires two iron–sulfur [Fe–S] clusters of the [4Fe–4S] type (Fig. 1a). These clusters are essentially involved in the *S*-adenosyl-methionine-dependent mechanism of cPMP formation (Hanzelmann et al. 2004; Hanzelmann and Schindelin 2004) and have to be preassembled in the mitochondria, from where mitochondrial as well as extramitochondrial [Fe–S] clusters originate (Balk and Lill 2004). In addition, both XDH and AO depend on iron in the form of [2Fe–2S] clusters, whereas NR requires iron in the form of iron-heme.

Another crosslink between iron and Mo has been found by investigating the *sta1* mutant of Arabidopsis, which carries a T-DNA insertion in the *ATM3* gene (Kushnir et al. 2001). ATM3 represents a mitochondrial ABC-type transporter and is a homolog of the well-studied Atm1p from baker's yeast, which mediates the export of compounds required for maturation of cytosolic [Fe–S] clusters (Kispal et al. 1999). Due to the mutation in the *ATM3* gene, *sta1* mutants are dwarf and chlorotic and were reported to accumulate free iron inside the mitochondria (Kushnir et al.

2001). Recently, Moco biosynthesis was also demonstrated to be affected in the *sta1* mutant as MPT/Moco amounts and Mo-enzyme activities were found to be significantly reduced (Teschner et al. 2009). Interestingly, this phenotype is accompanied by accumulation of cPMP, in particular inside the mitochondria. This observation led to the assumption of step 1 being located inside the mitochondria, which could be confirmed by immuno analysis using Cnx2- and Cnx3-specific antibodies. As steps 2 to 4 of Moco biosynthesis reside in the cytosol (Gehl, Hänsch, Mendel; unpublished), the mitochondrial localization of step 1 requires export of cPMP into the cytosol to allow conversion of cPMP into MPT and Moco. Due to the fact that the mutation in ATM3 resulted in accumulation of cPMP inside the mitochondria, it was concluded that ATM3 does not only export [Fe–S] cluster equivalents but also facilitates export of cPMP. However, as the substrates are neither known for ATM3 nor for its homologs from yeast (Atm1p) and humans (ABCb7), these need to be identified in future work in order to prove or disprove the capacity of these transporters to export different substrates. Nevertheless, it is obvious that ATM3 bridges [Fe–S] and Mo-metabolism as it is crucial for the export of a Moco biosynthesis-related molecule from mitochondria into the cytosol.

6 Conclusion

Within the last 15 years, we have gained detailed insight into the biological role and function of Mo in plants. Nevertheless, although most of the relevant genes were cloned and the basic functions of the respective proteins are known, we have to consider that a number of unknown genes and proteins are still there which hold crucial roles in Mo-metabolism such as the recently identified mARC proteins, the MoBPs, and the ABC transporter ATM3. Besides the identification of novel genes and proteins, future research also has to concentrate on both the detailed enzymology of Moco biosynthesis/allocation and on studying the regulation and structure–function relationships of Mo-enzymes. Questions that remain to be answered are for instance: How are mitochondrial and cytosolic steps of Moco biosynthesis bridged and how is the putative multienzyme complex for Moco biosynthesis organized? Where in the cell are the molybdate transporters localized and how are they organized in detail? What can MPT synthesis mechanism tell us about the state of copper bound to MPT, and what is the role of copper in Moco synthesis in general? How and in what order are Moco and other prosthetic groups inserted into apo-Mo-enzymes? How is Moco biosynthesis regulated to meet the changing demands of the cell for Moco? What is the physiological role of mARC and what is the precise function of ATM3 for Moco biosynthesis? – It will be a challenging issue for the following years to bring new insights into the regulatory connections of Moco-biosynthesis and Mo-enzymes, and to clarify the metabolic and physiological Mo-network of the cell, which starts with molybdate uptake and, at least presently, ends with five Mo-enzymes.

Acknowledgements We thank the many people who worked with us over the years on molybdenum. In particular we are grateful to Dr. Robert Hänsch for critical discussions. The research was consistently supported by the Deutsche Forschungsgemeinschaft (R.R.M., F.B.) which is gratefully acknowledged. Financial support came also from the European Union (R.R.M.) and the Fonds der Chemischen Industrie (R.R.M.).

References

Aguilar M, Cardenas J, Fernandez E (1992) Quantitation of molybdopterin oxidation product in wild-type and molybdenum cofactor deficient mutants of *Chlamydomonas reinhardtii*. Biochim Biophys Acta 1160:269–274

Akaba S, Seo M, Dohmae N, Takio K, Sekimoto H, Kamiya Y, Furuya N, Komano T, Koshiba T (1999) Production of homo- and hetero-dimeric isozymes from two aldehyde oxidase genes of *Arabidopsis thaliana*. J Biochem 126:395–401

Amaya Y, Yamazaki K, Sato M, Noda K, Nishino T, Nishino T (1990) Proteolytic conversion of xanthine dehydrogenase from the NAD-dependent type to the O_2-dependent type. Amino acid sequence of rat liver xanthine dehydrogenase and identification of the cleavage sites of the enzyme protein during irreversible conversion by trypsin. J Biol Chem 265:14170–14175

Ataya FS, Witte CP, Galvan A, Igeno MI, Fernandez E (2003) Mcp1 encodes the molybdenum cofactor carrier protein in *Chlamydomonas reinhardtii* and participates in protection, binding, and storage functions of the cofactor. J Biol Chem 278:10885–10890

Badwey JA, Robinson JM, Karnovsky MJ, Karnovsky ML (1981) Superoxide production by an unusual aldehyde oxidase in guinea pig granulocytes. Characterization and cytochemical localization. J Biol Chem 256:3479–3486

Balk J, Lill R (2004) The cell's cookbook for iron-sulfur clusters: recipes for fool's gold? Chembiochem 5:1044–1049

Baxter I, Muthukumar B, Park HC, Buchner P, Lahner B, Danku J, Zhao K, Lee J, Hawkesford MJ, Guerinot ML, Salt DE (2008) Variation in molybdenum content across broadly distributed populations of *Arabidopsis thaliana* is controlled by a mitochondrial molybdenum transporter (MOT1). PLoS Genet 4:e1000004

Beligni MV, Lamattina L (2000) Nitric oxide stimulates seed germination and de-etiolation, and inhibits hypocotyl elongation, three light-inducible responses in plants. Planta 210:215–221

Bittner F, Oreb M, Mendel RR (2001) ABA3 is a molybdenum cofactor sulfurase required for activation of aldehyde oxidase and xanthine dehydrogenase in *Arabidopsis thaliana*. J Biol Chem 276:40381–40384

Boland MJ, Schubert KR (1982) Purine biosynthesis and catabolism in soybean root nodules: incorporation of 14C from 14CO2 into xanthine. Arch Biochem Biophys 213:486–491

Bordas J, Bray RC, Garner CD, Gutteridge S, Hasnain SS (1980) X-ray absorption spectroscopy of xanthine oxidase. The molybdenum centres of the functional and the desulpho forms. Biochem J 191:499–508

Bortels H (1930) Molybdän als Katalysator bei der biologischen Stickstoffbindung. Arch Mikrobiol 1:333–342

Braaksma FJ, Feenstra WJ (1982) Isolation and characterization of nitrate reductase-deficient mutants of *Arabidopsis thaliana*. Theor Appl Genet 64:83–90

Brychkova G, Xia Z, Yang G, Yesbergenova Z, Zhang Z, Davydov O, Fluhr R, Sagi M (2007) Sulfite oxidase protects plants against sulfur dioxide toxicity. Plant J 50:696–709

Campbell WH, Kinghorn KR (1990) Functional domains of assimilatory nitrate reductases and nitrite reductases. Trends Biochem Sci 15:315–319

Campbell WH (2001) Structure and function of eukaryotic NAD(P)H:nitrate reductase. Cell Mol Life Sci 58:194–204

Corpas FJ, Palma JM, Sandalio LM, Valderrama R, Barroso JB, Del Rio LA (2008) Peroxisomal xanthine oxidoreductase: characterization of the enzyme from pea (*Pisum sativum* L.) leaves. J Plant Physiol 165:1319–1330

Crawford NM (1992) Study of chlorate resistant mutants of *Arabidopsis*: insights into nitrate assimilation and ion metabolism of plants. In: Setlow JK (ed) Genetic engineering, principles and methods, vol 14. Plenum Press, New York, pp 89–98

Da Cruz S, Xenarios I, Langridge J, Vilbois F, Parone P, Martinou J (2003) Proteomic analysis of the mouse liver mitochondrial inner membrane. J Biol Chem 278:41566–41571

Datta DB, Triplett EW, Newcomb EH (1991) Localization of xanthine dehydrogenase in cowpea root nodules: implications for the interaction between cellular compartments during ureide biogenesis. Proc Natl Acad Sci USA 88:4700–4702

Del Rio LA, Pastori GM, Palma JM, Sandalio LM, Sevilla F, Corpas FJ, Jimenez A, Lopez-Huertas E, Hernandez JA (1998) The activated oxygen role of peroxisomes in senescence. Plant Physiol 116:1195–1200

Dos Santos PC, Dean DR, Hu Y, Ribbe MW (2004) Formation and insertion of the nitrogenase iron-molybdenum cofactor. Chem Rev 104:1159–1173

Durner J, Klessig DF (1999) Nitric oxide as a signal in plants. Curr Opin Plant Biol 2:369–374

Eilers T, Schwarz G, Brinkmann H, Witt C, Richter T, Nieder J, Koch B, Hille R, Hansch R, Mendel RR (2001) Identification and biochemical characterization of *Arabidopsis thaliana* sulfite oxidase. A new player in plant sulfur metabolism. J Biol Chem 276:46989–46994

Fernandez E, Matagne RF (1986) In vivo complementation analysis of nitrate reductase-deficient mutants in *Chlamydomonas reinhardtii*. Curr Genet 10:397–403

Fischer K, Barbier GG, Hecht HJ, Mendel RR, Campbell WH, Schwarz G (2005) Structural basis of eukaryotic nitrate reduction: crystal structures of the nitrate reductase active site. Plant Cell 17:1167–1179

Fischer K, Llamas A, Tejada-Jimenez M, Schrader N, Kuper J, Ataya FS, Galvan A, Mendel RR, Fernandez E, Schwarz G (2006) Function and structure of the molybdenum cofactor carrier protein from *Chlamydomonas reinhardtii*. J Biol Chem 281:30186–30194

Gabard J, Pelsy F, Marion-Poll A, Caboche M, Saalbach I, Grafe R, Müller AJ (1988) Genetic analysis of nitrate reductase deficient mutants of *Nicotiana plumbaginifolia*: evidence for six complementation groups among 70 classified molybdenum cofactor deficient mutants. Mol Gen Genet 213:275–281

Garattini E, Mendel RR, Romao MJ, Wright R, Terao M (2003) Mammalian molybdo-flavoenzymes, an expanding family of proteins: structure, genetics, regulation, function and pathophysiology. Biochem J 372:15–32

González-Guzmán M, Abia D, Salinas J, Serrano R, Rodrigez PL (2004) Two new alleles of the abscisic aldehyde oxidase 3 gene reveal its role in abscisic acid biosynthesis in seeds. Plant Physiol 135:325–333

Gruenewald S, Wahl B, Bittner F, Hungeling H, Kanzow S, Kotthaus J, Schwering U, Mendel RR, Clement B (2008) The fourth molybdenum containing enzyme mARC: cloning and involvement in the activation of N-hydroxylated prodrugs. J Med Chem 51:8173–8177

Gutzke G, Fischer B, Mendel RR, Schwarz G (2001) Thiocarboxylation of molybdopterin synthase provides evidence for the mechanism of dithiolene formation in metal-binding pterins. J Biol Chem 276:36268–36274

Hänsch R, Lang C, Riebeseel E, Lindigkeit R, Gessler A, Rennenberg H, Mendel RR (2006) Plant sulfite oxidase as novel producer of H_2O_2: combination of enzyme catalysis with a subsequent nonenzymatic reaction step. J Biol Chem 281:6884–6888

Hanzelmann P, Hernandez HL, Menzel C, Garcia-Serres R, Huynh BH, Johnson MK, Mendel RR, Schindelin H (2004) Characterization of MOCS1A, an oxygen-sensitive iron-sulfur protein involved in human molybdenum cofactor biosynthesis. J Biol Chem 279:34721–34732

Hanzelmann P, Schindelin H (2004) Crystal structure of the S-adenosylmethionine-dependent enzyme MoaA and its implications for molybdenum cofactor deficiency in humans. Proc Natl Acad Sci USA 101:12870–12875

Havemeyer A, Bittner F, Wollers S, Mendel R, Kunze T, Clement B (2006) Identification of the missing component in the mitochondrial benzamidoxime prodrug-converting system as a novel molybdenum enzyme. J Biol Chem 281:34796–34802
Heidenreich T, Wollers S, Mendel RR, Bittner F (2005) Characterization of the NifS-like domain of ABA3 from *Arabidopsis thaliana* provides insight into the mechanism of molybdenum cofactor sulfuration. J Biol Chem 280:4213–4218
Hesberg C, Hansch R, Mendel RR, Bittner F (2004) Tandem orientation of duplicated xanthine dehydrogenase genes from *Arabidopsis thaliana*: differential gene expression and enzyme activities. J Biol Chem 279:13547–13554
Hille R, Nishino T (1995) Flavoprotein structure and mechanism. 4. Xanthine oxidase and xanthine dehydrogenase. Faseb J 9:995–1003
Hille R (1996) The mononuclear molybdenum enzymes. Chem Rev 96:2757–2816
Hille R (2002) Molybdenum enzymes containing the pyranopterin cofactor: an overview. Met Ions Biol Syst 39:187–226
Hille R (2005) Molybdenum-containing hydroxylases. Arch Biochem Biophys 433:107–116
Hoff T, Schnorr KM, Meyer C, Caboche M (1995) Isolation of two Arabidopsis cDNAs involved in early steps of molybdenum cofactor biosynthesis by functional complementation of *Escherichia coli* mutants. J Biol Chem 270:6100–6107
Ibdah M, Chen YT, Wilkerson CG, Pichersky E (2009) An aldehyde oxidase in developing seeds of *Arabidopsis* converts benzaldehyde to benzoic Acid. Plant Physiol 150:416–423
Ivanov NV, Hubalek F, Trani M, Edmondson DE (2003) Factors involved in the assembly of a functional molybdopyranopterin center in recombinant Comamonas acidovorans xanthine dehydrogenase. Eur J Biochem 270:4744–4754
Johnson JL, Hainline BE, Rajagopalan KV (1980) Characterization of the molybdenum cofactor of sulfite oxidase, xanthine, oxidase, and nitrate reductase. Identification of a pteridine as a structural component. J Biol Chem 255:1783–1786
Johnson JL, Hainline BE, Rajagopalan KV, Arison BH (1984) The pterin component of the molybdenum cofactor. Structural characterization of two fluorescent derivatives. J Biol Chem 259:5414–5422
Johnson JL, Duran M (2001) Molybdenum cofactor deficiency and isolated sulfite oxidase deficiency. In: Scriver C, Beaudet A, Sly W, Valle D (eds) The metabolic and molecular bases of inherited disease. McGraw-Hill, New York, pp 3163–3177
Kaiser WM, Huber SC (2001) Post-translational regulation of nitrate reductase: mechanism, physiological relevance and environmental triggers. J Exp Bot 52:1981–1989
Kisker C, Schindelin H, Pacheco A, Wehbi WA, Garrett RM, Rajagopalan KV, Enemark JH, Rees DC (1997a) Molecular basis of sulfite oxidase deficiency from the structure of sulfite oxidase. Cell 91:973–983
Kisker C, Schindelin H, Rees DC (1997b) Molybdenum-cofactor-containing enzymes: structure and mechanism. Annu Rev Biochem 66:233–267
Kispal G, Csere P, Prohl C, Lill R (1999) The mitochondrial proteins Atm1p and Nfs1p are essential for biogenesis of cytosolic Fe/S proteins. Embo J 18:3981–3989
Kleinhofs A, Warner RL, Lawrence JM, Jeter JM, Kudrna DA (1989) Molecular genetics of nitrate reductase in barley. In: Wray JL, Kinghorn JR (eds) Molecular and genetic aspects of nitrate assimilation. Oxford University Press, Oxford, pp 197–211
Koiwai H, Nakaminami K, Seo M, Mitsuhashi W, Toyomasu T, Koshiba T (2004) Tissue-specific localization of an abscisic acid biosynthetic enzyme, AAO3, in *Arabidopsis*. Plant Physiol 134:1697–1707
Kozmin SG, Leroy P, Pavlov YI, Schaaper RM (2008) YcbX and yiiM, two novel determinants for resistance of *Escherichia coli* to N-hydroxylated base analogues. Mol Microbiol 68:51–65
Kruse T, Geisler M, Lehrke M, Ringel P, Mendel RR (2009) Identification and biochemical characterization of molybdenum cofactor-binding proteins from *Arabidopsis thaliana*. J Biol Chem (unpublished)

Kuper J, Palmer T, Mendel RR, Schwarz G (2000) Mutations in the molybdenum cofactor biosynthetic protein Cnx1G from *Arabidopsis thaliana* define functions for molybdopterin binding. Mo-insertion and molybdenum cofactor stabilization. Proc Natl Acad Sci USA 97:6475–6480

Kuper J, Llamas A, Hecht HJ, Mendel RR, Schwarz G (2004) Structure of molybdopterin-bound Cnx1G domain links molybdenum and copper metabolism. Nature 430:806–806

Kushnir S, Babiychuk E, Storozhenko S, Davey MW, Papenbrock J, De Rycke R, Engler G, Stephan UW, Lange H, Kispal G, Lill R, Van Montagu M (2001) A mutation of the mitochondrial ABC transporter Sta1 leads to dwarfism and chlorosis in the *Arabidopsis* mutant starik. Plant Cell 13:89–100

Lang C, Popko J, Wirtz M, Hell R, Herschbach C, Kreuzwieser J, Rennenberg H, Mendel RR, Hänsch R (2007) Sulfite oxidase as key enzyme for protecting plants against sulfur dioxide. Plant Cell Environ 30:447–455

Leimkühler S, Klipp W (1999) Role of XDHC in Molybdenum cofactor insertion into xanthine dehydrogenase of *Rhodobacter capsulatus*. J Bacteriol 181:2745–2751

Leimkühler S, Wuebbens MM, Rajagopalan KV (2001) Characterization of *Escherichia coli* MoeB and its involvement in the activation of molybdopterin synthase for the biosynthesis of the molybdenum cofactor. J Biol Chem 276:34695–34701

Leon-Kloosterziel KM, Gil MA, Ruijs GJ, Jacobsen SE, Olszewski NE, Schwartz SH, Zeevaart JAD, Koorneef M (1996) Isolation and characterization of abscisic acid-deficient *Arabidopsis* mutants at two new loci. Plant J 10:655–661

Lewis NJ, Hurt P, Sealy-Lewis HM, Scazzocchio C (1978) The genetic control of the molybdoflavoproteins in *Aspergillus nidulans*. IV. A comparison between purine hydroxylase I and II. Eur J Biochem 91:311–316

Lill R, Muhlenhoff U (2006) Iron-sulfur protein biogenesis in eukaryotes: components and mechanisms. Annu Rev Cell Dev Biol 22:457–486

Llamas A, Mendel RR, Schwarz G (2004) Synthesis of adenylated molybdopterin: an essential step for molybdenum insertion. J Biol Chem 279:55241–55246

Llamas A, Otte T, Multhaup G, Mendel RR, Schwarz G (2006) The Mechanism of nucleotide-assisted molybdenum insertion into molybdopterin. A novel route toward metal cofactor assembly. J Biol Chem 281:18343–18350

Magalon A, Mendel RR. Chapter 3.6.3.13, Biosynthesis and Insertion of the Molybdenum Cofactor. In: Böck A, Curtiss III R, Kaper JB, Karp PD, Neidhardt FC, Nyström T, Slauch JM, Squires CL (eds) EcoSal–*Escherichia coli* and *Salmonella*: cellular and molecular biology. http://www.ecosal.org. ASM Press, Washington, D.C.

Massey V, Edmondson DE (1970) On the mechansim of inactivation of xanthine oxidase by cyanide. J Biol Chem 245:6595–6598

Matthies A, Rajagopalan KV, Mendel RR, Leimkuhler S (2004) Evidence for the physiological role of a rhodanese-like protein for the biosynthesis of the molybdenum cofactor in humans. Proc Natl Acad Sci USA 101:5946–5951

Matthies A, Nimtz M, Leimkuhler S (2005) Molybdenum cofactor biosynthesis in humans: identification of a persulfide group in the rhodanese-like domain of MOCS3 by mass spectrometry. Biochemistry 44:7912–7920

Mauch-Mani B, Mauch F (2005) The role of abscisic acid in plant-pathogen interactions. Curr Opin Plant Biol 8:409–414

Mendel RR (2009) Cell biology of molybdenum. BioFactors 35:429–434

Mendel RR, Müller AJ (1976) A common genetic determinant of xanthine dehydrogenase and nitrate reductase in *Nicotiana tabacum*. Biochem Physiol Pflanzen 170:538–541

Mendel RR (1992) The plant molybdenum cofactor (MoCo) – its biochemical and molecular genetics. In: Gresshoff PM (ed) Plant biotechnology and development – current topics in plant molecular biology, 1st edn. CRC Press, Boca Raton, pp 11–16

Mendel RR, Schwarz G (2002) Biosynthesis and molecular biology of the molybdenum cofactor (Moco). Met Ions Biol Syst 39:317–368

Mercer JF (2001) The molecular basis of copper-transport diseases. Trends Mol Med 7:64–69
Millar LJ, Heck IS, Sloan J, Kana'n GJ, Kinghorn JR, Unkles SE (2001) Deletion of the cnxE gene encoding the gephyrin-like protein involved in the final stages of molybdenum cofactor biosynthesis in *Aspergillus nidulans*. Mol Genet Genomics 266:445–453
Montalbini P (1992) Inhibition of hypersensitive response by allopurinol applied to the host in the incompatible relationship between *Phaseolus vulgaris* and *Uromyces phaseoli*. J Phytopathol 134:218–228
Montalbini P, Della Torre G (1996) Evidence of a two-fold mechanism responsible for the inhibition by allopurinol of the hypersensitive reponse induced in tobacco by tobacco necrosis virus. Phys Mol Plant Pathol 48:273–287
Montalbini P (1998) Purification and some properties of xanthine dehydrogenase from wheat leaves. Plant Sci 134:89–102
Montalbini P (2000) Xanthine dehydrogenase from leaves of leguminous plants: purification, characterization and properties of the enzyme. J Plant Phys 156:3–16
Moorhead GB, Meek SE, Douglas P, Bridges D, Smith CS, Morrice N, MacKintosh C (2003) Purification of a plant nucleotide pyrophosphatase as a protein that interferes with nitrate reductase and glutamine synthetase assays. Eur J Biochem 270:1356–1362
Müller AJ, Mendel RR (1989) Biochemical and somatic cell genetics of nitrate reductase in *Nicotiana*. In: Wray JL, Kinghorn JR (eds) Molecular and genetic aspects of nitrate assimilation. Oxford University Press, Oxford, pp 166–185
Nishino T, Nishino T (1997) The conversion from the dehydrogenase type to the oxidase type of rat liver xanthine dehydrogenase by modification of cysteine residues with fluorodinitrobenzene. J Biol Chem 272:29859–29864
Noritake T, Kawakita K, Doke N (1996) Nitric oxide induces phytoalexin accumulation in potato tuber tissues. Plant Cell Physiol 37:113–116
Nowak K, Luniak N, Witt C, Wustefeld Y, Wachter A, Mendel RR, Hansch R (2004) Peroxisomal localization of sulfite oxidase separates it from chloroplast-based sulfur assimilation. Plant Cell Physiol 45:1889–1894
Nussaume L, Vincentz M, Meyer C, Boutin JP, Caboche M (1995) Post-transcriptional regulation of nitrate reductase by light is abolished by an N-terminal deletion. Plant Cell 7:611–621
Pateman JA, Cove DJ, Rever BM, Roberts DB (1964) A common cofactor for nitrate reductase and xanthine dehydrogenase which also regulates the synthesis of nitrate reductase. Nature 201:58–60
Pau RN, Lawson DM (2002) Transport, homeostasis, regulation, and binding of molybdate and tungstate to proteins. Met Ions Biol Syst 39:3–74
Perez-Vicente R, Alamillo JM, Cardenas J, Pineda M (1992) Purification and substrate inactivation of xanthine dehydrogenase from *Chlamydomonas reinhardtii*. Biochim Biophys Acta 1117:159–166
Raab S, Drechsel G, Zarepour M, Hartung W, Koshiba T, Bittner F, Hoth S (2009) Identification of a novel E3 ubiquitin ligase that is required for suppression of premature senescence in *Arabidopsis*. Plant J 59:39–51
Rajagopalan KV, Johnson JL (1992) The pterin molybdenum cofactors. J Biol Chem 267: 10199–10202
Rajagopalan KV (1996) Biosynthesis of the molybdenum cofactor. In: Neidhardt FC (ed) *Escherichia coli* and *Salmonella typhimurium*. ASM Press, Washington, DC, pp 674–679
Rees DC, Akif Tezcan F, Haynes CA, Walton MY, Andrade S, Einsle O, Howard JB (2005) Structural basis of biological nitrogen fixation. Philos Transact A Math Phys Eng Sci 363: 971–984 discussion 1035-1040
Reiss J, Cohen N, Dorche C, Mandel H, Mendel RR, Stallmeyer B, Zabot MT, Dierks T (1998) Mutations in a polycistronic nuclear gene associated with molybdenum cofactor deficiency. Nat Genet 20:51–53
Reiss J, Johnson JL (2003) Mutations in the molybdenum cofactor biosynthetic genes MOCS1, MOCS2, and GEPH. Hum Mutat 21:569–576

Rieder C, Eisenreich W, O'Brien J, Richter G, Götze E, Boyle P, Blanchard S, Bacher A, Simon H (1998) Rearrangement reactions in the biosynthesis of molybdopterin – an NMR study with multiply $^{13}C/^{15}N$ labelled precursors. Eur J Biochem 255:24–36

Rockel P, Strube F, Rockel A, Wildt J, Kaiser WM (2002) Regulation of nitric oxide (NO) production by plant nitrate reductase in vivo and in vitro. J Exp Bot 53:103–110

Rodriguez-Trelles F, Tarrio R, Ayala FJ (2003) Convergent neofunctionalization by positive Darwinian selection after ancient recurrent duplications of the xanthine dehydrogenase gene. Proc Natl Acad Sci USA 100:13413–13417

Sagi M, Scazzocchio C, Fluhr R (2002) The absence of molybdenum cofactor sulfuration is the primary cause of the flacca phenotype in tomato plants. Plant J 31:305–317

Sandalio LM, Fernandez VM, Ruperez FL, del Rio LA (1988) Superoxide free radicals are produced in glyoxysomes. Plant Physiol 127:1–4

Santamaria-Araujo JA, Fischer B, Otte T, Nimtz M, Mendel RR, Wray V, Schwarz G (2004) The tetrahydropyranopterin structure of the sulfur-free and metal-free molybdenum cofactor precursor. J Biol Chem 279:15994–15999

Sauer P, Frebortova J, Sebela M, Galuszka P, Jacobsen S, Pec P, Frebort I (2002) Xanthine dehydrogenase of pea seedlings: a member of the plant molybdenum oxidoreductase family. Plant Physiol Biochem 40:393–400

Schrader N, Kim EY, Winking J, Paulukat J, Schindelin H, Schwarz G (2004) Biochemical characterization of the high affinity binding between the glycine receptor and gephyrin. J Biol Chem 279:18733–18741

Schwarz G, Boxer DH, Mendel RR (1997a) Molybdenum cofactor biosynthesis. The plant protein Cnx1 binds molybdopterin with high affinity. J Biol Chem 272:26811–26814

Schwarz G, Boxer DH, Mendel RR (1997b) The plant protein Cnx1 binds molybdopterin and is involved in the last step of molybdenum cofactor biosynthesis. In: Pfleiderer W, Rokos H (eds) Chemistry and biology of pteridines and folates. Blackwell Science, Berlin, pp 697–702

Schwarz G, Schulze J, Bittner F, Eilers T, Kuper J, Bollmann G, Nerlich A, Brinkmann H, Mendel RR (2000) The molybdenum cofactor biosynthetic protein Cnx1 complements molybdate-repairable mutants, transfers molybdenum to the metal binding pterin, and is associated with the cytoskeleton. Plant Cell 12:2455–2472

Seo M, Koiwai H, Akaba S, Komano T, Oritani T, Kamiya Y, Koshiba T (2000a) Abscisic aldehyde oxidase in leaves of *Arabidopsis thaliana*. Plant J 23:481–488

Seo M, Peeters AJ, Koiwai H, Oritani T, Marion-Poll A, Zeevaart JA, Koornneef M, Kamiya Y, Koshiba T (2000b) The *Arabidopsis* aldehyde oxidase 3 (AAO3) gene product catalyzes the final step in abscisic acid biosynthesis in leaves. Proc Natl Acad Sci USA 97:12908–12913

Seo M, Koshiba T (2002) The complex regulation of ABA biosynthesis in plants. Trends Plant Sci 7:41–48

Sigel A, Sigel H (2002) Molybdenum and tungsten. Their roles in biological processes. Marcel Dekker, New York

Skipper L, Campbell WH, Mertens JA, Lowe DJ (2001) Pre-steady-state kinetic analysis of recombinant *Arabidopsis* NADH:nitrate reductase: rate-limiting processes in catalysis. J Biol Chem 276:26995–27002

Sofia HJ, Chen G, Hetzler BG, Reyes-Spindola JF, Miller NE (2001) Radical SAM, a novel protein superfamily linking unresolved steps in familiar biosynthetic pathways with radical mechanisms: functional characterization using new analysis and information visualization methods. Nucleic Acids Res 29:1097–1106

Stallmeyer B, Drugeon G, Reiss J, Haenni AL, Mendel RR (1999a) Human molybdopterin synthase gene: identification of a bicistronic transcript with overlapping reading frames. Am J Hum Genet 64:698–705

Stallmeyer B, Schwarz G, Schulze J, Nerlich A, Reiss J, Kirsch J, Mendel RR (1999b) The neurotransmitter receptor-anchoring protein gephyrin reconstitutes molybdenum cofactor biosynthesis in bacteria, plants, and mammalian cells. Proc Natl Acad Sci USA 96: 1333–1338

Stiefel EI (2002) The biogeochemistry of molybdenum and tungsten. Met Ions Biol Syst 39:1–29

Tejada-Jimenez M, Llamas A, Sanz-Luque E, Galvan A, Fernandez E (2007) A high-affinity molybdate transporter in eukaryotes. Proc Natl Acad Sci USA 104:20126–20130

Terao M, Kurosaki M, Marini M, Vanoni MA, Saltini G, Bonetto V, Bastone A, Federico C, Saccone S, Fanelli R, Salmona M, Garattini E (2001) Purification of the aldehyde oxidase homolog 1 (AOH1) protein and cloning of the AOH1 and aldehyde oxidase homolog 2 (AOH2) genes. Identification of a novel molybdo-flavoprotein gene cluster on mouse chromosome 1. J Biol Chem 276:46347–46363

Teschner J, Lachmann N, Geisler M, Selbach K, Balk J, Mendel RR, Bittner F (2009) A novel role for the mitochondrial ABC transporter ATM3 from *Arabidopsis* in molybdenum cofactor biosynthesis. Plant Cell (under revision)

Tomatsu H, Takano J, Takahashi H, Watanabe-Takahashi A, Shibagaki N, Fujiwara T (2007) An *Arabidopsis thaliana* high-affinity molybdate transporter required for efficient uptake of molybdate from soil. Proc Natl Acad Sci USA 104:18807–18812

Turnlund JR (2002) Molybdenum metabolism and requirements in humans. Met Ions Biol Syst 39:727–739

Verslues PE, Zhu JK (2005) Before and beyond ABA: upstream sensing and internal signals that determine ABA accumulation and response under abiotic stress. Biochem Soc Trans 33:375–379

Wahl RC, Rajagopalan KV (1982) Evidence for the inorganic nature of the cyanolyzable sulfur of molybdenum hydroxylases. J Biol Chem 257:1354–1359

Warner CK, Finnerty V (1981) Molybdenum hydroxylases in *Drosophila*. II Molybdenum cofactor in xanthine dehydrogenase, aldehyde oxidase and pyridoxal oxidase. Mol Gen Genet 184:92–96

Witte CP, Igeno MI, Mendel RR, Schwarz G, Fernandez E (1998) The Chlamydomonas reinhardtii MoCo carrier protein is multimeric and stabilizes molybdopterin cofactor in a molybdate charged form. FEBS Lett 431:205–209

Wollers S, Heidenreich T, Zarepour M, Zachmann D, Kraft C, Zhao Y, Mendel RR, Bittner F (2008) Binding of sulfurated molybdenum cofactor to the C-terminal domain of ABA3 from *Arabidopsis thaliana* provides insight into the mechanism of molybdenum cofactor sulfuration. J Biol Chem 283:9642–9650

Wuebbens MM, Rajagopalan KV (1993) Structural characterization of a molybdopterin precursor. J Biol Chem 268:13493–13498

Wuebbens MM, Rajagopalan KV (1995) Investigation of the early steps of molybdopterin biosynthesis in *Escherichia coli* through the use of in vivo labeling studies. J Biol Chem 270:1082–1087

Wuebbens MM, Rajagopalan KV (2003) Mechanistic and mutational studies of *Escherichia coli* molybdopterin synthase clarify the final step of molybdopterin biosynthesis. J Biol Chem 278:14523–14532

Xiong L, Ishitani M, Lee H, Zhu JK (2001) The arabidopsis los5/aba3 locus encodes a molybdenum cofactor sulfurase and modulates cold stress- and osmotic stress-responsive gene expression. Plant Cell 13:2063–2083

Yamasaki H, Sakihama Y (2000) Simultaneous production of nitric oxide and peroxynitrite by plant nitrate reductase: in vitro evidence for the NR-dependent formation of active nitrogen species. FEBS Lett 468:89–92

Yesbergenova Z, Yang G, Oron E, Soffer D, Fluhr R, Sagi M (2005) The plant Mo-hydroxylases aldehyde oxidase and xanthine dehydrogenase have distinct reactive oxygen species signatures and are induced by drought and abscisic acid. Plant J 42:862–876

Zhang Y, Gladyshev VN (2008) Molybdoproteomes and evolution of molybdenum utilization. J Mol Biol 379:881–899

Cellular Biology of Nitrogen Metabolism and Signaling

Françoise Daniel-Vedele, Anne Krapp, and Werner M. Kaiser

Abstract This chapter summarizes major aspects of N-nutrition in plants. N distribution within a plant varies widely according to the organ, the development stage, and mostly to the environmental conditions. Within the cell, the different N forms are stored in different compartments and the pool sizes are controlled in contrasting manner. Plants can take up nitrate, ammonium, urea, and other organic N forms. Various transporters for these compounds have been characterized, and the localization and properties of these proteins give rise to a complex pattern of N fluxes within the plant. The further assimilation of nitrate is well described, but the *in planta* role of all proteins, as for example GS1 and GDH, is far from being evident. Some are involved in N remobilization which is an important N source for example during seed filling.

Regulation of N assimilation occurs at the transcriptional and post-transcriptional levels, and regulation of the different steps is highly coordinated. However, only very few molecular players are known. As a special case in N-signaling, NO, a side product of N assimilation, is considered in some detail.

1 Introduction

Nitrogen is the mineral nutrient required in highest amounts by plants and is most frequently limiting growth and yield. Inorganic or organic N forms participate to plant nutrition in a variable extent depending on plant parameters as well as soil

F. Daniel-Vedele and A. Krapp
Institut Jean-Pierre Bourgin (IJPB, UMR 1318) Institut National de Recherche Agronomique (INRA), Route de St Cyr, 78026 Versailles, France
e-mail: vedele@versailles.inra.fr, krapp@versailles.inra.fr

W.M. Kaiser (✉)
Julius-von-Sachs-Institut für Biowissenschaften Lehrstuhl für Botanik I, Julius-von-Sachs-Platz 2, 97082 Würzburg, Germany
e-mail: kaiser@botanik.uni-wuerzburg.de

R. Hell and R.-R. Mendel (eds.), *Cell Biology of Metals and Nutrients*, Plant Cell Monographs 17,
DOI 10.1007/978-3-642-10613-2_7,

characteristics. In temperate climatic conditions, inorganic N forms are predominant, and fertilizers are often supplied as nitrate, ammonium, or urea (http://www.fertilizer.org/ifa/). However, the soil solution may contain different organic N forms such as soluble proteins or amino acids derived from proteolytic processes. A variety of plant species are able to use organic N forms in artic, boreal, temperate, Mediterranean shrub-land, or alpine natural ecosystems (reviewed in Näsholm et al. 2009). In legume plants, atmospheric N_2 is fixed in the nodule, a specialized organ resulting from the interaction between bacteria and roots (Gordon et al. 2001). In the same manner, nutrient use efficiency is increased by symbioses between fungi and plants, the mycorrhizal system being involved in nutrient uptake and the plant partner providing reduced carbon to the fungus (Martin et al. 2001).

Although such symbioses are important in natural ecosystems, this chapter describes only direct N uptake by root cells. We give an overview of (1) N distribution within the plant and more precisely within a plant cell, (2) the molecular elements involved in different fluxes or in assimilatory steps, and (3) the regulatory mechanisms that control these processes. N metabolites, such as nitrate, ammonium, and glutamate act as signal molecules as well. However, this is out of the scope of this chapter and has been reviewed recently (Walch-Liu et al. 2005); instead, we extend this chapter by (4) a detailed description of the synthesis and mode of action of NO.

2 Distribution of N Forms in Plant Cells

2.1 N in Different Tissues

The N forms and N quantities within a plant vary widely according to the organ, the development stage, and the environmental conditions. The root is obviously the predominant organ where large exchanges of a variety of N forms occur between root cells and the soil solution. The differential expression and localization of channel- or transporter proteins (see below) led to a complex picture of the root cellular organization, with specialized uptake functions for lateral root caps or epidermis/cortex, and horizontal transport toward the vasculature for endodermis/pericycle and stele lines (see below 1.2.2 ammonium transport). Inorganic N forms can then enter the xylem to be transported to the shoots. N assimilation and remobilization take place in roots and shoots, and organic N forms are then distributed to sinks organs (Brouquisse et al. 2001).

2.2 N Cellular Distribution

Within the cell, N forms are stored in different compartments (Fig. 1).

Cytoplasmic ammonium pools originate not only from ammonium uptake across the plasma membranes but also from amino acid catabolism occurring during

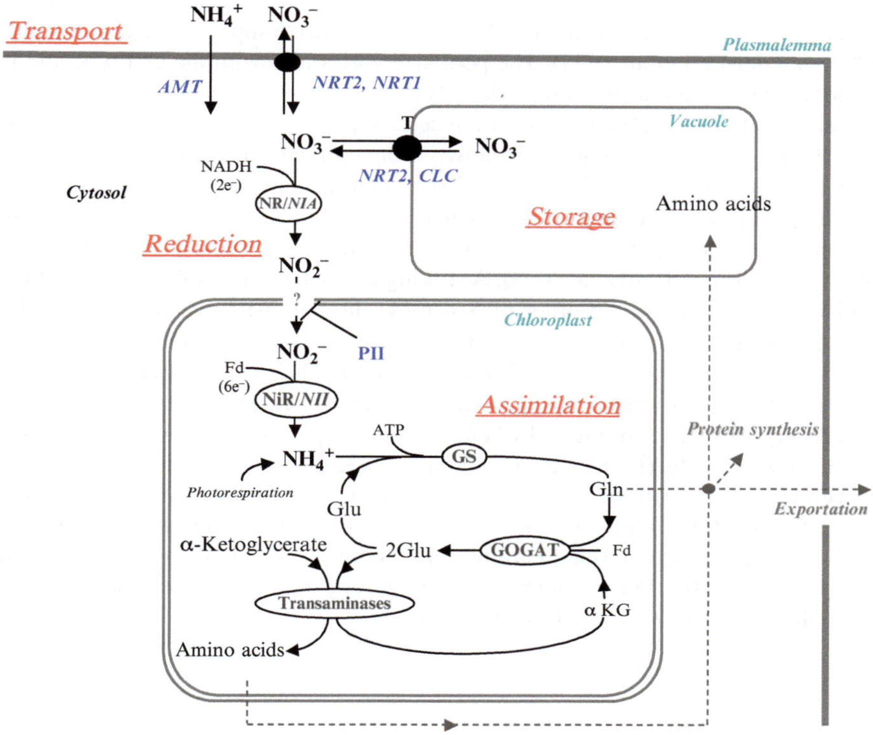

Fig. 1 *N Storage in different compartments*. Nitrate and ammonium enter the cell and can be either stored in the vacuole, transported to other tissues, or assimilated in the cytosol and the chloroplast. Small italic letters: nitrogen assimilation steps. Capital letters: enzymes. Capital italic letters: genes. NR: Nitrate reductase. NiR: nitrite reductase/ GS: Glutamine synthethase. GOGAT: glutamate synthase, AMT: ammonium transporter. NRT: nitrate transporter; CLC: chloride channel

photorespiration in illuminated leaves (Leegood et al. 1996) or in senescent tissues (Matsson and Schjoerring 2003). Ammonium concentrations have been measured using analysis of ^{1}H-coupled ^{14}N-NMR signals (review in Mesnard and Ratcliffe 2005) or with ammonium-selective microelectrodes (Wells and Miller 2000). In both cases, the cytoplasmic ammonium concentrations were no more than a few millimolar (8–15 mM), but this concentration could be increased in maize roots when ammonium assimilation was blocked (Lee and Ratcliffe 1991). In vacuoles, ammonium concentrations vary between 1 and 45 mM in nonstressed plants (Miller et al. 2001), indicating a possible role of this compartment for the storage of ammonium.

The global nitrate concentrations in leaves or roots are highly dependent on external N supply, and nitrate, among all N-compounds, disappears most quickly in response to N starvation (Richard-Molard et al. 2008). The pool of nitrate associated with purified chloroplasts remains remarkably constant under various conditions (Schröppel-Maier and Kaiser 1988). In contrast, the vacuolar nitrate pools show a positive correlation with the external nitrate supply (Miller and Smith 2007; van der Leij et al. 1998). The pool size varies also with the cell type and is higher in epidermal than in mesophyll cells in barley leaves (Karley et al. 2000) and higher in cortical compared to epidermal cells in barley roots. In roots, remobilization of vacuolar nitrate occurs more slowly from cortical cells than from epidermal cells (van der Leij et al. 1998). This tissue heterogeneity revealed by single-cell techniques implies that knowledge obtained for vacuoles from one type of tissue cannot be necessarily transferred to vacuoles from other tissues, as also shown for gene expression (Gifford et al 2008). A striking characteristic of the cytosolic NO_3^- pool is its low size (in the order of 3–4 mM). In contrast to vacuoles, cytosolic nitrate is maintained at a remarkably stable value that is independent of changes in the external nitrate concentration (Miller and Smith 2007; van der Leij et al. 1998).

The global amino acid contents in leaves depend on external N supplies and can vary from 150 to 45 nmol/mgDM when *Arabidopsis* plants are fed with 10 or 3 mM nitrate, respectively (Loudet et al. 2003). Subcellular volumes and amino acid concentrations have been analyzed using non-aqueous fractionation in spinach (Winter et al. 1994), barley (Winter et al. 1993), or potato (Leidreiter et al. 1995). In all cases, the concentration of amino acids is much lower in the vacuoles than in the cytosol (1.7/40 mM for glutamate in barley, for example). These concentrations are quite similar between cytosol and stroma.

3 N Fluxes Within a Plant Cell

3.1 Nitrate and Nitrite Fluxes

Two nitrate transport systems have been shown to co-exist in plants and act co-ordinately to take up nitrate from the soil solution and distribute nitrate within the whole plant (Fig. 2) (review in Daniel-Vedele et al. 1998; Tsay et al. 2007).

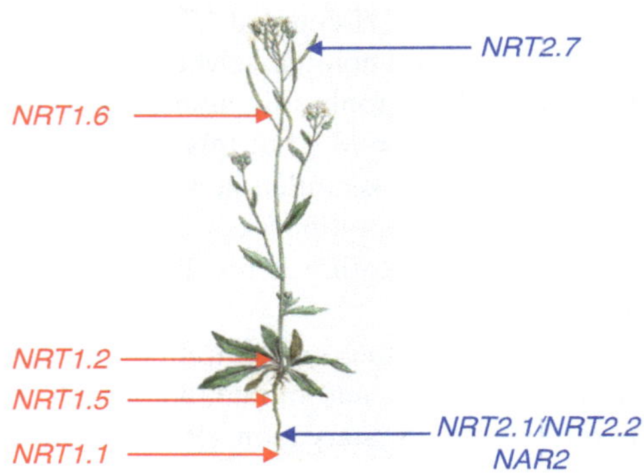

Fig. 2 *Schematic presentation of the known localization of NRT1 and NRT2 genes in Arabidopsis.* Two nitrate transport systems have been shown to co-exist in plants and act co-ordinately to take up nitrate from the soil solution and distribute nitrate within the whole plant

It is generally assumed that the NRT1 gene family mediates the root low-affinity transport system (LATS), with the exception of the At*NRT1.1*, which is a dual affinity transporter (Wang et al. 1998; Liu et al. 1999). In *Arabidopsis*, 53 genes belong to the NRT1 family. Among them 51 genes are expressed and exhibit different tissue expression patterns in the whole plant (Tsay et al. 2007), suggesting a specialized and unique function for at least some of them. The most extensively studied gene is the first one isolated, *AtNRT1.1* (formerly *Chl1*; Tsay et al. 1993). The gene is expressed in epidermis of the root tips and in the cortex and endodermis in the more mature part of the root (Huang et al. 1996) but also accumulates in nascent organs (Guo et al. 2001). *AtNRT1.1* is also considered as a nitrate sensor that could regulate other processes like regulation of other components of nitrate uptake (Krouk et al. 2006), stomatal opening (Guo et al. 2001), relieving of seed dormancy (Alboresi et al. 2005), or stimulation of root proliferation by nitrate (Remans et al. 2006a). Beside this gene, the *AtNRT1.2* gene is constitutively expressed only in the root epidermis and belongs to the constitutive low-affinity system (Huang et al. 1999). *AtNRT1.5* is located on the plasma membrane of root pericycle cells close to the xylem. The protein is a low-affinity, pH-dependent bidirectional nitrate transporter and is involved in long distance transport of nitrate from the root to the shoot (Lin et al. 2008). The *AtNRT1.4* gene is only expressed in leaf petioles, and the nitrate content is twice lower in the petiole of the mutant compared to that of the wild type (Chiu et al. 2004). Recently, *AtNRT1.6* was shown to be involved in embryo development. The gene is expressed in the vascular tissue of the silique. Expression in oocytes and mutant phenotypes suggest that the protein could deliver nitrate from maternal tissue to the developing embryo (Almagro et al. 2008). A striking particularity of the NRT1 family is that certain members belonging to the group II (reviewed in Tsay et al. 2007) are able to transport not only nitrate but also di or tripeptides in heterologous systems, while OPT proteins transport tetra/pentapeptides.

The high-affinity transport system (HATS), acting when the external nitrate concentration is low, relies on the activity of the so called NRT2 family (reviewed in Williams and Miller 2001). *AtNRT2.1* is a major component of the iHATS in *Arabidopsis*, as shown by the fact that a mutant disrupted for the *AtNRT2.1* gene has lost up to 75% of the inducible high-affinity NO_3^- uptake activity and showed a lower leaf nitrate content (Cerezo et al. 2001; Filleur et al. 2001). As a consequence, growth of these mutants is severely impaired at low NO_3^- concentration (Orsel et al. 2004; Orsel et al. 2006). Li and coworkers showed that the *AtNRT2.2* makes only a small contribution to iHATS under normal growth conditions (Li et al. 2007).

Nitrate can also be exported from the cytosolic pool by an efflux mechanism. Segonsac and co-workers have identified an *Arabidopsis* excretion transporter, localized at the plasma membrane of cortical root cells and encoded by the *NAXT1* gene belonging to the *NRT1* family (Segonzac et al. 2007).

Regarding vacuolar nitrate pools, classical experiments using indirect assay of H^+ transport provided evidence for the presence of a NO_3^-/H^+ antiporter in the tonoplast (Schumaker and Sze 1987). Recently, De Angeli et al. (2006)

demonstrated that the AtCLCa protein, localized in the vacuolar membrane, behaves as a NO_3^-/H^+ exchanger, allowing the accumulation of nitrate within the vacuole. Residues important for nitrate/proton coupling have been identified in plant and mammalian CLC transporters (Eun-Yeong et al. 2009; Zifarelli and Pusch 2009). Insertion mutants within the *AtCLCa* gene exhibit normal development but show a reduced capacity to store nitrate but not other anions (Geelen et al. 2000). This phenotype was also recently found when the expression of the vacuole-located nitrate transporter AtNRT2.7 was affected. This *AtNRT2* gene is expressed in aerial organs and also highly induced in dry seeds. In two allelic *atnrt2.7* mutants, less nitrate is accumulated in the seed. In contrast, seeds from plants overexpressing the *AtNRT2.7* coding region accumulate more nitrate, and as a consequence they are less dormant than the corresponding wild type seeds (Chopin et al. 2007).

Finally, little is known on potential channels or transporters that could be involved in fluxes towards the chloroplast (reviewed in Weber et al. 2005). Fusion proteins with the GFP marker revealed the chloroplastic subcellular localization of the AtCLCe protein. The *atclce* mutants display a phenotype linked both to photosynthesis (Marmagne et al. 2007) and nitrate content (Monachello et al. 2009). The flux of nitrite, the product of nitrate reduction in the cytosol, into the chloroplast could also play a role in the flux of nitrate towards the chloroplast and thus in the homeostasis of cytosolic nitrate. A nitrite transporter belonging to the NRT1 family has been recently identified in cucumber and *Arabidopsis* (Sugiura et al. 2007).

3.2 Ammonium Fluxes

Since the cloning of the first gene involved in ammonium transport (Ninnemann et al. 1994), five other genes belonging to the same family were found in *Arabidopsis* (Gazzarrini et al. 1999; Sohlenkamp et al. 2000), ten in rice (Sonoda et al. 2003), a species adapted to ammonium nutrition, and 14 in poplar (Couturier et al. 2007). Focusing on the results obtained in *Arabidopsis*, kinetics properties of the AMT proteins expressed in oocytes showed Km values ranging from 34 mM for *AMT1;1* (Wood et al. 2006) to 140 mM for *AMT1;2* (Neuhäuser et al. 2007). Among the six genes, *AMT1;1*, *AMT1;2*, *AMT1;3*, and *AMT2;1* are highly expressed in roots (Loqué and von Wirén 2004) and encode proteins that are located in the plasma membranes (Loqué et al. 2006; Yuan et al. 2007). In order to analyze the function of each of this genes separately *in planta*, physiological and ammonium influx studies were carried out on single, double, triple, and quadruple mutants (Yuan et al. 2007). Additive contribution of AMT1;1 and AMT1;3 was shown, while a second saturable transport is thought to be coded by the *AMT1;5* gene. A complex picture is now emerging from these studies (Fig. 3). There is a spatial organization of AMT1 proteins, the transporters possessing the highest ammonium affinities being located in outer root cells or root hairs where they can uptake ammonium from the soil solution (*AMT1;1*, *AMT1;3*, *AMT1;5*). The lower

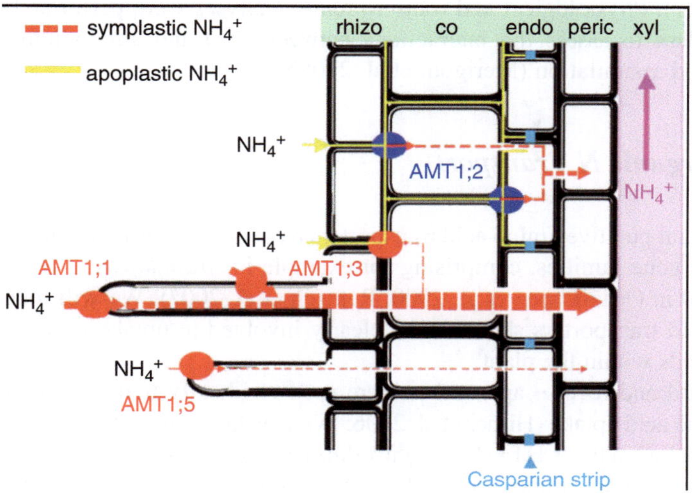

Fig. 3 *Model summarizing the functions of AMT1-type transporters in high-affinity ammonium uptake in Arabidopsis roots* (from Yuan et al. 2007). This schematic representation shows the contribution to ammonium uptake and spatial expression in root tissues of AMT1;1, AMT1;3, AMT1;5 (all in *red*), and AMT1;2 (*blue*) under nitrogen deficiency. AMT-dependent ammonium influx is proportionally represented by the size of their arrows. rhizo, rhizodermis; co, cortex; endo, endodermis; peric, pericycle; xyl, xylem

affinity of *AMT1;2* and its location in the endodermis along the root hair zone suggest a function in the retrieval of ammonium that is released from the cortex, or that enters the root via the apoplastic route.

The electrochemical gradient between vacuole and cytosol would drive NH_3 import to and NH_4^+ export out of the vacuole. Indeed, tonoplast intrinsic proteins of the TIP family were shown to play a role in NH_3 transport into the vacuole (Loqué et al. 2005). Vacuolar loading with NH_4^+ should require an electrogenic ammonium transporter, which has not yet been identified.

3.3 Urea Transport

Although urea is the major nitrogen form supplied as fertilizer in agricultural plant production, its uptake by plant roots or leaves before its hydrolysis has been a matter of debate for a long time. However, studies in crop plants (Merigout et al. 2008a) and *Arabidopsis* (review in Kojima et al. 2006) showed the uptake of urea. The identification of the high-affinity urea transporter *AtDUR3* by Liu et al. (2003a) and of the *AtTIP* urea permeases (Liu et al. 2003b) led to new insights regarding the molecular basis of urea uptake in plants. Growth of mutant lines carrying T-DNA insertions in *AtDUR3* is impaired when urea is the sole nitrogen source. (Kojima et al. 2007). Physiological and transcriptomic analyses were performed in *Arabidopsis* plant to assess the interactions between urea and ammonium or nitrate uptake and assimilation (Merigout et al. 2008b).

3.4 Organic N Transport

So far, plant putative amino acid transporters have been identified as members of at least five gene families, comprising for example in *Arabidopsis* at least 67 genes (reviewed in Ortiz-Lopez et al. 2000; Rentsch et al. 2007). We will focus here on amino acid transporters shown to be clearly involved in uptake or distribution of amino acids within the plant.

Forward and reverse approaches were used to identify transporters involved in root amino acid uptake (Hirner et al. 2006; Svennerstam et al. 2007). Both studies led to the conclusion that LHT1 (Lysine/histidine transporter) is crucial for root uptake of acidic and neutral amino acids. The AAP1 protein was also shown to transport uncharged amino acids, but only when they are supplied at high concentrations in the external medium (Lee et al. 2007b). Uptake of cationic amino acids like L-Lys or L-Arg is mediated by AAP5 within the concentration range relevant for field conditions (Svennerstam et al. 2008). Näsholm et al. (2009) suggests a hypothetic mode of root amino acid uptake in nonmycorrhizal plants. Although expression of many seed amino acid transporters precedes storage protein synthesis during seed maturation, only a few organic N transporters, among them AtOPT3, have been shown to be essential for seed loading or development (Stacey et al. 2002).

Intracellular transport is expected to be important particularly in the case of amino acid transport. Indeed, plastids are key compartments for amino acid biosynthesis, some of them being exclusively synthesized there (phenylalanine, tyrosine, tryptophan, and lysine) whereas others (glutamine, aspartate, and serine) are produced in multiple compartments. Strikingly, only one protein, Dit2.1, is so far clearly localized at the inner envelope membrane and functions as a glutamate/malate exchanger, essential for the photorespiratory pathway (Renné et al. 2003). Similarly, only transporters for basic amino acids have been localized in the mitochondrial membrane (Catoni et al. 2003; Hoyos et al. 2003). Some transporters have been localized at the tonoplast and their function remains to be demonstrated. The concentration of amino acids in the vacuole is lower than in the cytosol, but so far a vacuolar export system has been shown only in *Chara* vacuoles (Martinoia et al. 2000).

4 N Assimilation Pathways

As described before, the main nitrogen sources taken up by higher plants are nitrate or ammonium as inorganic N sources, and eventually amino acids under particular conditions. Here, we will briefly describe the main steps of nitrate or ammonium assimilation in growing cells and summarize recent results obtained for source organs when N is remobilized.

4.1 N Assimilation

A global overview of N assimilation in plants is given in Fig. 1. Nitrogen assimilation requires the reduction of nitrate to ammonium, followed by ammonium assimilation into amino acids.

Nitrate reduction into nitrite is catalysed in the cytosol by the enzyme nitrate reductase (NR). This enzyme is a homodimer, each monomer being associated with three prosthetic groups: flavin adenine dinucleotide, a haem, and a molybdenum cofactor (MoCo). Characterization of mutants resistant to chlorate, which can be reduced into toxic chlorite by NR, identified two classes of genes, the *NIA* genes encoding the NR apoenzyme and the *CNX* genes encoding the MoCo cofactor (Pelsy and Caboche 1992; Crawford and Arst 1993). Since 1993, a lot of work has been done to characterize the NR in different species (reviewed in Meyer and Stitt 2001). Although the NR enzyme is thought to be localized in the cytosol (Solomonson and Barber 1990), an association with the plasma membrane (PM-NR) has been found in some species like in corn roots (Chen and Wang 1995) or barley roots (Ward et al. 1989). The structural characteristics and the potential role of this PM-NR have been intensively studied in *Chlorella* by Tischner and collaborators (reviewed in Tischner 2000). Nitrite is then translocated to the cytosol where it is reduced to ammonium by the second enzyme of the pathway, nitrite reductase (NiR). The *NII* genes encoding the NiR enzyme have been cloned from various species, the number of genes varying from one to two copies (Meyer and Stöhr 2002).

Ammonium, originating from nitrate reduction, photorespiration, or amino acid catabolism, is assimilated in the chloroplast by the so-called GS/GOGAT cycle (Lea and Miflin 2004). The glutamine synthetase fixes ammonium on a glutamate molecule to form glutamine. This glutamine reacts subsequently with 2-oxoglutarate to form two molecules of glutamate, this step being catalysed by the glutamine 2-oxoglutarate amino transferase (or glutamate synthase GOGAT). Two classes of genes code for GS: the *GS2* gene, present as a single nuclear gene in all species studied so far, codes for a chloroplastic GS, involved in the assimilation of ammonium stemming from nitrate reduction or photorespiration. Conversely, the *GS1* nuclear gene family codes for cytosolic GS isoforms, present in different organs such as roots or stems and thought to be involved in ammonium recycling during particular developmental steps such as grain filling or leaf senescence (reviewed in Hirel and Lea 2001; Corruzzi 2003). Two different forms of glutamate synthase are present in plants: the Fd-GOGAT and NADH-GOGAT use ferredoxin and NADH as electron donors, respectively. Fd-GOGAT is predominantly localized in leaf chloroplasts, while NADH-GOGAT is primarily located in plastids of non-photosynthetic tissues, such as roots or etiolated leaf tissues. The structural, mechanistic, and regulatory properties of GOGAT enzymes and their role in amino-acid metabolism have been recently reviewed by Suzuki and Knaff (2005).

4.2 N Remobilization

Although nitrogen uptake still operates at the reproductive stage (Gallais et al. 2007), it is generally assumed that seeds receive a large part of nitrogen from remobilization of different N forms present in source organs (Feller and Keist 1986). During senescence, a re-distribution of amino acids, free or produced by proteolysis of proteins (Patrick and Offler 2001) leads to an increase of asparagine in pea (Rochat and Boutin 1991) and an increase in glutamine in other species, in the phloem sap (Herrera-Rodriguez et al. 2006; Masclaux-Daubresse et al. 2006). Some amino acid transporters of the AAP family are putatively involved in phloem loading (see above). During these particular developmental stages, specific enzymes related to N metabolism are activated (reviewed in Masclaux-Daubresse et al. 2008). Induction of cytosolic GS1 as well as induction of glutamate dehydrogenase appears in a large variety of plants. The latter, catalysing glutamate de-amination as well as glutamate synthesis, carried out the de-amination reaction in source leaves (Masclaux-Daubresse et al. 2006). This N remobilization during senescence is also triggered in response to environmental factors such as drought, nutrient limitation, or pathogen attack (Pageau et al. 2006).

5 Regulation of N Uptake and Metabolism

N uptake by the roots and N assimilation are integrated to match the nutrient demand of the whole organism. Regulatory mechanisms that modulate the expression and/or the activity of transport systems and enzymes, according to the nutritional status of the plant and to external stimuli or stresses, ensure both rapid adjustments of metabolism and long term adaptations (Fig. 4).

5.1 Regulation at the mRNA Level

Patterns for changes in mRNA abundance of many components of N uptake and N assimilation have been observed, which allow coordinated regulation of N metabolism. Two main metabolic cues operate in the control of N uptake and assimilation.

The first mechanism includes the induction by substrates and repression by endogenous N assimilates, mediating a negative feedback regulation by the N status of the whole plant (Gazzarrini et al. 1999; Cerezo et al. 2001). This results in up regulation when N is low and down regulation when N is high. Accordingly,

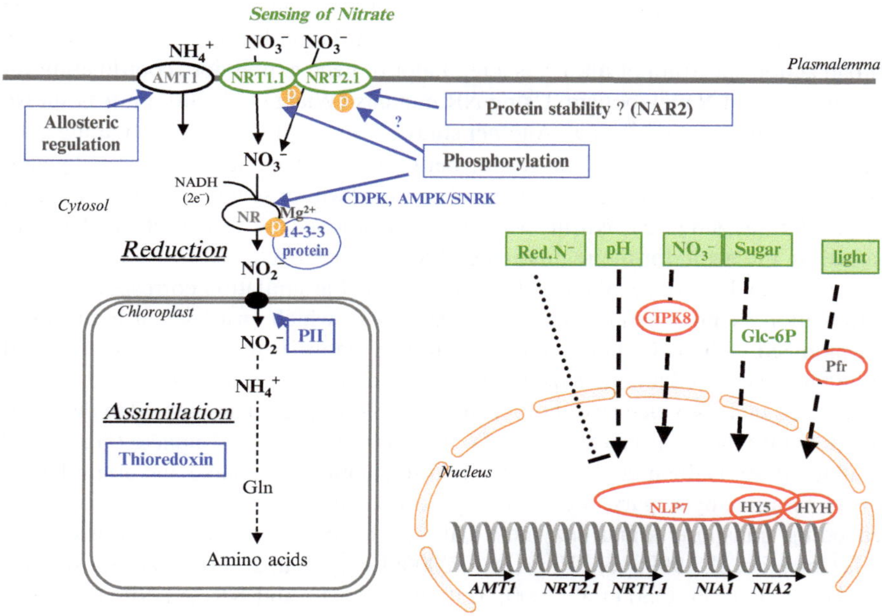

Fig. 4 *Regulatory mechanisms that modulate the expression and/or the activity of transport systems and enzymes.* Components acting on protein level are given in blue, transcriptional regulators in red, and signals and sensors in green (for abbreviation see 1.4)

several *NRT2* and *AMT1* transporters as well as *NIA* and *NII* were found to be repressed at the mRNA level by N metabolites such as amino acids (Tsay et al. 2007). Further studies using the glutamate synthase inhibitor, AZA, or exposure to NH_4^+ or various amino acids suggested that glutamine plays an important role in the down regulation of *NRT2.1* (Nazoa et al. 2003, Zhuo et al. 1999).

In response to N deprivation, *AMT1.1*, *AMT1.3,* and *AMT1.5* (Gazzarrini et al. 1999; Gansel et al. 2001; Loqué et al. 2006), as well as *AtNRT2.1*, *AtNRT2.2,* and *AtDUR3,* are induced (Lejay et al 1999; Scheible et al. 2004). Interestingly, two genes of the NRT2 family are slowly but steadily induced by starvation (Orsel and Krapp, unpublished data). Resupply of nitrate re-induces *NRT2.1,* and *NRT2.2* as well as *NIA* and *NII* expression after long term starvation (Scheible et al. 2004), whereas expression of *NRT2.4* and *NRT2.5* is repressed by the resupply of any N source (Okamoto et al. 2003).

Transcriptional regulation of genes involved in LATS for NH_4^+ and NO_3^- is less documented. *NRT1.1* shares many regulatory features with *NRT2.1*. *NRT1.1* is rapidly induced by nitrate and by starvation but less subjected to regulation by N metabolites (Tsay et al. 1993), while *AtNRT1.2* is constitutively expressed (Huang et al. 1999); *AtNRT1.5* is much more slowly induced by nitrate, and is in addition regulated by potassium. *AtNRT1.1* and *AtNRT1.5* are both regulated by pH (Tsay et al. 1993; Lin et al. 2008).

Global transcriptome studies (Wang et al. 2003; Scheible et al. 2004) confirmed transcriptional regulation of N uptake and assimilation by nitrate and showed a broad action spectrum of nitrate as a regulator of gene expression, coordinating for example C and N metabolism. Using NR mutants (Wang et al. 2004) it was shown that nitrate itself acts as signal. Another study (Wang et al. 2007) investigating gene regulation by nitrite showed an overlap between nitrate and nitrite regulated genes. Nevertheless, specific regulation by nitrite was shown for several genes of N uptake (e.g. *NRT2.5, AMT1.3*). Nitrite was already discussed by Loqué et al (2006) as signaling molecule for the regulation of *NRT1.1* and *NIA1*.

The second major regulation of N uptake and N assimilation corresponds to the stimulation by photosynthesis (Lejay et al. 2003), which ensures that N uptake is harmonized with the C status. A major common feature is the diurnal fluctuation of N uptake and N reduction. This control has often been attributed to the regulatory action of sugars produced by photosynthesis and transported downward to the roots, as shown by the positive effect of CO_2 concentration on NO_3^- uptake (Gastal and Saugier 1989; Delhon et al. 1996) Diurnal fluctuations in uptake and assimilation, or stimulation by sugars, are generally correlated with the expression of genes encoding transporters and enzymes. This has been shown for NH_4^+ transporters (Gazzarrini et al. 1999; von Wirén et al. 2000; Lejay et al. 2003), NO_3^- transporters (Lejay et al. 1999; Ono et al. 2000; Matt et al. 2001), and NR and NiR (Vincentz et al. 1993). In *Arabidopsis*, genes tested by Lejay et al. (2003, *AtNRT2.1* and *AtNRT1.1)*, showed 5–10 times higher expression during the light period compared with the dark period. Nitrate uptake, measured using $^{15}NO_3^-$ also increased after the onset of light. The increase was approx. two-fold during the photoperiod. The decrease in *AtNRT2.1* and *AtNRT1.1* mRNA levels and nitrate uptake during the dark period was prevented by supply of 1% sucrose to the roots, which is a further indication for the role of sugars during diurnal regulation. This regulation seems to be independent of the known sugar regulation pathways, such as hexokinase signaling (Lejay et al. 2003). Recently Lejay et al. (2008) showed that up-regulation of nitrate transporters (*AtNRT2.1* and *AtNRT1.1*) was related to the concentration of glucose 6-phosphate. Contrary to that of the transporters, the diurnal regulation of *NIA* transcripts is not only governed by sugars but also by light regulation via phytochrome (Rajasekhar et al. 1988). In addition, *NIA* expression is controlled by signals from photosynthetic electron flow, which adds to the picture of intracellular cross-talk between chloroplasts and the nucleus (Sherameti et al. 2002).

Despite the very important regulation of transcript abundance by external and internal factors, information about the molecular players such as transcription factors, miRNA, etc. is still rather rare. Lately two bZIP (basic leucine zipper) transcription factors have been discovered as being involved in the light regulation of N metabolism (Jonassen et al. 2008): HY5 and its homolog HYH are essential for phytochrome dependent light-activated expression of NR. ChIPchip analyses showed a binding site for HY5 in the *NIA2* promoter (Lee et al. 2007a). Interestingly also the *NRT1.1* promoter has three binding sites for HY5, but HY5 has a negative effect on transcription in this case (Lillo 2009). However, not all light regulation of N metabolism is governed by the HY5/HYH system (Lillo 2009).

Camargo et al. (2007) identified CrNIT2 as a main regulator of *NIA* expression in *Chlamydomonas,* and Castaings et al. (2009) showed that *Arabidopsis* mutants in a homologous gene (NLP7) are defective in the nitrate induction of *N*IA genes, *NRT2.1* and *NRT2.2*. Both proteins belong to a class of putative transcription factors homologous to a protein first identified in *Medicago* and essential for nodulation (NIN = nodulation inception). The CrNIT2 protein has been shown to bind to multiple sites of the *NIA* promoter, but no target genes are yet known for the AtNLP7 protein. Interestingly, mutants in the *CIPK8* gene which encode a protein kinase (Hu et al. 2009), are also unable to fully induce expression of several genes by nitrate, such as the *NIA* genes, *NRT2.1, NRT1.1,* and several others. It is tempting to speculate that CIPK8 might be involved in the same regulation pathway than NLP7. *NLP7* belongs to a gene family with nine different members, but the functions of the other NLP proteins are still unknown.

5.2 Regulation at the Protein Level

N metabolism has to respond fast to external stimuli. This can be achieved by rapid post-translational protein modification.

The best studied case of post-translational regulation in N metabolism is the regulation of higher plant NR. NR is inactivated via a two step process that involves phosphorylation of ser^{543} in spinach and the subsequent magnesium-dependent binding of an inhibitory 14-3-3 protein to NR (Bachmann et al. 1996; Moorhead et al. 1996). This activation/inactivation process is linked to the production of C assimilates that thus control NR activity (De Cires et al. 1993; Kaiser and Huber 2001). Both CDPK (calcium-dependent protein kinases) and AMPK/SNRK (SNF1-related kinase)-related protein kinases are able to phosphorylate NR at least in vitro (McMichael et al. 1995; Douglas et al. 1997; Sugden et al. 1999; Ikeda et al. 2000). The inactive phosphorylated form is re-activated by dephosphorylation probably by PP2A (MacKintosh 1992).

Protein phosphorylation may act as a trigger for protein degradation, as well as for binding of the inhibitory 14-3-3 proteins. When a modified form of NR with a truncated N-terminus that was not susceptible to post-translational dark inactivation was overexpressed, the resulting protein did not decline in the second part of the photoperiod (Nussaume et al. 1995). There is also a correlation between the phosphorylation state or the activation state of NR and the rate at which NR protein decreases (Geiger et al. 1998; Kaiser and Huber 1997; Scheible et al. 1997; Weiner and Kaiser 1999).

Post-translational regulation of nitrate transporters has recently been described. The nitrate transporter NRT1.1 is regulated by phosphorylation. When phosphorylated, AtNRT1.1 functions as a high affinity transporter, whereas it is active in the low affinity range when dephosphorylated (Liu and Tsay 2003). Recent data show that NRT1.1 acts not only as a transporter, but is also involved in N signaling (Remans et al. 2006; Walch-Liu and Forde 2008). Interestingly in one case, only the phosphorylated form is an active signaling component (Walch-Liu and Forde

2008). Nitrate transporters from the NRT2 family are also subjected to post-transcriptional regulation. First indications of putative phosphorylation of NRT2 proteins came from their amino acid sequences (Forde 2000). In addition, several of the NRT2 proteins have been identified in global phosphoprotein studies (Benschop et al. 2007). Such a post-transcriptional regulation may explain why high affinity NO_3^- influx is down-regulated by NH_4^+ in transgenic plants expressing *NpNRT2.1* cDNA under the control of a constitutive, root specific promoter (Fraisier et al. 2000). Recently, Wirth et al (2007) showed that despite strict transcriptional regulation of *AtNRT2.1* , NRT2.1 protein levels are rather constant in response to light, sucrose, or nitrogen treatments that strongly affect both *NRT2.1* mRNA level and HATS activity. Again post-translational regulation processes are required to explain these observations. One such mechanism could correspond to the cleavage of NRT2.1 C terminus, which results in the presence of both intact and truncated proteins in the plasma membrane (Wirth et al. 2007). Several forms of the protein seem to co-exist in cell membranes (the monomer and at least one higher molecular weight complex). However, the monomer is the most abundant form of NRT2.1, and seems to be the one involved in $NO3^-$ transport (Wirth et al. 2007). Interestingly, AtNRT2.1 is only present and active at the plasma membrane in the presence of AtNAR2.1 (Orsel et al. 2006; Wirth et al. 2007). The mechanism by which NAR2.1 affects NRT2.1 is so far unknown, but might open a new level of regulation by protein stability or protein transport.

A different form of post-translational regulation has been revealed for ammonium transporters allowing rapid shut-off in order to avoid toxic accumulation of ammonium. Loqué et al (2007) showed that the soluble carboxy terminus of the oligomeric AtAMT1 serves as an allosteric regulator essential for function. It is suggested that this C terminus interacts physically with cytosolic loops in the neighboring subunit with phosphorylation as a regulating mechanism.

Less is known about nitrite transport and its regulation. In *E. coli*, the PII protein regulates nitrite transport. This regulation seems to be conserved in plants. The chloroplastic PII protein might be involved in the regulation of nitrite uptake by chloroplast as mutants affected in the gene exhibit a nitrite sensitive phenotype (Ferrario-Méry et al. 2005). This hypothesis was re-enforced by the increased nitrite uptake by chloroplasts isolated from PII mutants (Ferrario-Méry et al. 2008).

Several chloroplastic enzymes of nitrogen assimilation such as NIR, GS2, and Fd-GOGAT are redox regulated through the thioredoxin system (Lemaire et al. 2007; Lichter and Häberlein 1998). In addition NR is also regulated by NO, a by-product of its own activity. NO production and the broad mode of action are described in the following paragraph.

6 N- Signaling: Nitric Oxide – A Special Case

Nitrate and other low molecular weight intermediates of nitrogen metabolism are not only substrates, but also act as signals regulating the interaction between metabolic pathways of growth and differentiation, or plant interactions with the environment. Among these nitrogen signals, nitric oxide has gained specific attention during the last decade. Therefore, the role of this N-compound will be considered in more detail in context with N-metabolism.

6.1 Sources for NO in Plants

NO (+2) may be formed either by reduction of higher N-oxidation states, preferentially nitrite, or by oxidation of more reduced N-forms (for review see del Rio et al. 2004). Figure 5 summarizes pathways for NO production.

Reductive NO formation: Nitrate reduction appears always linked to the production of trace amounts of NO, originating from a one-electron reduction of nitrite. The reduction can be mediated by NR, or, at least in non-green plant tissues, by mitochondrial electron transport (Planchet et al. 2005; Gupta et al. 2005). In both cases, nitrite competes with the "normal" substrates (e.g. nitrate in the case of NR or oxygen in the case of mitochondrial ET), and therefore rather high nitrite concentrations are required for appreciable rates of NO production. Cytosolic nitrite concentrations are usually low (10–20 μM). Nevertheless, nitrate-fertilized plants emit NO into NO- free air at rates that can be detected and quantified by sensitive analytical methods such as gas-phase chemiluminescence. For example, with illuminated tobacco leaves, NO emission was 0.3 nmoles/g FW h (Rockel et al. 2002). Rates were lower in the dark, because NR activity is down regulated. As NO is rather reactive, real NO production rates inside leaf cells could be much higher, but this is not known with certainty. NR is activated by light or by anoxia in the dark, whereas nitrite reduction becomes very low under anoxia in the dark, presumably because NADPH production via oxidative pentose phosphate cycle ceases. In consequence, nitrite accumulates in anoxic cells and tissues to millimolar concentrations, and therefore anoxic NO emission can become 1,000-fold higher than in air (Rockel et al. 2002; Planchet et al. 2005). In NiR-deficient tobacco mutant leaves, which always accumulate nitrite even in air (*light*), NO emission was as high in air (*light*) as in nitrogen (*dark*). In NR-free *nia1nia2* double mutants, NO emission in air and in nitrogen was absent (Planchet et al. 2005). The oxygen-dependent NOS reaction appeared not to contribute to this normal "bulk"-NO emission from leaves.

Plants possess yet another PM-bound NR plus a nitrite::NO reductase, which together can also produce NO (Stöhr and Stremlau 2006). No genes for these two enzymes have been identified so far, and their physiological role is still under investigation.

NO generation from nitrite may also occur non-enzymatically in acidic compartments at pH-values below 5. Such compartments might be either the mesophyll apoplast or vacuoles. While apoplastic NO formation has been localized by DAF-2 fluorescence (Bethke et al. 2004), no vacuolar NO production has been reported so far, which is actually astonishing.

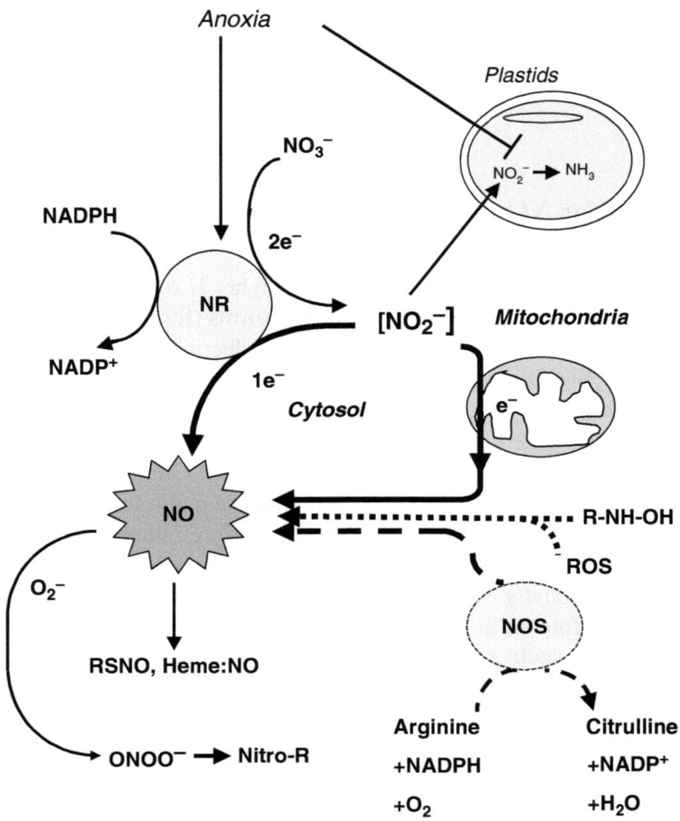

Fig. 5 *Pathways of nitric oxide (NO) synthesis, and basic reactions of NO with different targets.* NO can be synthesized by nitrite reduction, mediated either by NR itself of by mitochondrial electron transport (the latter only in roots). Nitrite to NO reduction requires high nitrite concentrations, which can become especially high under anoxia, when NR is highly active (dephosphorylated), and nitrite reduction is impaired. Also shown are the two oxidative pathways for NO synthesis; one is the (probably non-enzymatic) oxidation of hydroxylamines by reactive oxygen, the other one is the oxidation of L-arginine by a NOS-like activity. NO may either react directly with heme groups of enzymes forming Fe:NO adducts, or it may react with thiol groups to form nitrosothiols. At least in theory, NO may also react with superoxide radicals to form the highly reactive peroxynitrite, which may nitrosate aromatic amino acids

Oxidative NO formation: In animals the major NO source is L-arginine, which is oxidized to NO and L-citrulline in a complex process catalyzed by the family of NOS-enzymes (nitric oxide synthases), using NADPH and O_2 as further substrates. No gene homolog to the animal NOS family has been detected so far in *Arabidopsis*. Nevertheless, there are numerous indirect hints on NOS-like activities in plants, on the basis of effects of NOS-inhibitors, and also of enzyme activity measurements using NO measurement by EPR, of nitrite + nitrate production, or of conversion of labeled L-arginine into L-citrulline (for review see del Rio et al. 2004). Recently, an enzyme converting L-arginine to citrulline and NO has been purified from *Arabidopsis* shoots. The activity depends on the typical NOS-cofactors BH4 and Calmodulin (R. Tischner, pers. communication). Sequence information on that preparation may give a first insight into the nature of plant NOS.

Another substrate for oxidative NO formation are hydroxylamines, which can be oxidized by plant cells to NO, probably using superoxide and/or H_2O_2 as oxidants (Rümer et al. 2009). Although it is not yet clear whether the reaction is physiologically relevant, there is little doubt that plants are able to produce NO not only via nitrite reduction, but also via oxidation of amine-N.

Concentrations of NO and its "bioavailability" in cells will also depend on NO consumption (Vanin et al. 2004). NO oxidation involving reactive oxygen species (ROS), or O_2-dependent oxidation catalyzed by hemoglobins (Dordas et al. 2003) should be among the most important reactions consuming NO. In addition, reversible binding of NO to thiols may be an important aspect regulating cellular NO levels (see below).

6.2 *Mechanisms Through Which NO Affects Targets*

In the complex cellular environment, NO may undergo various oxidation and/or dismutation reactions, yielding compounds like NO_2, N_2O_3, the nitrosonium cation (NO^+), or the nitroxyl anion (NO^-). Some of these products may rapidly and reversibly nitrosylate protein- or non-protein thiols, or form nitrosyl-iron complexes with metal ions, e.g. in heme-proteins. Peroxynitrite ($ONOO^-$) may be formed from the reaction of NO with superoxide anions. However, it is not clear to what extend the reaction occurs under natural conditions in vivo. Peroxynitrite may serve as a substrate for oxidation or nitration of aromatic amino acids. Nitration appears less easily reversible than nitrosylation. Because 3-tyrosine nitration occurs on the same position (3) that is also the site for phosphorylation, it can be assumed that tyrosine nitration has important consequence for regulation mediated via tyrosine protein kinases/phosphatases.

Cysteine-*S*-nitrosylation (also called nitrosation) appears as the most widespread way in which proteins are post-translationally modulated by NO (Fig. 5). More than 100 redox-sensitive proteins were identified in *Arabidopsis* as putative candidates for cysteine *S*-nitrosylation (Lindermayr et al. 2005). In animals, NO was shown to regulate by *S*-nitrosylation signaling-related proteins including soluble guanylate cyclase, the GTP-binding protein p21ras, Ca^{2+} permeable channels, and protein kinases (for review see Courtois et al. 2008, and literature cited). Already a decade before, Stamler et al. (1997) had suggested a general "nitrosylation motiv" consisting of three or four basic or acidic amino acids surrounding the regulatory cysteine, which would permit an acid-base-autocatalyzed *S*-nitrosylation and denitrosylation. In general, the actual nitrosylating agent appears to be the nitrosonium cation NO^+, and hence *S*-nitrosylation would require an electron acceptor.

Glutathion in its reduced form is major cellular antioxidant. It reacts readily with NO to form the acid-stable *S*-nitrosoglutathione (GSNO), which may act as a NO donor to other cellular thiols. Such transnitrosation would include transfer of NO^+ to another reduced thiol (Dutton et al. 2005), or RSNO may be homolytically cleaved to release free NO and disulfide (Singh et al. 1996). GSNO can be metabolized by *S*-nitrosoglutathione reductase (GSNOR), yielding, e.g. GSSG, hydroxylamine, and NH_3 (Jensen et al. 1998). Hydroxylamine can be oxidized back to NO, probably involving ROS (Rümer et al. 2009). The relevance of GSNOR and GSNO levels for stress tolerance was recently demonstrated. Transgenic plants *Arabidopsis* with decreased GSNOR levels showed enhanced resistance against *Peronospora parasitica* correlated with higher intracellular GSNO levels (Rustérucci et al. 2007). The *Arabidopsis* HOT5 encoding a mutated GSNOR was unable to acquire thermotolerance and also had other important developmental defects (Lee et al. 2008).

NO also induces complex changes in the expression of many genes involved, e.g. in defense and cell death, transport, basic metabolism, and ROS production or degradation. Here again, *S*-nitrosylation of proteins acting as transcription factors might be the way for transcriptional control by NO. Seven families of transcription factor binding sites, among them WRKY-, GBOX-, and OCSE-elements, have been identified, which are preferentially located in the promoter regions of NO regulated genes, and co-expression of many genes can be explained by the cooperation of a set of such transcription factors (Palmieri et al. 2008).

As NO may be too short lived to diffuse via longer distances within tissues or even within single cells, it has been suggested that NO production (preferentially by NOS) and NO reception may be organized within supra molecular structures in which NO signaling occurs within highly localized environments and with minimal diffusion of free NO (Kone et al. 2003). Although this is an attractive idea, today there is no experimental evidence in context with NO that such supra molecular structures would exist and function in plants.

NO-regulated reactions in plants. The list of physiological processes in plants that are (probably) regulated by NO includes the induction of the hypersensitive response in resistance to incompatible pathogens, ABA-induced stomatal closure, seed germination and breakage of seed dormancy, iron homeostasis, flowering induction, and response to abiotic stresses such as drought, UV-B, salinity, chilling, or high temperatures (for recent reviews see Hong et al. 2008; Courtois et al. 2008; Neill et al. 2008). In spite of these many putative NO-regulated processes, today only few plant enzymes have been proven experimentally to be regulated by *S*-nitrosylation, hemoglobin 1, GAP-dehydrogenase, *S*-adenosyl synthethase, metacaspase, and potassium channels in guard cells being among them (summarized by Palmieri et al. 2008).

As mentioned, "regulatory" NO is either stemming from a NOS-like reaction or from nitrite to NO reduction. Involvement of nitrate metabolism in production of regulatory NO has been evidenced in a few cases only. For example, ABA-induced stomatal closure in *Arabidopsis* is impaired in the *nia* double mutant. Tungstate, which prevents synthesis of functional NR, also inhibited stomatal closure, whereas nitrite addition induced stomatal closure (Bright et al. 2006; Neill et al. 2008, and literature cited). Similarily, Chitosan-induced stomatal closure in *Pisum sativum*, which may prohibit easy entry of pathogens into the leaf, was impaired by tungstate treatment, which would again suggest some role for nitrite-dependent NO (together with NOS-derived NO) (Srivastava et al. 2009). ABA-induced stomatal closure was also reduced in a nia1::DS deletion mutant, indicating that only *NIA1*, but not *NIA2* was required for effective ABA signal transduction (Bright et al. 2006). This is surprising, as *NIA1* is thought to contribute only about 10% to total nitrite production (Wilkinson and Crawford 1991), and because a specific response to *NIA1* would require a mechanism by which cells can distinguish between nitrite and NO derived from one or the other protein.

Another connection between nitrate reduction, NO production, and a physiological response seems to exist for the induction of the HR in *Arabidopsis* by incompatible strains of *Pseudomonas syringae*. Here, the HR was impaired in the *nia1nia2* mutant compared to WT, and was restored by addition of nitrite (Modolo et al. 2006). However, the *nia* mutants had significantly lower arginine contents compared to WT, which might limit their NOS activity. Thus, it appeared possible that this was an indirect response to the low arginine and not directly related to the lack of nitrite.

Recently it was suggested that NO produced from nitrite would enhance NR activity in roots of *Brassica chinensis* L, thereby forming a positive feedback loop. The conclusion was based on the observation that treatment of roots with NO gas, NO donors, or NO scavengers modified extractable NR activity in the roots. In addition, treatment of purified NR or of NR in root extracts of tomato with NO in vitro also increased NR activity (Du et al. 2008; Jin et al. 2009), suggesting a direct interaction of NO with NR. It is not known yet in detail how NO modifies NR, i.e. whether NO interacts with heme-iron of the cytochrome domain or whether it forms a nitrosothiol. One consequence appears to be an increase in V_{max}, of all partial reactions of NR.

7 Conclusion

Plants use a multitude of N forms, and their uptake, transport in the plant, and assimilation are taken care of by numerous transporters and enzymes. Their quantity, localization, and the regulation of their activity enable plants to adapt quickly and finely their N acquisition and utilization strategies to developmental and environmental changes. The availability of full genome sequences, in addition to new tools and resources for functional genomics, allows the use of systems biology in the last decade to give an entire view of this important metabolic pathway in plants. Still some effort is needed to reach a virtual plant. The *in planta* function of many of the proteins is still to discover and the actors implicated in the regulation on mRNA and protein levels are just about to emerge. The N metabolite NO is implicated in many regulatory processes, but its synthesis pathways and their control, as well as the exact mode of interaction of NO with multiple targets, still need to be elucidated.

References

Alboresi A, Gestin C, Leydecker MT, Bedu M, Meyer C, Truong HN (2005) Nitrate, a signal relieving seed dormancy in *Arabidopsis*. Plant Cell Environ 28:500–512

Almagro A, Lin SH, Tsay YF (2008) Chrarcterization of the *Arabidopsis* nitrate transporterNRT1.6 reveals a role of nitrate in early embryo development. Plant Cell 20:3289–3299

Bachmann M, Shiraishi N, Campbell WH, Yoo BC, Harmon AC, Huber SC (1996) Identification of ser-543 as the major regulatory phosphorylation site in spinach leaf nitrate reductase. Plant Cell 8:505–517

Benschop JJ, Mohammed S, O'Flaherty M, Heck AJR, Slijiper M, Menke FL (2007) Quantitative phosphoproteomics of early elicitor signalling in *Arabidopsis*. Mol Cell Proteomics 65:1198–1214

Bethke PC, Badger MR, Jones RL (2004) Apoplastic synthesis of nitric oxide by plant tissues. Plant Cell 16:332–341

Bright J, Desikan R, Hancock JT, Weir IS, Neill JS (2006) ABA-induced NO generation and stomatal closure in *Arabidopsis* are dependent on H_2O_2 synthesis. Plant J 45:113–122

Brouquisse R, Masclaux C, Feller U, Raymond P (2001) Protein hydrolysis and nitrogen remobilization in plant life. In: Lea P, Morot-Gaudry JF (eds) Plant nitrogen. Springer, Berlin, pp 275–294

Camargo A, Llamas A, Scnell RA, Higuera JJ, Gonzales-Ballester D, Lefebvre PA, Fernandez E, Galvan A (2007) Nitrate signalling by the regulatory gene NIT2 in Chlamydomonas. Plant Cell 19:3491–3503

Castaings L, Camargo A, Pocholle D, Gaudon V, Texier Y, Boutet-Mercey S, Taconnat L, Renou JP, Daniel-Vedele F, Fernandez E, Meyer C, Krapp A (2009) The nodule inception-like protein 7 modulates nitrate sensing and metabolism in *Arabidopsis*. Plant J 57:426–435

Catoni E, Desimone M, Hilpert M, Wipf D, Kunzae R, Schneider A, Flugge UI, Schumacher K, Frommer WB (2003) Expression pattern of a nuclear encoded mitochondrial arginine-ornithine translocator gene from *Arabidopsis*. BMC Plant Biol 3:1

Cerezo M, Tillard P, Filleur S, Munos S, Daniel-Vedele F, Gojon A (2001) Major alterations of the regulation of root NO_3^- uptake are associated with the mutation of *Nrt2.1* and *Nrt2.2* genes in *Arabidopsis*. Plant Physiol 127:262–271

Chen J, Wang X (1995) Existence and characteristics of nitrate reductase in plamam membrane of maize roots. Sci China Ser B Chem Life Sci Earth Sci 38:564–572

Chiu C, Lin CS, Hsia AP, Su RC, Lin HL, Tsay YF (2004) Mutation of a nitrate transporter, AtNRT1:4, results in a reduced petiole nitrate content and altered leaf development. Plant Cell Physiol 45:1139–1148

Chopin F, Orsel M, Dorbe MF, Chardon F, Truong HN, Miller AJ, Krapp A, Daniel-Vedele F (2007) The *Arabidopsis* ATNRT2.7 nitrate transporter controls nitrate content in seeds. Plant Cell 19:1590–1602

Couturier J, Montanini B, Martin F, Brun A, Blaudez D, Chalot M (2007) The expanded family of ammonium transporters in the perennial poplar plant. New Phytol 174:137–150

Corruzzi G M (2003) Primary N-assimilation into aminoacid in *Arabidopsis*. I. The *Arabidopsis* book, Rockville, MD: American Society of Plant Biologists. doi: 10.1199/tab.0111, http://www.aspb.org/publications/arabidopsis/

Courtois C, Besson A, Dahan J, Bourque S, Dobrowolska G, Pugin A, Wendehenne D (2008) Nitric oxide signaling in plants: interplays with Ca^{2+} and proteinkinases. J Exp Bot 59:155–164

Crawford NM, Arst HN Jr (1993) The molecular genetics of nitrate assimilation in fungi and plants. Annu Rev Genet 27:115–146

Daniel-Vedele F, Filleur S, Caboche M (1998) Nitrate transport: a key step in nitrate assimilation. Curr Opin Plant Biol 1:235–239

De Angeli A, Monachello D, Ephritikhine G, Frachisse JM, Thomine S, Gambale F, Barbier-Brygoo H (2006) The nitrate/proton antiporter AtCLCa mediates nitrate accumulation in plant vacuoles. Nature 442:939–943

De Cires A, De la Torre A, Delgado B, Lara C (1993) Role of light and CO2 fixation in the control of nitrate-reductase activity in barley leaves. Planta 190:277–283

Dehlon P, Gojon A, Tillard P, Passama L (1996) Diurnal regulation of NO3- uptake in soybean plants. 4. Dependence on current photosynthesis and sugar availability to the roots. J Exp Bot 47:893–900

del Rio LA, Corpas FJ, Barroso JB (2004) Nitric oxide and nitric oxide synthase activity in plants. Phytochemistry 65:783–792

Dordas C, Hasinoff BB, Igamberdiev AU, Manac'h N, Rivoal J, Hill RD (2003) Expression of a stress-induced hemoglobin affects NO levels produced by alfalfa under hypoxic stress. Plant J 35:763–770

Douglas P, Pigaglio E, Ferrer A, Halford NG, MacKintosh C (1997) Three spinach leaf nitrate reductase-3-hydroxy-3-methylglutaryl-CoA reductase kinases that are regulated by reversible phosphorylation and/or Ca2+ ions. Biochem J 325:101–109

Du S, Zhang Y, Lin X, Wang Y, Tang C (2008) Regulation of nitrate reductase by nitric oxide in Chinese cabbage pakchoi (*Brassica chinensis* L.). Plant Cell Environ 31:195–204

Dutton AS, Fukuto JM, Houk KN (2005) Quantum mechanical determinations of reaction mechanisms, acid base, and redox properties of nitrogen oxides and their donors. In: Packer L, Cadenas E (eds) Methods in enzymology, 396th edn. Elsevier, Amsterdam, pp 26–44

Eun-Yeong B, Zdebik AA, Jentsch TJ (2009) Residues important for nitrate/proton coupling in plant and mammalian CLC transporters. J Biol Chem. doi:10.1074/jbc.M901170200

Feller U, Keist M (1986) Senescence and nitrogen metabolism in annual plants. In: Lambers H, Neetson JJ, Stulen I (eds) Fundamental, ecological and agricultural aspects of nitrogen metabolism. Martinus Nijhoff Publishers Dordrecht, the Netherlands, pp 219–234

Ferrario-Méry S, Boutet M, Leleu O, Savino G, Hodges M, Meyer C (2005) Physiological characterization of *Arabidopsis* mutants affected in the expression of the putative regulatory protein PII. Planta 223:28–39

Ferrario-Méry S, Meyer C, Hodges M (2008) Chloroplast nitrite uptake is enhanced in *Arabidopsis* PII mutants. FEBS Lett 582:1061–1066

Filleur S, Dorbe MF, Cerezo M, Orsel M, Granier F, Gojon A, Daniel-Vedele F (2001) An *Arabidopsis* T-DNA mutant affected in Nrt2 genes is impaired in nitrate uptake. FEBS Lett 489:220–224

Forde BG (2000) Nitrate transporters in plants: structure, function and regulation. Biochim Biophys Acta Biomembr 1465:219–235

Fraisier V, Gojon A, Tillard P, Daniel-Vedele F (2000) Constitutive expression of a putative high-affinity nitrate transporter in *Nicotiana plumbaginifolia*: evidence for post-transcriptional regulation by a reduced nitrogen source. Plant J 23:489–496

Gallais A, Coque M, Quilléré I, Le Gouis J, Prioul JL, Hirel B (2007) Estimating proportions of N remobilization and of post-silking N uptake allocated to maize kernels by 15N labelling. Crop Sci 47:685–691

Gastal F, Saugier B (1989) Relationship between nitrogen uptake and carbon assimilation in whole plant of tall fescue. Plant Cell Environ 12:407–416

Gansel X, Munos S, Tillard P, Gojon A (2001) Differential regulation of the NO3- and NH4+ trasnporter genes AtNRT2.1 and AtAMT1.1 in *Arabidopsis*: relation with long distance and local controls by N status of the plant. Plant J 26:143–155

Gazzarrini S, Lejay L, Gojon A, Ninnemann O, Frommer WB, von Wirén N (1999) Three functional transporters for constitutive, diurnally regulated, and starvation-induced uptake of ammonium into *Arabidopsis* roots. Plant Cell 11:937–947

Geelen D, Lurin C, Bouchez D, Frachisse JM, Lelievre F, Courtial B, Barbier-Brygoo H, Maurel C (2000) Disruption of putative anion channel gene *AtCLC-a* in *Arabidopsis* suggests a role in the regulation of nitrate content. Plant J 21:259–269

Geiger M, Walch-Liu P, Engels C, Harnecker J, Schulze ED, Ludewig SU, Scheible WR, Stitt M (1998) Enhanced carbon dioxide leads to a modified diurnal rhythm of nitrate reductase activity in older plants, and a large stimulation of nitrate reductase activity and higher levels of amino acids in young tobacco plants. Plant Cell Environ 21:253–268

Gifford LM, Dean A, Gutierrez RA, Coruzzi GM, Birnbaum KD (2008) Cell-specific nitrogen responses mediate developmental plasticity. Proc Natl Acad Sci USA 105:803–808

Gordon T, Lea PJ, Rosenberg C, Trinchnat JC (2001) Nodule formation and function. In: Lea P, Morot-Gaudry JF (eds) Plant nitrogen. Springer, Berlin, pp 101–146

Gupta KJ, Stoimenova M, Kaiser WM (2005) In higher plants, only root mitochondria, but not leaf mitochondria reduce nitrite to NO, in vitro and in situ. J Exp Bot 56:2601–2609

Guo FQ, Wang R, Chen M, Crawford NM (2001) The *Arabidopsis* dual-affinity nitrate transporter gene AtNRT1.1 (CHL1) is activated and functions in nascent organ development during vegetative and reproductive growth. Plant Cell 13:1761–17677

Herrera-Rodriguez MB, Maldonado JM, Perez-Vicente R (2006) Role of asparagine and asparagine synthetase genes in sunflower (*Helianthus annuus*) germination and natural senescence. J Plant Physiol 163:11061–11070

Hirel B, Lea PJ (2001) Ammonia assimilation. In: Lea P, Morot-Gaudry JF (eds) Plant nitrogen. Springer, Berlin, pp 79–100

Hirner A, Ladwig F, Stransky H, Okumoto S, Keinath M, Harms A, Frommer WB, Koch W (2006) *Arabidopsis* LHT1 is a high-affinity transporter for cellular amino acid uptake in both root epidermis and leaf mesophyll. Plant Cell 18:1931–1934

Hong JK, Yun BW, Kang J-G, Raja MU, Kwon E, Sorhagen K, Chu C, Wang Y, Loake GJ (2008) Nitric oxide function and signaling in plant disease resistance. J Exp Bot 59:147–154

Hoyos ME, Palmieri L, Wertin T, Arrigoni R, Polacco JC, Palmieri F (2003) Identification of a mitochondrial transporter for basic amino acids in *Arabidopsis thaliana* by functional reconstitution into liposomes and complementation in yeast. Plant J 33:1027–1035

Hu HC, Wang YY, Tsay YF (2009) AtCIPK8, a CBL-interating protein kinase, regulates the low-affinity phase of the primary nitrate response. Plant J 57:264–278

Huang NC, Chiang CS, Crawford NM, Tsay YF (1996) *CHL1* encodes a component of the low-affinity nitrate uptake system in *Arabidopsis* and shows cell type-specific expression in roots. Plant Cell 8:2183–2191

Huang NC, Liu KH, Lo HJ, Tsay YF (1999) Cloning and functionnal characterization of an *Arabidopsis* nitrate transporter gene that encodes a constitutive component of a low-affiniy uptake. Plant Cell 11:1381–1392

Ikeda Y, Koizumi N, Kusano T, Sano H (2000) Specific binding of a 14–13–3 protein to autophosphorylateld WPK4, and SNF1-related wheat protein kinase, and to WPK4-phosphorylated nitrate reductase. J Biol Chem 275:31695–31700

Jensen DE, Belka GK, Du Bois GC (1998) S-nitrosoglutathione is a substrate for rat alcohol dehydrogenase class III isoenzyme. Biochem J 331:659–668

Jin CW, Du ST, Zhang YS, Lin XY, Tang CX (2009) Differential regulatory role of nitric oxide in mediating nitrate reductase activity in roots of tomata (*Solanum lycocarpum*). Ann Bot. DOI:10.1093/mcp087

Jonassen EM, Lea US, Lillo C (2008) HY5 and HYH are positive regulators of nitrate reductase in seedlings and rosette stage plants. Planta 227:559–564

Kaiser WM, Huber SC (1997) Correlation between apparent activation state of nitrate reductase (NR), NR hysteresis and degradation of NR protein. J Exp Bot 48:1367–1374

Kaiser WM, Huber SC (2001) Post-translational regulation of nitrate reductase: mechanism, physiological relevance and environmental triggers. J Exp Bot 52:1981–1989

Karley AJ, Leigh RA, Sanders D (2000) Differential ion accumulation and ion fluxes in the mesophyll and epidermis of barley. Plant Physiol 122:835–844

Kojima S, Bohner A, von Wirèn N (2006) Molecular mechanisms of urea transport in plants. J Membr Biol 212:83–91

Kojima S, Bohner A, Gassert B, Yuan L, von Wirén N (2007) AtDUR3 represents the major transporter for high-affinity urea transport across the plasma membrane of nitrogen-deficient *Arabidopsis* roots. Plant J 52:30–40

Kone BC, Kuncewicz T, Zhang W, Yu ZY (2003) Protein interactions with nitric oxide synthases: controlling the right time, the right place and the right amount of nitric oxide. Am J Renal Physiol 285:178–190

Krouk G, Tillard P, Gojon A (2006) Regulation of the high-affinity $NO3^-$ uptake system by the NRT1.1 mediated $NO3^-$ demand signalling in *Arabidopsis*. Plant Physiol 142:1075–1086

Lea PJ, Miflin BJ (2004) Glutamate synthase and synthesis of glutamate in plants. Plant Physiol Biochem 41:555–564

Lee RB, Ratcliffe RG (1991) Observations on the subcellular distribution of the ammonium ion in maize root tissue using in-vivo 14N-nuclear magnetic resonnance spectroscopy. Planta 183:359–367

Lee J, He K, Stolc V, Lee H, Figueroa P, Gao Y, Tongprasit W, Zhao H, Lee I, Deng XW (2007a) Analysis of transcription factor HY5 genomic binding sites revealed its hierarchical role in light regulation of development. Plant Cell 19:731–749

Lee YH, Foster J, Chen J, Voll LM, Weber APM, Tegeder M (2007b) AAP1transports uncharged amino acids into roots of *Arabidopsis*. Plant J 50:305–319

Lee U, Wie C, Fernandez BO, Feelisch M, Vierling E (2008) Modulation of nitrosative stress by S-nitrosoglutathione reductase is critical for thermotolerance and plant growth in *Arabidopsis*. The Plant Cell 20:786–802

Leegood RC, Lea PJ, Hauser RE (1996) Use of barley mutants to study the control of photo-respiratory metabolism. Biochem Soc Trans 24:757–761

Leidreiter K, Kruse A, Heineke D, Robinson DG, Heldt HW (1995) Subcellular volumes and metabolites concentrations in potato (*Solanum tuberosum* cv Désirée) leaves. Botanica Acta 108:403–406

Lejay L, Tillard P, Lepetit M, Olive F, Filleur S, Daniel-Vedele F, Gojon A (1999) Molecular and functional regulation of two NO_3^- uptake systems by N- and C-status of *Arabidopsis* plants. Plant J 18:509–519

Lejay L, Gansel X, Cerezo M, Tillard P, Muller C, Krapp A, von Wirén N, Daniel-Vedele F, Gojon A (2003) Regulation of root ion transporters by photosynthesis: functional importance and relation with hexokinase. Plant Cell 15:2218–2232

Lejay L, Wirth J, Pervent M, Cross JM, Tillard P, Gojon A (2008) Oxidative pentose phosphate pathway-dependent sugar sensing as a mechanism for regulation of root ion transporters by photosynthesis. Plant Physiol 146:2036–2053

Lemaire SD, Michelet L, Zaffagnini M, Massot V, Issakidis-Bourguet E (2007) Thioredoxins in chloroplasts. Curr Genet 51:343–365

Li W, Wang Y, Okamoto M, Crawford N, Siddiqi MY, Glass ADM (2007) Dissection of the *AtNRT2.1:AtNRT2.2* inducible high-affinity nitrate transporter gene cluster. Plant Physiol 143:425–433

Lichter A, Häberlein I (1998) A light-dependent redox signal participates in the regulation of ammonia fixation in chloroplasts of higher plants – ferredoxin: glutamate synthase is a thioredoxin dependent enzyme. J Plant Physiol 153:83–90

Lillo C (2009) Signalling cascades integrating light-enhanced nitrate metabolism. Biochem J 415:11–19

Lin SH, Kuo HF, Canivenc G, Lin CS, Lepetit M, Hsu PK, Tillard P, Lin HL, Wang YY, Tsai CB, Gojon A, Tsay YF (2008) Mutation of the *Arabidopsis* NRT1.5 nitrate transporter causes defective root-to-shoot nitrate transport. Plant Cell 20:2514–2528

Lindermayr C, Saalbach G, Durner J (2005) Proteomic identification of S-nitrosylated proteins in *Arabidopsis*. Plant Physiol 137:921–930

Liu KH, Huang CY, Tsay YF (1999) CHL1 is a dual-affinity nitrate transporter of *Arabidopsis* involved in multiple phases of nitrate uptake. Plant Cell 11:865–874

Liu KH, Tsay YF (2003) Switching between the two action modes of the dual-affinity nitrate transporter CHL1 by phosphorylation. EMBO J 22:1005–1013

Liu LH, Ludewig U, Frommer WB, von Wirén N (2003a) AtDUR3 encodes a new type of high-affinity urea/H+ symporter in *Arabidopsis*. Plant Cell 15:790–800

Liu LH, Ludewig U, Gassert B, Frommer WB, von Wirén N (2003b) Urea transport by nitrogen-regulated tonoplast intrinsic proteins in *Arabidopsis*. Plant Physiol 133:1220–1228

Loqué D, von Wirén N (2004) Regulatory levels for the transport of ammonium in plant roots. J Exp Bot 55:1293–1305

Loqué D, Ludewig U, Yuan L, von Wirén N (2005) Tonoplast intrinsic proteins AtTIP2; 1and AtTIP2;3 facilitate NH_3 transport of ammonium in plant roots. Plant Phys 137:671–680

Loqué D, Yuan L, Kojima S, Gojon A, Wirth J, Gazzarini S, Ishiyama K, Takahashi H, von Wirèn N (2006) Additive contribution of AMT1;1 and AMT1;3 to high affinity ammonium uptake across the plasma membrane of nitrogen deficient *Arabidopsis* roots. Plant J 48:522–534

Loqué D, Lalonde S, Looger LL, von Wiren N, Frommer WB (2007) A cytosolic trans-activation domain is essential for ammonium uptake. Nature 446:195–198

Loudet O, Chaillou S, Merigout P, Talbotec J, Daniel-Vedele F (2003) Quantitative trait loci analysis of nitrogen use efficiency in *Arabidopsis*. Plant Physiol 131:345–358

MacKintosh C (1992) Regulation of spinach-leaf nitrate reductase by reversible phosphorylation. Biochim Biophys Acta 1137:121–126

McMichael RW Jr, Bachmann M, Huber SC (1995) Spinach leaf sucrose-phosphate synthase and nitrate reductase are phosphorylated/inactivated by multiple protein kinases in vitro. Plant Physiol 108:1077–1082

Marmagne A, Vinauger-Douard M, Monachello D, Falcon de Longevialle A, Charon C, Allot M, Rappaport F, Wollman F, Barbier-Brygoo H, Ephritikhine G (2007) Two members of the *Arabidopsis* CLC (Chloride Channels) family, AtCLCe and AtCLCf, are associated with thylakoid and Golgi membranes respectively. J Exp Bot 12:3385–3393

Martinoia E, Massonneau A, Frangne N (2000) Transport processes of solutes across the vacuolar membrane of higher plants. Plant Cell Phsyiol 41:1175–1186

Martin F, Cliquet JB, Stewart G (2001) Nitrogen acquisition and assimilation in mycorrhizal symbioses. In: Lea P, Morot-Gaudry JF (eds) Plant nitrogen. Springer, Berlin, pp 147–166

Masclaux-Daubresse C, Reisdorf-Cren M, Pageau K, Lelandais M, Grandjean O, Kronenberger J, Valadier MH, Feraud M, Jouglet T, Suzuki A (2006) Glutamine synthetase-glutamate synthase pathway and glutamate dehydrogenase play distinct roles in the sink-source nitrogen cycle in tobacco. Plant Physiol 140:444–456

Masclaux-Daubresse C, Reisdorf-Cren M, Orsel M (2008) Leaf nitrogen remobilisation for plant development and grain filling. Plant Biol 10:23–36

Matsson M, Schjoerring JK (2003) Senescence-induced changes in apoplastic and bulk tissue ammonia concentrations of ryegrass leaves. New Phytol 160:489–499

Matt P, Geiger M, Walch-Liu P, Engels C, Stitt M (2001) Elevated carbon dioxide increases nitrate uptake and nitrate reductase activity when tobacco is growing on nitrate, but increases ammonium uptake and inhibits nitrate reductase activity when tobacco is growing on ammonium nitrate. Plant Cell Environ 24:1119–1137

Merigout P, Gaudon V, Quilleré I, Briand X, Daniel-Vedele F (2008a) Urea use efficiency of hydroponically grown maize and wheat plants. J Plant Nutr 31:427–443

Merigout P, Lelandais M, Bitton F, Renou JP, Briand X, Meyer C, Daniel-Vedele F (2008b) Physiological and transcriptomic aspects of urea utptake and assimilation in *Arabidopsis* plants. Plant physiol 147:1225–1238

Mesnard F, Ratcliffe RG (2005) NMR analysis of plant nitrogen metabolism. New phytol 83:163–180

Meyer C, Stitt M (2001) Nitrate reduction and signalling. In: Lea P, Morot-Gaudry JF (eds) Plant nitrogen. Springer, Berlin, pp 37–59

Meyer C, Stöhr C (2002) Soluble and plasma membrane-bound enzymes involved in nitrate and nitrite metabolism. In: Foyer C, Noctor G (eds) Photosynthetic nitrogen assimilation and associated carbon and respiratory metabolism. Kluwer Academic, Dordrecht, pp 49–62

Miller AJ, Cookson SJ, Smith SJ, Wells DM (2001) The use of microelectrodes to investigate compartmentation and the transport of metabolized inorganic ions in plants. J Exp Bot 52:541–549

Miller AJ, Smith SJ (2008) Cytosolic nitrate ion homeostasis: could it have a role in sensing nitrogen status? Annals Bot 101:485–489

Modolo LV, Augusto O, Almeida IMG, Pinto-Maglio CAF, Oliveira HC, Seligman K, Salgado I (2006) Decreased arginine and nitrite levels in nitrate reductase-deficient *Arabidopsis thaliana* plants impair nitric oxide synthesis and the hypersensitive response to *Pseudomonas syringae*. Plant Sci 171:34–40

Monachello D, Allot M, Oliva S, Krapp A, Daniel-Vedele F, Barbier-Brygoo H, Ephritikine G (2009) The txo anions transporters AtCLCa and AtCLCe fulfil interconnecting but not redundant roles in nitrate assimilation pathways. New Phytol, in press

Moorhead G, Douglas P, Morrice N, Scarabel M, Aitken A, MacKintosh C (1996) Phosphorylated nitrate reductase from spinach leaves is inhibited by 14–3–3 proteins and activated by fusicoccin. Curr Biol 6:1104–1113

Nazoa P, Videmar J, Tranbarger TJ, Mouline K, Damiani I, Tillard P, Glass ADM, Touraine B (2003) Regulation of the nitrate transporter gene *AtNRT2.1* in *Arabidopsis thaliana*: responses to nitrate, amino acids and developmental stage. Plant Mol Biol 52:689–703

Näsholm T, Kielland K, Ganeteg U (2009) Uptake of organic nitrogen by plants. New Phytol 182:31–48

Neill S, Barros R, Bright J, Desikan R, Hancock J, Harrison J, Morris P, Ribeiro D, Wilson I (2008) Nitric oxide, stomatal closure and abiotic stress. J Exp Bot 59:165–176

Neuhäuser B, Dynowski M, Mayer M, Ludewig U (2007) Regulation of $NH4^+$ transport by essential cross talk between AMT monomers through carboxy tails. Plant Phys 143: 1651–1659

Ninnemann O, Jauniaux J, Frommer WB (1994) Identification of a high affinity NH_4^+ transporter from plants. EMBO J 13:3464–3471

Nussaume L, Vincentz M, Meyer C, Boutin JP, Caboche M (1995) Posttranscriptional regulation of nitrate reductase by light is abolished by an N-terminal deletion. Plant Cell 7:611–621

Okamoto M, Vidmar JJ, Glass ADM (2003) Regulation of *NRT1* and *NRT2* Gene Families of *Arabidopsis thaliana*: responses to nitrate provision. Plant Cell Physiol 44:304–317

Ono F, Frommer WB, von Wirén N (2000) Coordinated diurnal regulation of low-and high-affinity nitrate transporters in tomato. Plant Biol 2:17–23

Orsel M, Eulenburg K, Krapp A, Daniel-Vedele F (2004) Disruption of the nitrate transporter genes AtNRT2.1 and AtNRT2.2 restricts growth at low external nitrate concentration. Planta 219:714–721

Orsel M, Chopin F, Leleu O, Smith SJ, Krapp A, Daniel-Vedele F, Miller AJ (2006) Characterization of a two-component high-affinity nitrate uptake system in *Arabidopsis*: physiology and protein-protein interaction. Plant Physiol 142:1304–1317

Ortiz-Lopez A, Chang HC, Bush DR (2000) Amino acid transporters in plants. Biochem Biophys Acta Biomembr 1465:275–280

Pageau K, Reisdorf-Cren M, Morot-Gaudry JF, Masclaux-Daubresse C (2006) The two senescence-related markers, GS1 (cytosolic glutamine synthetase) and GDH (glutamate dehydrogenase), involved in nitrogen mobilization, are differentially regulated during pathogen attack and by stress hormones and reactive oxygen species in Nicotiana tabacum L. leaves. J Exp Bot 57:547–557

Palmieri MC, Sell S, Hunag X, Scherf M, Werner T, Durner J, Lindermayr C (2008) Nitric-oxide responsive genes and promoters in *Arabidopsis thaliana*: a bioinformatics approach. J Exp Bot 59:177–186

Patrick JW, Offler CE (2001) Compartmentation of transport and transfer events in developing seeds. J Exp Bot 52:551–564

Pelsy F, Caboche M (1992) Molecular genetics of nitrate reductase in higher plants. In: Scandalios JG, Wright TRF (eds) Advances in genetics, Academic Press Inc, San Diego, CA, pp 1–40

Planchet E, Gupta KJ, Sonoda M, Kaiser WM (2005) Nitric oxide emission from tobacco leaves and cell suspensions: rate limiting factors and evidence for the involvement of mitochondrial electron transport. Plant J 41:732–743

Rajasekhar VK, Gowri G, Campbell WH (1988) Phytochrome-mediated light regulation of nitrate reductase expression in squash cotyledons. Plant Physiol 88:242–244

Remans T, Pervent M, Filleur S, Diatloff E, Mounier E, Tillard P, Forde BG, Gojon A (2006) The *Arabidopsis* NRT1.1 transporter participates in the signaling pathway triggering root colonization of nitrate-rich patches. Proc Natl Acad Sci USA 103:19206–19211

Renné P, Dressen U, Hebbeker U, Hille D, Flugge UI, Westhoff P, Weber APM (2003) The *Arabidopsis* mutant dct is deficient in the plastidic glutamate/malate translocator DiT2. Plant J 35:316–331

Rentsch D, Schmidt S, Tegeder M (2007) Transporters for uptake and allocation of organic nitrogen compounds in plants. FEBS Lett 581:2281–2289

Richard-Molard C, Krapp A, Brun F, Ney B, Daniel-Vedele F, Chaillou S (2008) Plant response to nitrate starvation is determined by N storage capacity matched by nitrate uptake capacity in two *Arabidopsis* genotypes. J Exp Bot 59:779–791

Rochat C, Boutin J-P (1991) Metabolism of phloem-borne amino acids in maternal tissues of fruit of nodulated or nitrate-fed pea plants. J Exp Bot 42:207–214

Rockel P, Strube F, Rockel A, Wildt J, Kaiser WM (2002) Regulation of nitric oxide (NO) production in vivo and in vitro. J Exp Bot 53:103–110

Rümer S, Kapuganti JG, Kaiser WM (2009) Plants cells oxidize hydroxylamines to NO. J Exp Bot 60:2065–2072

Rustérucci C, Espunya MC, Díaz M, Chabannes M, Martínez CM (2007) S-nitrosoglutathione reductase affords protection against pathogens in *Arabidopsis*, both locally and systemically. Plant Physiol 143:1282–1292

Schröppel-Maier G, Kaiser WM (1988) Ion homeostasis in chloroplasts under salinity and mineral deficiency. Plant Physiol 87:828–832

Schumaker KS, Sze H (1987) Decrease of pH gradients in tnonplast vesicles by NO_3^- and Cl^-: evidence for H^+-coupled anion transport. Plant Physiol 83:7490–7496

Scheible WR, Gonzalez-Fontes A, Morcuende R, Lauerer M, Geiger M, Glaab MJ, Gojon A, Schulze ED, Stitt M (1997) Tobacco mutants with a decreased number of functional NIA genes compensate by modifying the diurnal regulation of transcription, post-translational modification and turnover of nitrate reductase. Planta 203:304–319

Scheible WR, Morcuende R, Czechowski T, Fritz C, Osuna D, Placios-Rojas N, Schindelasch D, Thimm O, Udvardi K, Stitt M (2004) Genome-wide reprogramming of primary and secondary metabolism; protein synthesis, cellular growth processes and the regulatory infrastructure of Arabidopsis in response to nitrogen. Plant Physiol 136:2483–2499

Sherameti I, Sopory SK, Trebicka A, Pfannschmidt T, Oelmuller R (2002) Photosynthetic electron transport determines nitrate reductase gene expression and activity in higher plants. J Biol Chem 277:46594–46600

Segonzac C, Boyer JC, Ipotesi E, Szponarski W, Tillard P, Touraine B, Sommerer N, Rossignol M, Gibrat R (2007) Nitrate efflux at the root plasmam membrane: identification of an *Arabidopsis* excretion transporter. Plant Cell 19:3760–3777

Singh RJ, Hogg N, Joseph J, Kalyanraman B (1996) Mechanism of nitric oxide release from S-nitrosothiols. J Biol Chem 271:18596–18603

Sohlenkamp C, Shelden M, Howitt S, Udvardi M (2000) Characterization of *Arabidopsis* AtAMT2, a novel ammonium transporter in plants. FEBS Lett 467:273–278

Solomonson LP, Barber MJ (1990) Assimilatory nitrate reductase: functionel properties and regulation. Ann Rev Plant Physiol Plant Mol Biol 41:225–253

Sonoda Y, Ikeda A, Saiki S, von Wirén N, Yamaya T, Yamaguchi J (2003) Distinct expression and function of three ammonium transporter genes (OsAMT1;1–1;3) in rice. Plant Cell Physiol 44:726–734

Srivastava N, Gonugunta VK, Puli MR, Raghavendra AS (2009) Nitric oxide production occurs downstream of reactive oxygen species in guard cells during stomatal closure induced by chitosan in abaxial epidermis of *Pisum sativum*. Planta 229:757–765

Stacey MG, Koh S, Becker J, Stacey G (2002) AtOPT3, a member of the oligopeptide transporter family, is essential for embryo development in *Arabidopsis*. Plant Cell 14:2799–2811

Stamler JS, Toone EJ, Lipton SA, Sucher NJ (1997) (S)NO signals: translocation, regulation, and a consensus motif. Neuron 18:691–696

Stöhr C, Stremlau S (2006) Formation and possible roles of nitric oxide in plant roots. J Exp Bot 57:463–470

Sugden C, Donaghy PG, Halford NG, Hardie DG (1999) Two SNF1-related protein kinases from spinach leaf phosphorylate and inactivate 3-hydroxy-3- methylglutaryl Coenzyme A reductase, nitrate reductase, and sucrose phosphate synthase in vitro. Plant Physiol 120:257–274

Sugiura M, Georgescu MN, Takahashi M (2007) A nitrite transporter associated with nitrite uptake by higher plants chloroplasts. Plant Cell Physiol 48:1022–1035

Suzuki A, Knaff DB (2005) Glutamate synthase: strucutral, mechanistic and regulatory properties, and role in the amino acid metabolism. Photosynth Res 83:191–217

Svennerstam H, Ganeteg U, Bellini C, Näsholm T (2007) Comprehensive screening of Arabidopsis mutants suggests the lysine histidine transporter 1 to be involved in plant uptake of amino acids. Plant Physiol 143:1853–1860

Svennerstam H, Ganeteg U, Näsholm T (2008) Root uptake of cationic amino acids by *Arabidopsis* depends on functional expression of amino acid permease. New Phytol 180:620–630

Tischner R (2000) Nitrate uptake and reduction in higher and lower plants. Plant Cell Environ 23:1005–1024

Tsay YF, Schroeder JI, Feldmann KA, Crawford NM (1993) The herbicide sensitivity gene *CHL1* of *Arabidopsis* encodes a nitrate-inducible nitrate transporter. Cell 72:705–713

Tsay YF, Chiu CC, Tsai CB, Ho CH, Hsu PK (2007) Nitrate transporters and peptide transporters. FEBS Lett 58:2290–2300

van der Leij M, Smith SJ, Miller AJ (1998) Remobilisation of vacuolar stored nitrate in barley root cells. Planta 205:64–72

Vanin AF, Svistunenko DA, Mikoyan VD, Serezhenkov VA, Fryers MJ, Baker NR, Coopers CE (2004) Endogenous superoxide production and the nitrite/nitrate ratio control the concentration of bioavailable free nitric oxide in leaves. J Biol Chem 279:24100–24107

Vincentz M, Moureaux T, Leydecker MT, Vaucheret H, Caboche M (1993) Regulation of nitrate and nitrite reductase expression in *Nicotiana plumbaginifolia* leaves by nitrogen and carbon metabolites. Plant J 3:315–324

von Wirén N, Lauter FR, Ninnemann O, Gillissen B, Walch-Liu P, Engels C, Jost W, Frommer WB (2000) Differential regulation of three functional ammonium transporter genes by nitrogen in root hairs and by light in leaves of tomato. Plant J 21:167–175

Walch-Liu P, Filleur S, Gan YB, Forde BG (2005) Signaling mechanisms integrating root and shoot responses to changes in the nitrogen supply. Photosynth Res 83:239–250

Walch-Liu P, Forde BG (2008) Nitrate signalling mediated by the NRT1.1 nitrate transporter antagonises L-Glutamate-induced changes in root architecture. Plant J 54:820–828

Wang R, Liu D, Crawford NM (1998) The *Arabidopsis* CHL1 protein plays a major role in high-affinity nitrate uptake. Proc Natl Acad Sci USA 95:15134–15139

Wang R, Okamoto M, Xing X, Crawford NM (2003) Microarray analysis of the nitrate response in *Arabidopsis* roots and shoots reveals over 1,000 rapidly responding genes and new linkages to glucose, trehalose-6-phosphate, iron, and sulfate metabolism. Plant Physiol 132:556–567

Wang R, Tischner R, Gutiérrez RA, Hoffman M, Xing X, Chen M, Coruzzi G, Crawford NM (2004) Genomic analysis of the nitrate response using a nitrate reductase-null mutant of *Arabidopsis*. Plant Physiol 136:2512–2522

Wang R, Xing X, Crawford N (2007) Nitrite acts as a transcriptome signal at micromolar concentrations in *Arabidopsis* roots. Plant Physiol 145:1735–1745

Ward MR, Grimes HD, Huffaker RC (1989) Latent nitrate reductase-activity associated with the plasma membrane of corn roots. Planta 177:470–475

Weber A, Schwacke R, Flügge UI (2005) Solute transporters of the plastid envelope membrane. Ann Rev Plant Biol 56:133–164

Weiner H, Kaiser WM (1999) 14–3–3 proteins control proteolysis of nitrate reductase in spinach leaves. FEBS Lett 455:75–78

Wells DM, Miller AJ (2000) Intracellular measurements of ammonium in *Chara corallina* using ion-selective microelectrodes. Plant Soil 221:103–106

Wilkinson J, Crawford NM (1991) Identification of the *Arabidopsis* CHL3 gene as the nitrate reductase structural gene NIA2. Plant Cell 3:461–471

Williams LE, Miller AJ (2001) Transporters responsible for the uptake and partitioning of nitrogenous solutes. Ann Rev Plant Physiol Plant Mol Biol 52:659–688

Winter H, Robinson DG, Heldt HW (1993) Subcellular volumes and metabolite concentrations in barley leaves. Planta 191:180–190

Winter H, Robinson DG, Heldt HW (1994) Subcellular volumes and metabolite concentrations in spinach leaves. Planta 193:530–535

Wirth J, Chopin F, Santoni V, Viennois G, Tillard P, Krapp A, Lejay L, Daniel-Vedele F, Gojon A (2007) Regulation of root nitrate uptake at the NRT2.1 protein level in *Arabidopsis thaliana*. J Biol Chem 282:23541–23552

Wood CC, Poree F, Dreyer I, Koelher GJ, Udvardi MK (2006) Mechanisms of ammonium transport, accumulation and retention in oocytes and yeast cells expressing *Arabidopsis* AtAMT1;1. FEBS Lett 580:3931–3936

Yuan L, Loqué D, Kojima S, Rauch S, Ishiyama K, Inoue E, Takahashi H, von Wirén N (2007) The organization of high-affinity ammonium uptake in *Arabidopsis* roots depend on the spatial arrangement and biochemical properties of AmMT1-type transporters. Plant Cell 19:2636–2652

Zhuo D, Okamoto M, Vidmar JJ, Glass ADM (1999) Regulation of a putative high-affinity nitrate transporter (Nrt2;1At) in roots of *Arabidopsis thaliana*. Plant J 17:63–568

Zifarelli G, Pusch M (2009) Conversion of the 2 $Cl^-/1H^+$ antiporter CLC-5 in a NO_3^-/H^+ antiporter by a single point mutation. EMBO J 28:175–182

Phosphorus: Plant Strategies to Cope with its Scarcity

Lenin Sánchez-Calderón, Alejandra Chacon-López, Claudia-Anahí Pérez-Torres, and Luis Herrera-Estrella

Abstract Low phosphorus (P) availability is considered a major constraint for plant growth and crop productivity; therefore, the ability of plant roots to acquire P from soil and the mechanisms that regulate phosphorous homeostasis in the plant are topics of great interest. Low P availability elicits a Pi-starvation response that includes morphological, metabolic, and physiological changes oriented to increase P availability and the efficiency of P uptake and usage by the plant. Recent advances in the study of the plant responses to low P availability allowed the identification of several key molecular components of the P rescue system. However, the complete signaling pathways, as well as the putative phosphate receptors, remain largely unknown. In this chapter, we review current research aimed at dissecting the components of the biochemical, molecular, and physiological adaptations associated with the plant responses to P starvation and its relation to the efficiency and effectiveness of P uptake and assimilation from rhizosphere.

1 Introduction

1.1 Phosphorus is Needed to Sustain Life

Phosphorus (P) is required for practically all cellular metabolic processes. During development, reproduction, and environmental adaptation of all organisms, P plays

L. Sánchez-Calderón, C-A. Pérez-Torres, and L. Herrera-Estrella (✉)
Laboratorio Nacional de Genómica para la Biodiversidad, Centro de Investigación y de Estudios Avanzados del Instituto Politécnico Nacional, Irapuato, Guanajuato, México
e-mail: lherrera@ira.cinvestav.mx

A. Chacon-López
Departamento de Ingeniería Genética de Plantas, Centro de Investigación y de Estudios Avanzados del Instituto Politécnico Nacional, Irapuato, Guanajuato, México

R. Hell and R.-R. Mendel (eds.), *Cell Biology of Metals and Nutrients*, Plant Cell Monographs 17,
DOI 10.1007/978-3-642-10613-2_8,

a crucial role in the post-translational regulation of enzymes and the control of signal transduction cascades participating in the phosphorylation and dephosphorylation cycles of signaling proteins. P also takes part in physiological events such as photosynthesis, oxidative respiration, carbon and nitrogen assimilation, energy conservation, and lipid metabolism. Additionally, P is a structural component of several biomolecules such as nucleic acids, membrane phospholipids, and high-energy compounds such as ATP and NADPH. Therefore, P is indispensable for growth, development, fertility, and reproduction of all the living beings (Abelson 1999; Ticconi and Abel 2004). In all the ecosystems worldwide, the primary production, structure, and diversity of species are strongly limited by P availability (Tiessen 2008).

The aim of this chapter is to provide a review on the role of phosphorus as an essential element during the plant life cycle, and how plants have adapted to changes in phosphorus availability in their ecosystems. We also focus on presenting recent findings on molecular mechanisms regulating the plant phosphate starvation response. We begin with a discussion about the phosphorous paradox and continue with a compilation of the responses that allow plants to survive better during phosphate starvation. Finally, we summarize the current knowledge about the components involved in P homeostasis.

2 The Phosphorus Paradox

2.1 Phosphorus is Necessary for Plant Welfare

As phosphorus is necessary for plant growth and reproduction, it is considered as an essential macronutrient. The typical symptomatology of P deprivation in plants is first evidenced as a gradual change in foliage coloration, starting from dark green in young leaves to purple in the older ones, as a consequence of anthocyanin accumulation. P-deprived plants also show reduced growth due to changes in the rate of cell division and elongation, a reduction in size and number of flowers and seeds, and changes in their phenology and dormancy, decreasing to a great extent plant productivity (Bould et al. 1986; Benton 1998).

2.2 Phosphorus: Its Limited Availability

P is the 11th most abundant element in the earth's crust. Plants acquire P from soil solution by the root system mainly in the form of $H_2PO_4^-$ and to a lesser extent as HPO_4^{-2}. Although the total amount of P in the soil may be high, it is often present in chemical forms not readily available for plant uptake (Schachtman et al. 1998; White and Hammond 2008). Its availability is limited in the majority of soils around the world, including the sandy semiarid, volcanic, and alkaline soils of subtropical areas, and acid and humid soils rich in organic compounds of tropical and subtropical lands (Tiessen 2008; Oberson et al. 2001). Arable soils with problems of P availability are calculated to have an extension from 2 to 5.7 billion hectares (Hinsinger 2001; Cakmak 2002; Oberson et al. 2001). The reason behind low P availability is in part due to its physico-chemical properties that affect its rate of sorption/desorption and water solubility, as well as its high conversion into organic forms and mineralization by organisms present in the soil (Holford 1997; Schachtman et al. 1998).

2.3 Phosphorus in Soil

Almost all naturally occurring P (over 99%) is present as phosphates, either as inorganic phosphates (inorganic pool) or as organic phosphate esters (organic pool). The inorganic one (Pi) is the form readily available for plant uptake. Pi is negatively charged, making it highly reactive with positive charged cations on the surface of soil particles, restricting its mobility in the soil (Bieleski 1973; Tiessen 2008). Typically, in alkaline soils Pi is largely bonded to calcium (Ca) and magnesium (Mg), and in acid soils to aluminum (Al) and iron (Fe), having, as a result, a strong reduction of its water solubility causing its fixation in the inorganic pool. Pi in the soil is also present, forming organic compounds such as nucleic acids, phospholipids, and mainly phytic acid (inositol hexaphosphate) that are not available for plant uptake. Pi in the organic fixed-pool may constitute between 20%–80% of the total phosphorus found in the uppermost layer of the soil. These Pi fixed pools must be solubilized, desorbed, or degraded (mineralized) to release soluble Pi for plant nutrition (Schachtman et al. 1998; White and Hammond 2008).

2.4 Phosphorus Availability: Economical and Environmental Problems

In most agricultural systems, the amount of Pi released in the soil is not sufficient to support the high growth rates of crop species. To sustain high productivity levels in soils with low pi availability, a continuous Pi input in the form of fertilizers is required. Recovery of the Pi fertilizer applied by crop plants is very low because of the fact that more than 80% becomes immobile and unavailable for plant uptake (Holford 1997; Schachtman et al. 1998). To compensate the low fertilizer capture by plants, farmers frequently apply Pi fertilizers in excessive amounts to ensure crop production (Goldstein 1992). However, the excessive use of fertilizers has had severe environmental consequences in terrestrial freshwater and near-shore marine ecosystems due to leakage of Pi from cropping areas. Moreover, since the P cycle is very slow and the use of Pi fertilizers keeps increasing, the source of low-cost Pi-rock is being quickly depleted.Thus, it is necessary to develop a more efficient program of P fertilization, as well as understand how plants respond to Pi starvation in order to design more effective breeding programs to produce plant varieties with enhanced Pi uptake and use efficiency in order to develop more sustainable agricultural practices.

3 Pi Uptake and Transport by Plants

3.1 Pi Uptake and Translocation in Whole Plant

Available forms of Pi in soil are rapidly acquired by plant roots, generating in their surrounding area "depletion zones" with little bioavailable Pi, that will only slowly recharge through diffusion and mineralization (Hinsinger 2001) Pi enters by epidermal cell membranes and is transported by the symplastic pathway (Fig. 1a, b) in which Pi transport through the cell membrane is performed against a concentration gradient, because the soil concentration rarely surpasses 10 μM and inside the cell it reaches millimolar concentrations. The energy required for Pi transport against a concentration gradient is provided by ATPases that generate a proton gradient in the cell membrane. This gradient is used to co-transport Pi into the cell. It has

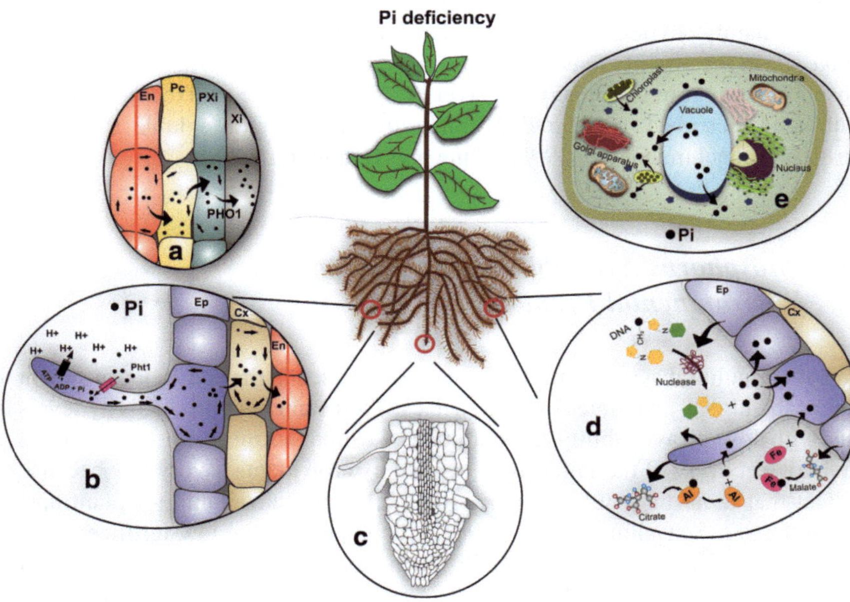

Fig. 1 Strategies shown by plants to cope with phosphate starvation. (**a**, **b**) Pi uptake by roots. Pi movement from soil to xylem. (**c**) Root meristem differentiation. (**d**) Root exudates to increase Pi mobilization from organic and inorganic Pi soil pool. (**e**) Pi cell remobilization from vacuole and plastids. *Ep* epiderm, *Cx* cortex, *En* endoderm, *Pc* pro-cambium, *Pxi* pro-xylem, *Xi* xylem

been proposed that there exists a dual mechanism for phosphate uptake that is characterized by the constitutive activity of low-affinity Pi transporters operating at high (millimolar) Pi concentrations and the activation of genes encoding high-affinity phosphate transporters that operate at low (micromolar) Pi concentrations (Schachtman et al. 1998; Raghothama 1999).

The next step is Pi translocation. As Pi is easily mobile, it can move through the symplasm to xylem parenchyma cells in order to upload Pi into xylem cells and then be transported toward sink tissues (Fig. 1a). It has been postulated that the PHOSPHATE1 (PHO1) protein (which is not a Pi transporter) is implicated in modulating Pi loading into the xylem. PHO1 is expressed in vascular root tissue and the *pho1* mutant over-accumulates Pi in the root because of its incapacity to load phosphate into the xylem (Poirier et al. 1991; Hamburger et al. 2002). Finally, Pi transported to leaves exits the xylem and is distributed to the remaining tissues. In senescence and Pi starvation conditions, Pi is remobilized from older to younger leaves and other sink. In this process, PHO2 (PHOSPHATE2) and RNA mir399 are key regulators (Delhaize and Randall 1995; Fujii et al. 2005; Chiou 2007), both of which will be described in more detail below.

Once inside cells, Pi would follow several different pathways: a) to be used in phospholipid, DNA, and RNA biosynthetic routes; b) transported to plastids or/and mitochondria to be used in metabolic processes, c) transported to vacuoles to be stored, or d) to be transported to other cells (Rausch and Bucher 2002). The final or transient destination of Pi will depend on the kind of tissue and the metabolic requirements of cells. Under Pi starvation conditions (Fig. 1e), the Pi stored in vacuole and present in plastids and membranes is recycled in order to reprioritize its usage (Schachtman et al. 1998).

3.2 Pi Transporters in Plants

Pi uptake, transport, and translocation within the plant are carried out by phosphate transporter (Pht). There are four phosphate transporter families, namely, Pht1, Pht2, Pht3, and Pht4 reported to date. The phosphate transport 1 (*Pht1*) gene family encodes high-affinity proton/Pi co-transporters that mediate Pi uptake across the plasma membrane. In the *Arabidopsis* genome, there are nine members of the *Pht1* gene family, *Pht1;1 to 9*. The expression promoter of four members of the *Pht1* gene family (*Pht1;1*, *Pht1;2*, *Pht1;3*, and *Pht1;4*) is induced during Pi-deprivation in a root-specific manner, suggesting their participation in Pi uptake from the soil under Pi-limiting conditions. Members of *Pth1* gene family also could have an important role in Pi translocation within the plant, because their expression are detected also in flowers, hyadothodes of cotyledons, pollen, leaf vasculature, and shoot buds (Karthikeyan et al. 2002; Mudge et al. 2002). The phosphate transporter *Pht2* and *Pht3* families encode organelle-associated Pi transporters. Pht2;1, the only one member of *Pht2* family, is a low-affinity Pi transporter located in the chloroplast membrane (Daram et al. 1999; Versaw and Harrison 2002). The *Pht3* family, having three members, encodes proteins located in the mitochondria. The *Pht4* gene family comprises six members that share similarity to the SLC17/type 1 animal Pi transporters. *Pht4* genes are expressed both in roots and leaves, five of which are plastid located and one in the Golgi apparatus and are probably involved in the transport of Pi between the cytosol and plastids and the Golgi apparatus (Guo et al. 2008). Although it has been functionally shown that a low-affinity Pi transport system exists in plant roots, the genes encoding these transporters remain to be identified.

4 The Plant Phosphate Starvation Response

Plants have developed elaborate mechanisms to adapt to extended periods of Pi deficiency. This includes a complex array of biochemical, physiological, and morphological changes collectively known as the Pi-starvation response, which ultimately enable plants to better cope with Pi limiting conditions.

4.1 Biochemical Adaptations of Phosphate Starved Plants

In Arabidopsis grown under high phosphate, by far the largest amount of phosphate is stored as inorganic phosphate, and smaller amounts in lipids, esters, and nucleic acids. During low phosphate conditions, phosphate is mobilized from the inorganic phosphate pool and from phospholipids, while nucleic acids and esters are hardly affected (Poirier et al. 1991). In order to provide additional Pi to maintain their metabolism, plants activate two major biochemical responses which permit them to enhance Pi uptake: (1) increase of endogenous and soil Pi availability, and (2) increase of Pi mobilization and recycling activity within the plant (Raghothama 1999; Vance et al. 2003). The processes leading to an increased Pi uptake include root-induced modifications in the chemistry of the rhizosphere and mobilization of various inorganic and organic P forms for plant uptake. The general biochemical response of plants to Pi starvation includes increased production of phosphate-scavenging proteins such as nucleases and acid phosphatases, replacement of phospholipids with nonphosphorus lipids, expression of high-affinity phosphate transporters genes, and increased proton release and organic acid exudation.

4.1.1 Gathering and Recycling Phosphorus from Organic Pi Pool

In order to reprioritize internal Pi use and to maximize external Pi acquisition, plants produce and exude acid phosphatases (APases) during Pi stress (Fig. 1d, e).

The function of intracellular (vacuolar) or extracellular (secreted) APases is to provide Pi to the plant from external and internal expendable organic phosphate pools (Ticconi and Abel 2004). These APases catalyze the hydrolysis of a wide spectrum of orthophosphate-monoesters and anhydrides. The study of several genes that encode APases has revealed the differential expression of these genes in both a temporal and tissue-specific fashion (Wu et al. 2003; Zimmermann et al. 2004; Amtmann et al. 2006; Bozzo et al. 2006).

Nucleic acids present in decaying organic matter represent an important source of extracellular Pi (Fig. 1d) that may be exploited by plants under Pi deficiency conditions. The induction of genes encoding ribonucleases by Pi deficiency is a rapid and reversible process that is sensitive to changes in Pi concentration (Bariola et al. 1994; Köck et al. 1995; Köck et al. 1998; Chen et al. 2008). Thus, remobilization of Pi from intra- and extracellular nucleic acid substrates collaborates to maintain constant Pi cytoplasmic concentrations (Abel et al. 2000). In cell membranes, Pi is present as a component of phospholipids, which constitute one of the largest phosphorus pools in plants. When plants grow under P starvation conditions, the relative phospholipids abundance is reduced and the abundance of nonphosphorus lipids such as sulphoquinovosyldiacylglycerol (SQDG), digalatosyldiacyglycerol (DGDG), and monogalatosyldiacyglycerol (MGDG) increases. (Essigmann et al. 1998; Härtel et al. 2000; Dörmann and Benning 2002; Andersson et al. 2003; Jouhet et al. 2004; Andersson et al. 2005; Benning and Ohta 2005; Kobayashi et al. 2006; Li et al. 2006). Phospholipases DZ 1 and 2 (PLDZ1 and PLDZ2) and phosphoinositide phospholipase C 5 (PLC5), participate in the hydrolysis of membrane phospholipids in order to provide Pi for cell metabolism (Cruz-Ramirez et al. 2006; Li et al. 2006; Gaude et al. 2008). During this tightly controlled process, the expression of genes encoding enzymes that break down phospholipids is regulated by Pi availability. In Pi-deprived plants, phospholipids are replaced by galactolipids synthesized by DGDG and MGDG synthases and sulfolipids synthesized by SQD1 and SQD2 enzymes (Essigmann et al. 1998; Härtel et al. 2000; Sanda et al. 2001; Yu et al. 2002; Kobayashi et al. 2009).

4.1.2 Solubilizing Phosphorus from Inorganic Pi Pool

During Pi starvation, in several plant species, the exudation of particular organic acids (Fig. 1d) to mobilize different types of Pi-associated compounds, such as $FePO_4$, $AlPO_4$, or $Ca_3(PO_4)_2$, has been reported (Zhang et al. 1997). Malate and citrate are the main components released by roots under Pi deficiency, facilitating the chelation of cations commonly associated with Pi in soil (Bar-Yosef 1991; Jones and Darrah 1994; Lan et al. 1995; Jones 1998). High rates of organic acids exudation are associated with a large capacity for P-mobilization in a variety of plant species including pigeon pea (*Cajanus cajan*), radish (*Raphanus sativus*), turnip (*Brassica napus*), barley (*Hordeum vulgare*), and *Oryza sativa* (Otani et al.

1996; Zhang et al. 1997; Kirk et al. 1999; Gahoonia et al. 2000). In transgenic tobacco (*Nicotiana tabacum*) plants that over-express a bacterial citrate synthase gene, the increased citrate efflux from their roots enable the transgenic lines to more effectively access P from Ca–P (López-Bucio et al. 2000).

4.1.3 Increasing the Pi Uptake Ratio and Translocation

The transport of phosphate (Pi) between subcellular compartments is central in metabolic regulation. Phosphate transporters are integral membrane proteins responsible for acquiring and mobilizing Pi to different tissues and organs. High-affinity Pi transporters (Pht1 family), work in low Pi concentration, and expression of their genes is modulated by Pi status (Karthikeyan et al. 2002; Raghothama and Karthikeyan 2005). *Pht1* genes have been found in many different plant species such as rice, barley, maize (*Zea mays*), potato (*Solanum tuberosum*), tomato (*Lycopersicon esculentum*), and tobacco (Kai et al. 2002; Paszkowski et al. 2002; Schunmann et al. 2004; Nagy et al. 2006). The expression pattern of *Pht1* members is differential, which indicates that *Pht1* genes are regulated by distinct mechanisms conferring tissue-specific expression promoting Pi translocation in response to Pi deprivation or internal cues (Chen et al. 2008).

4.2 Physiological Changes

Plants respond to tissue Pi status, enabling the efficient use of C, N, S, and P resources within the plant (White et al. 2005; Amtmann et al. 2006; Hammond and White 2008). During Pi starvation, when the available vacuole Pi reserves are depleted, many of the plant responses appear to be initiated or modulated by a decrease in the delivery of Pi to the shoot (Jeschke et al. 1997) with a consequent reduction in metabolism, impacting directly photosynthesis, glycolysis, and respiration (Jeschke et al. 1997; Hammond et al. 2003; Plaxton 2004; Hammond and White 2008). The reduction in photosynthesis is evident by down-regulation in Pi-deprived plants of many genes-encoding proteins involved in this process, including photosystem subunits, small subunits of RuBisCo, and enzymes needed for chlorophyll synthesis (Wu et al. 2003; Morcuende et al. 2007). Phosphorus deficiency also increases the export of triose-Pi from the chloroplast, which is subsequently converted to starch. The expression of several genes encoding enzymes involved in starch synthesis and degradation is known to be altered during Pi starvation (Nielsen et al. 1998; Wu et al. 2003; Morcuende et al. 2007; Müller et al. 2007). Changes in the synthesis, translocation, and degradation of sucrose as well as transcripts from genes encoding invertases, such as sucrose synthases, sucrose-phosphate synthases, and sucrose-phosphate phosphatases have been shown to be differentially expressed during Pi starvation. In addition, the expression patterns of several carbohydrate transporters are modified in the shoots and roots (Hammond et al. 2003; Wu et al. 2003; Hammond et al. 2005; Misson et al. 2005; Müller et al. 2007). Likewise, in Pi deficient plants, the reduction in cellular ADP and Pi result in decrease in the efficiency of respiration by inhibiting the cytochrome pathway of electron transport. It has been proposed that pyruvate accumulation in phosphate-deficient roots and nonphosphorylating pathways that bypass energy-requiring steps, including the activation of alternative oxidase in respiration are important aspects of plant metabolic adaptations to Pi limitation. This may play a role reducing oxidative stress maintaining cellular metabolic integrity by phosphate deficiency (Vance et al. 2003).

4.3 *Morphological Adaptations of Phosphate-Starved Plants*

4.3.1 Pi Can Modify Post-Embrionary Root Development

To cope with phosphate deficiency, plants modify their post-embryonic root developmental program towards a mechanism by which the exploratory capacity of the root and the absorptive surface area increase (Fig. 1). The modification of the root developmental program is reflected in changes of the root system architecture as a consequence of alterations in cell growth and expansion, root apical meristem activity, root hair (RH) formation, and elongation and formation and development of lateral, cluster, or proteoid roots (Ma et al. 2001; Schmidt and Schikora 2001; Williamson et al. 2001; López-Bucio et al. 2002; López-Bucio et al. 2003); however, these changes are not universal and may vary between plant species.

4.3.2 RH Formation

RHs are subcellular protrusions of epidermal cells in the root system with a major role in water and nutrient acquisition. RHs contribute up to 80% of the surface contact area of the root, and are the main site of Pi uptake in species that do not establish associations with mycorrhizae (Jungk 2001). The increase in RH length and number is an adaptation response to Pi stress (Bates and Lynch 2001). Species that develop more and/or longer RHs are far more efficient at accessing Pi from soils (Gahoonia et al. 2000) favoring the symbiotic association between RH and mycorrhizal fungi.

Some studies have suggested that the formation of ectopic RH is caused by alterations in the positional arrangement of epidermal cells. Pi starvation affects radial patterning of the root, making cortical cells smaller and more numerous. This results in a disorder in the position of epidermal cells adjacent to cortical cells generating more epidermal cells that finally form RHs (Ma et al. 2001; Zhang et al. 2003).

4.3.3 Root System Architecture

Diverse plant species present modifications in their root system architecture in response to Pi deficiency; in general, the root system is more branched. This change in root architecture helps to increase the capacity of root systems for soil exploration and interaction with beneficial soil microorganisms. In *Arabidopsis* the changes in root architecture include arrest of root tip growth and an increase in the initiation and elongation of lateral roots (LR) (Williamson et al. 2001; Linkohr et al. 2002; Lopez-Bucio et al. 2002; Al-Ghazi et al. 2003; Nacry et al. 2005; Jain et al. 2007; Pérez-Torres et al. 2008). Architectural modifications observed in common bean (*Phaseolus vulgaris*) include enhanced adventitious rooting and greater dispersion of LR (Lynch and van Beem 1993), and similar responses in LR development have been reported in maize (Zhu and Lynch 2004) and wheat (Manske et al. 2000; Liao et al. 2006).

In *Arabidopsis,* Pi deficiency induces a determinate growth program, which starts with alterations of cell division and elongation that later lead to an irreversible shift from an indeterminate to a determinate root growth program (Fig. 1c). In this process, the quiescent center plays a central role, showing an inhibition of cell elongation followed by the progressive loss of meristematic cells in a response mediated by *LPR* multicopper oxidase genes (Sánchez-Calderón et al. 2005; Svistoonoff et al. 2007).

Other species, such as white lupinus (*Lupinus albus*), respond to Pi deficiency developing specialized root structures known as proteoid roots (Keerthisinghe et al. 1998; Vance et al. 2003; Dinkelaker et al. 2005; Lambers et al. 2006), which are root clusters consisting of massive numbers of secondary and tertiary roots covered by RH (Kirkby and Johnston 2008). Proteoid roots are specialized in exuding organic acids, protons, acid phosphatases, and iron chelate reductase that release Pi from organic and inorganic compounds making it available for plant uptake (Gardner et al. 1982; Vance et al. 2003; Dinkelaker et al. 2005).

4.4 Interaction with Other Organisms

There is a wide range of bacterial and fungal species that have the ability to solubilize various forms of mineral Pi. It is believed that the establishment of mycorrhizal (fungus–root) symbioses in soil is one of the most successful strategies to maximize the access of plant roots to available Pi. The extra radical fungal mycelium has the ability to function as additional absorptive surface for the plant, thus increasing its capacity to forage for nutrients beyond the Pi depletion zone surrounding the roots (Barea et al. 2008; Vance 2008). Even though the vast majority (82%) of higher plant species have the capacity to form mycorrhizal symbioses, there are many plant families with species that either do not form or rarely form this kind of association. Those plants have developed alternative adaptations such as the development of active and extensive root systems or specialized root clusters (such as proteoid or dauciform clusters) for maximizing

Pi uptake from the soil solution. However, it should be noted that these specialized adaptations are not restricted to nonmycorrhizal species, there are several mycorrhizal species that also have the capacity to produce root clusters (Lambers et al. 2006). Moreover, most if not all, plant species that form mycorrizal associations harbor the mechanisms to respond to Pi deprivation.

5 Regulation and Signaling Mechanisms of Phosphate Starvation

5.1 Phosphate Starvation Response, a Coordinate Mechanism

5.1.1 Is There a Plant Pho Regulon?

The physiological and morphological responses in Pi starvation have been extensively studied and well understood, but so far the molecular mechanisms of how the plants perceive the presence or absence of Pi and how this signal evokes the adaptive responses to this deficient nutrition still remain largely unknown. Signaling mechanisms of Pi in bacteria and yeast are well established. It is known that in these organisms a Pi status-dependent signaling mechanism is activated, via the denominated phosphate (pho) regulon, which is integrated by several genes involved in Pi sensing and adaptation to limiting Pi environments. Pho regulon activation is dependent on a two-component system, which many times function as a sensor of environmental stimulus (Wanner 1996). In contrast, in vascular plants little is known about the underlying molecular mechanisms of the Pi starvation response; however, some components of the Pi signaling pathway have been identified that suggest the existence of a mechanism of regulation and signaling similar to that of bacteria and yeast. In the last decade, significant progresses have been made toward the understanding of Pi signaling pathways. In this sense, several important signaling components have been identified (Fig. 2). The majority of information for understanding the response mechanisms to Pi deprivation in plants comes from experimentation with *Arabidopsis thaliana*. Nevertheless, the research in other species such as legumes, rice, and tomato has generated valuable information. Because some regulating common components are present in these species, it is suggested that at least some regulatory mechanisms came from an ancestor and have been been conserved in vascular plants (Abel et al. 2002).

5.1.2 Sensing Pi Status

A molecular sensor that directly interacts with Pi to perceive its status has not yet been identified in plants. Nevertheless, analysis of several mutants affected in the

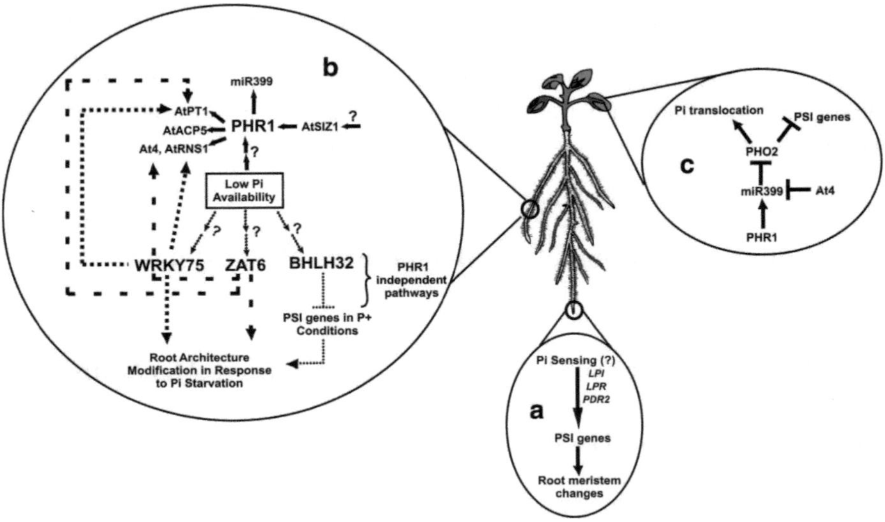

Fig. 2 Model *of Arabidopsis thaliana* Pi starvation signaling pathways. (**a**) Pi perception in root meristems by an unknown Pi starvation sensor (?) which triggers the pathways involved in changes in root architecture and in a subset of induced phosphate starvation (PSI) genes. (**b**) Unknown signal molecule (?) is activated by Pi status in dependent manner, transduces the signal to each transcription factor (*in bold*), and each TF is regulated independently. (**c**) PHO2 takes part in the translocation of Pi from shoot to the roots through xylem, PHO2 acts as inhibitor of some PSI genes and is downregulated by miRNA399. Additionally, miR399 activity is negatively regulated by *At4* through a target mimicry mechanism described only in plants, The signaling pathway directed by each TF is represented as PHR1 (——), WRKY75 (- - - -), ZAT6 (– – –), and BHLH32 (·· ·· ·· ···). *Arrows* indicate positive regulation and bars indicate negative regulation

response to Pi starvation, such as *pdr2* (*phosphate deficiency response 2*), *lpi* (*low phosphorus insensitive*), and *lpr* (*low phosphate root*); suggests that the meristem is the site where the external Pi concentration is being sensed (Fig. 2a) (Abel et al. 2002; Ticconi et al. 2004; Sanchez-Calderon et al. 2006; Svistoonoff et al. 2007). Although the mechanism by which the perception of Pi triggers all the response to Pi starvation remains to be determined, in the experiments made by Svistoonoff et al. (2007), they observed that the primary root arrested its growth when the root tip was in contact with the low Pi medium, even when the leaves were in contact with high Pi medium.

Through microarrays technology, a broad number of genes with differential expression during Pi stress have been detected (Hammond et al. 2003; Wu et al. 2003; Misson et al. 2005; Morcuende et al. 2007; Müller et al. 2007). Despite the large number of Pi-responsive genes that have been identified, the function of just a few genes is known. Among the genes that could be regulating the response to Pi starvation, transcription factors (TFs) genes are very important because of their capacity to activate signaling cascades.

5.1.3 Transcriptional Factors Involved in Phosphate Starvation

PHR1 (PHOSPHATE STARVATION RESPONSE1), BHLH32 (BASIC HELIX_ LOOP_HELIX 32), WRKY75 (which contains a conserved WRKYGQK sequence), and ZAT6 (ZINC FINGER OF *A. thaliana* 6) are so far the currently described TFs related in Pi starvation response (Fig. 2b). The first transcriptional factor identified that regulates Pi deprivation responses was PHR1 (Rubio et al. 2001). This TF was identified using transgenic *Arabidopsis* plants harboring a reporter gene specifically induced by Pi starvation and searching for mutants unable to activate the expression of the reporter gene.

PHR1 encodes an MYB transcription factor that is homolog to PSR1 (PHOSPHATE STARVATION RESPONSE 1) a TF involved in Pi sensing in *Chlamydomonas reinhardtii*. This suggests an ancestral mechanism of Pi signal sensing in unicellular photosynthetic ancestors that is present in vascular plants. PHR1 is not involved in modulating all phosphate starvation responses; for instance, this TF does not regulate the induction of genes involved in membrane lipid changes (Gaude et al 2008). However, the expression of several P starvation-induced genes is reduced in *phr1* mutant (Rubio et al. 2001; Bari et al. 2006), suggesting that PHR1 regulates a large subset of the Pi-responsive genes in *Arabidopsis*. The MYB-like domain of PHR1 binds to a GNATATNC DNA motif, named P1BS (Rubio et al. 2001). This imperfect palyndromic sequence is present in the promoter region of many Pi starvation-induced genes (Franco-Zorrilla et al. 2004; Misson et al. 2005; Müller et al. 2007). As the transcription of PHR1 is not altered by the Pi status in the plant, it was postulated that PHR1 regulation occurs postranslationally. Subsequently, it was found that PHR1 is a target of SIZ1 [SAP (SCAFFOLD ATTACHMENT FACTOR, acinus, protein inhibitor of activated signal transducer and activator of transcription and MIZ1 (Msx2-INTERACTING ZINC FINGER)], a SUMO E3 ligase (Miura et al. 2005), sumoylation being one of the control mechanisms activated in the Pi starvation response. The *siz1* mutant shows hypersensitivity and altered expression patterns of Pi starvation-inducible genes, including *AtIPS1* (*A. thaliana* induced by phosphate starvation 1) and *AtRNS1* (*A. thaliana* ribonuclease 1), which are genes regulated by PHR1. SIZ1 is considered also as a positive regulator of PHR1.

The BHLH32 TF has been recently shown to be involved in Pi starvation response. It is known that Pi-deprived plants increase their RH density and length, and present an elevated production of anthocyanins when compared to plants grown under Pi sufficient conditions (Williamson et al. 2001; Lopez-Bucio et al. 2002). The mutant *bhlh32* has increased anthocyanins accumulation, high Pi content, and increased RH number when compared to wild type, even when the plants are grown in abundance of Pi, suggesting that BHLH32 acts as a negative regulator of a set of genes that participate in low Pi-response and are repressed in high Pi conditions (Chen et al. 2007).

The expression of the TF WRKY75 is strongly induced during Pi depletion (Misson et al. 2005). WRKY75 belongs to the WRKY superfamily whose members have been implicated in the regulation of genes involved in the response to different biotic and abiotic stresses. This TF is a positive regulator of Pi deficiency responses, as WRKY75 RNAi silenced lines show accelerated Pi deficiency symptoms and a reduced expression of several genes induced in Pi starvation condition, including genes encoding phosphatases, high-affinity Pi transporters, and signaling and allocation mechanism components (Devaiah et al. 2007a). It is known that the action of WRKY TFs is primarily through their binding to conserved W box elements present in the promoters of specific genes (Ulker and Somssich 2004; Devaiah et al. 2007a). It has been shown that genes that are completely suppressed in WRKY75 RNAi plants, such as *AtPS2-1* (*A. thaliana* phosphate starvation-induced; representing an *acid phosphatase2-1*) and *At4* (*A. thaliana* 4), harbor multiple W boxes on their promoters, and those that did not have any W box motifs like *AtACP5* (*A. thaliana acid phosphatase 5*), *AtPS2-3* and *PLDZ2* (phospholipase DZ) were not affected. Genes that are partially suppressed, such as *AtPS2-2* and *AtIPS1*, had only one predicted W box with the TTGACT motif. The *Pht1;1* and *Pht1;4* phosphate transporters genes have multiple predicted W boxes, and they are partially suppressed. This could potentially lead to competition with the apparently more active TTGACC motif (Devaiah et al. 2007a).

A TF belonging to the Cys-2/His2 (C2H2) zinc finger family that has been shown to be involved in the response to Pi stress is ZAT6. Members of the zinc finger proteins family have been implicated in different plant processes, including stress responses. The suppression of ZAT6 through RNAi-mediated silencing resulted in lethality, whereas the overexpression of ZAT6 in seedlings resulted in retarded root growth, short primary roots, reduced Pi acquisition, and increased anthocyanin accumulation (Devaiah et al. 2007b). In contrast, older ZAT6 overexpressing plants showed a significant decrease in primary root length, enhanced LR length, and higher Pi content in shoots and roots, independently of their Pi regimen. Also, the expression of several Pi starvation-induced genes is reduced. Its role in Pi homeostasis seems to be linked to the positive regulation of genes related with Pi distribution and acquisition (*AtIPS1*, *AtPht1;4*), and to negatively controlling root development. A reduced expression of ZAT6 in auxin-resistant mutants *aux1-7*, *axr1-3*, *axr2-1*, and *ax4-1*, under Pi starvation suggests that this transcriptional factor could be an intermediary Pi-responsive regulator to the auxin-mediated changes in RSA (Devaiah et al. 2007b).

5.2 *Phosphate Homeostasis*

5.2.1 Signaling Pathway of Phosphate Starvation Dependent of PHR1, PHO2, and MicroRNA399

In order to advance the understanding of phosphate starvation signaling, the study of several phosphate starvation response mutants has been carried out. In particular, the study of *pho2* has revealed important knowledge about Pi homeostasis in plants.

PHO2 is an ubiquitin-conjugating E2 enzyme (UBC24) and the *pho2* mutant shows Pi over-accumulation in the shoot, in both Pi sufficient and Pi deficient conditions (Delhaize and Randall 1995; Aung et al. 2006; Bari et al. 2006). PHO2 is conserved among angiosperms and orthologs have been reported for rice, *Medicago truncatula*, and *Populus trichocarpa* (Bari et al. 2006). PHO2 is postranscriptionally regulated by microRNA 399 (miR399). The expression of miR399 increases rapidly during Pi depletion and is reversible by Pi re-supply. Its expression is typically induced under Pi deprivation and is partially inhibited in the *phr1* mutants, suggesting that PHR1 is required for miR399 expression and thus upstream of PHO2 in Pi signaling. The results of several investigations indicate that a regulatory pathway is active in Pi starvation, connecting PHR1 to PHO2 through the abundance of miR399 (Fig. 2c) (Bari et al. 2006; Chiou et al. 2006). The miR399-binding sites are located in the 5′ untranslated region of its target, leading to degradation of the PHO2 transcript. Promoter–reporter analysis demonstrated that UBC24/PHO2 and miR399 were co-localized to the vascular cylinder (Aung et al. 2006). The coordinated regulation of these genes supports the role of UBC24/PHO2 and miR399 in the regulation of Pi translocation and remobilization (Aung et al. 2006; Bari et al. 2006; Chiou et al. 2006; Chiou 2007). In addition to expression in the vascular system, miR399 expression was also found in root tips and mesophyll cells, suggesting the possibility of additional, unknown targets of miR399 in these tissues. Micrografting experiments with miR399-overexpressing and wild-type plants showed that miR399 is able to move from shoots to roots through the phloem and acts as a systemic signal (Lin et al. 2008; Pant et al. 2008).

The *IPS1/At4* gene family predicted to contain multiple short open-reading frames is known for being induced under low-Pi condition. At4 is implicated in internal allocation of Pi between shoot and roots during Pi stress, and it has been suggested that transcript levels may be adjusted at postranscriptional level by the activity of a miRNA (Shin et al. 2006). In an elegant work published by Franco-Zorrilla et al. (2007), the authors demonstrated that these non-coding transcripts attenuate the response to Pi deprivation through a process denominated "target mimicry". In this process, when *IPS1* is expressed, it sequesters miR399 through complementary interaction, resulting in an accumulation of the miR399 target PHO2 mRNA, because of a reduced degradation. These results suggest that Pi-starvation responses are tightly regulated and show that systemic Pi homeostasis is strongly regulated by the control of protein stability mediated by PHO2, opening new perspectives for the systemic role of miRNAs in the regulation of nutrient homeostasis. Several homologous components of this systemic Pi homeostasis (Fig. 2c) process have been identified in common bean and rice (Doerner 2008; Valdés-López and Hernández 2008; Zhou et al. 2008), showing that systemic Pi homeostasis is conserved in vascular plants.

Recently, in the response to Pi starvation, the role of *Arabidopsis* and rice genes encoding proteins that contain the SPX domain has been examined (Duan et al. 2008; Wang et al. 2009). SPX is a conserved domain present in the N-terminal region of the yeast SYG1 and PHO81, and human XPR1 proteins. SYG1 and XPR1 have been involved in signal transduction via association with G proteins and

PHO81 from *S. cerevisiae* and its homolog from *N. crassa* NUC2 are involved in Pi homeostasis through the control of the PHO regulon (Lenburg and O'Shea 1996; Poleg et al. 1996). In the *Arabidopsis* genome, there are 20 genes encoding proteins with a SPX domain, of which 16 besides the SPX domain contain an additional conserved domain and four (*AtSPX1-AtSPX4*) contain a single SPX domain (Duan et al. 2008). Most of the identified plant *SPX* gene products are involved in responses to environmental cues or internal regulation of nutrition homeostasis. In barley, *IDS4* (*iron-deficiency specific clone 4*) contains part of the SPX domain and is preferentially expressed in Fe-deficient roots (Nakanishi et al. 1993). Recently, it has been suggested that *AtSPX1–AtPSX4* have a potential role in the plant response to Pi signaling. They found that the expression of the *AtSPX1* and *AtSPX3* genes is induced by Pi starvation, and proposed that *AtSPX1* may be involved in transcriptional activation of genes, such as *AtACP5*, *PAP2* (*PURPLE ACID PHOSPHATASE2*), and *AtRNS1* that are related to Pi mobilization and scavenging of active oxygen species during Pi starvation conditions and that *AtSPX3* could have a positive role in plant adaptation to Pi deprivation. Moreover, it was shown that the expression of the Pi-responsive members of the *AtSPX* family depends of PHR1 (Duan et al. 2008). In rice, Wang et al. (2009) showed that *OsPSX1*, ortholog of *AtSPX1*, is specifically induced by Pi starvation and acts downstream of *OsPHR2* (*O. sativa phosphate starvation response transcription factor*, an ortholog of *Arabidopsis* PHR1) and *OsPHO2*. In the same line, overexpression of SPX1 caused suppression of several phosphate starvation induced (*PSI*) genes, suggesting that OsSPX1 acts as a negative feedback factor of the *PSI* signaling pathway in rice.

5.3 The Role of Sugars in Phosphate Starvation

Sugars are considered as signal molecules that regulate plant metabolic and developmental process including those in response to environmental stresses. In different sugar metabolic reactions, Pi has a pivotal role acting as a substrate or a product (Rolland and Sheen 2005). The results from some investigations indicate a relationship between sugar signaling and Pi starvation response. In soybean, several genes, such as a *vacuolar glycoprotein acid phosphatase* (*VspB*), *lipoxygenase A* (*LoxA*), a *proteinase inhibitor II* (*PinII*), and a chalcone synthase (*Chs*) respond to both stimulus, are induced by sugar, and inhibited by Pi in a dosage-dependent manner (Sadka et al. 1994). In tobacco seedlings, the expression of the *ADP glucose pyrophosphorylase* (*AGPase*) encoding gene is induced by Pi starvation and its induction is more prominent when sucrose is added (Nielsen et al. 1998). In *Arabidopsis*, the Pi transporter Pht1;4, was induced by sucrose supply (Lejay et al. 2003). Moreover, *Arabidopsis* mutants impaired in sugar transporting or signaling have a reduced induction of Pi responsive genes, and the phenotype in these mutants has been rescued or partially rescued by sugar addition (Karthikeyan et al. 2006). Originally the mutant *pho3* was associated with Pi response because of its altered acid phosphatase activity and a phenotype that apparently related with phosphate starvation response (Zakhleniuk et al. 2001). Later it was discovered that the *pho3* mutant is defective in the *SUC2* gene (Lloyd and Zakhleniuk 2004) that encodes for a sucrose-proton symporter which is important for phloem loading of sucrose (Gottwald et al. 2000).

Although the molecular mechanism by which sugar signaling and Pi status are interconnected is not well known, there are some clues that strongly suggest a cross talk between the two signaling pathways. Through microarray analysis, it has been observed that sugar-inducible genes respond to Pi and also the sugar affects the expression of several Pi starvation-inducible genes, revelating a close interaction between Pi and sugar dependent genes regulation (Müller et al. 2007). Additionally, microarray expression analysis of Pi-starved plants showed that the expression of genes involved in carbohydrate metabolism is altered under Pi stress (Hammond et al., 2003; Wu et al. 2003).

5.4 The Role of Plant Hormones in the Regulation of Phosphate Starvation Response

The adaptation of plants to diverse environmental conditions like Pi shortage is a process that involves biochemical and developmental changes to improve the acquisition of this essential nutrient. In Pi deprivation conditions, the plant has to perceive changes in environmental condition to activate the mechanisms that mediate these changes. Hormones act as chemical messengers in the regulation of physiological, biochemical, and molecular processes underlying growth and development. In order to survive, plants rely heavily on the proper physiological and developmental adjustments that determine their ability to secure edaphic resources. Therefore, hormones probably serve as essential integrators of developmental processes with environmental signals. The plant hormones more extensively studied so far in relation to Pi signaling are auxins, cytokinins, and ethylene.

In plants, auxins play a key regulator role in several growth and developmental processes, including cell division, cell elongation, cell differentiation, seed germination, flowering, and senescence (Paciorek and Friml 2006). In Pi starvation conditions, root architecture is modified in order to make Pi uptake more efficient. Plants grown under Pi deficiency show an increase in both length and density of LR or adventitious roots and RH. In *Arabidopsis*, auxins play an important role in mediating the Pi starvation effects on root system architecture (Lopez-Bucio et al. 2002; Al-Ghazi et al. 2003; Nacry et al. 2005; Jain et al. 2007), because of its central role in LR and RH formation and elongation (Stals and Inzé 2001). In White Lupin and *Arabidopsis* plants treated with the auxin transport inhibitors 2,3, 5-triiodobenzoic acid and *N*-1-naphthylphthalamic acid (NPA), the formation of proteoid or LR induced by the lack of Pi, respectively, is inhibited (Gilbert et al. 2000; López-Bucio et al. 2005). Nevertheless, it had not been determined whether auxin concentration, transport, and sensitivity were key factors modified by nutrient availability. Recently, Pérez-Torres et al. (2008) showed that in Pi starvation conditions the sensitivity to auxins is increased. In experiments made with *Arabidopsis* these authors observed that seedlings grown with the auxin transport inhibitor NPA and under Pi starvation regime, were able to form LR primordia, in comparison with seedlings grown with NPA and sufficient Pi that were not. In the same sense, seedlings grown under P deprivation and transferred to NPA-containing medium and exogenous auxin had greater formation of LR primordia compared with seedlings grown under P abundance and transferred to P-containing medium supplemented with the same concentration of NPA and auxin (Pérez-Torres et al. 2008). Based on the prior observations, it is clear that the pericycle cells are more sensitive to auxin provoking the proliferation of LR in plants grown under Pi starvation. The authors showed that this increased auxin sensitivity, is at least in part due to the induction of auxin-receptor TIR1 (TRANSPORT INHIBITOR RESPONSE 1) under Pi stress.

In the same fashion, cytokinins (CTK) also have been considered as molecular signals with important roles in growth and development. In the case of Pi limitation, exogenous application of CTK has a repressive effect on starvation-response genes expression (Martín et al. 2000). In the mutant *cre1* (*cytokinin response 1*), which is impaired in one cytokinin receptor, the repression of low-Pi-regulated genes is not observed, and because of this, the CTK two component signaling has been considered as a negative regulation system of Pi starvation responses (Franco-Zorrilla et al. 2002). Nevertheless, the action of plant hormones has not a straightforward effect as diverse studies reveal a complex network interacting with other factors to respond to specific environment conditions. *cre1* and *histidine kinase 4* (*ahk4*) mutant show increased sugar sensitivity and enhanced expression of both Pi-starvation and sugar-responsive genes in shoots of high sugar-grown plants (Franco-Zorrilla et al. 2005). These results suggest the existence of a multidirectional interaction between Pi starvation, sugar, and CTK. It has been reported also that the process of cell membrane phospholipids modifications is strictly regulated by Pi signaling and auxin/cytokinin cross-talk (Kobayashi et al. 2006).

Microarray analysis in rice carried out to investigate the influence of exogenous cytokinin on gene expression under Pi-deficient and -sufficient conditions has shown a combined effect of CTK and Pi-starvation signals in a global reduction of Pi-starvation signaling triggered by exogenous CTK with a remarkable increase of the cellular Pi concentration (Wang et al. 2006). This suggests that the increase of cellular Pi level caused by exogenous CTK treatment is likely to be one of the mechanisms for CTK repression of systemic Pi-starvation signaling.

It has also been proposed that ethylene has an important role in signal integration, by determining hierarchies between responses to multiple environmental challenges (Pierik et al. 2007). Ethylene has been connected with Pi-starvation signaling because the phenotype of *Arabidopsis* plants grown in the presence of ethylene precursor 1-amino-1-cyclopropane carboxylic acid (ACC) show a similar phenotype to low-Pi grown plants, both presenting a decrease in primary root length and an increase of RH formation. It is believed that ethylene has a role in mediating Pi nutrient deficiency response because of the similarities in RH response to low phosphorus and this hormone (Zhang et al. 2003).

Ethylene is known to be involved in cell expansion and hair root development (Dolan 2001). The production of ethylene in *P. vulgaris* roots grown in Pi starvation was twofold higher than that produced by the roots of plants grown under Pi-sufficient conditions. Nevertheless, a dual effect from ethylene has been observed depending of Pi-status. Experiments in beans (Borch et al. 1999) and *Arabidopsis* (Ma et al. 2001) revealed that in Pi sufficiency conditions ethylene inhibits primary root elongation but in Pi deprivation conditions it maintains root growth, suggesting that the root system acclimates to Pi deficiency by changing the signal transduction pathway connecting ethylene levels to root growth (Ma et al. 2003). The interaction of ethylene with auxin, establishes a complex interaction between these signaling pathways and their effect on plant development. However, the effect of ethylene on root growth is probably mediated through auxin, because ethylene stimulates auxin biosynthesis and the basipetal transport of auxin toward the root elongation zone (Ruzicka et al. 2007; Stepanova et al. 2007; Swarup et al. 2007). Under phosphate starvation, auxin biosynthesis is not increased (Pérez-Torres et al. 2008), but it is quite probable that there is a tight relation between these two hormones in order to modify the architecture of the root system in response to Pi starvation.

6 Conclusions

In the last few decades, significant progress has been made in understanding the Pi dynamics in the soil. Now we know that even if the total amount of Pi in most soils is enough to sustain plant productivity, it is generally not readily available to be taken up by plants. Pi availability depends on several biotic and abiotic soil processes such as precipitation/dissolution, sorption/desorption, immobilization, and mineralization. This availability partially determines the ecosystem layout around the world. To cope with low P availability, plants display a wide array of strategies that improve Pi acquisition and recycling, including root architectural traits that enhance topsoil foraging and metabolic adjustment. Most of these strategies have been well described in recent years and the molecular mechanisms underlying these adaptation strategies are starting to be understood.

Low Pi availability is a widespread nutritional problem greatly affecting crop production around the world and the source of Pi fertilizers tends to be exhausted; it is necessary to develop new agricultural strategies to reduce the Pi-fertilizer inputs without compromising yield or quality (White and Hammond 2008). It is also necessary to improve the study of phosphate starvation response in plants in order to identify the putative Pi sensors molecules as well as the many missing elements involved in the signaling pathways of this integrative program. The challenge over the next few years will be to identify and integrate these elements in a meaningful fashion to understand the different components of the Pi deprivation response in plants and develop the strategies to use this information to generate novel plant varieties with increased Pi uptake and use efficiency either by conventional breeding or by transgenic strategies.

References

Abel S, Nurnberger T, Ahnert V, Krauss G, Glund K (2000) Induction of an extracellular cyclic nucleotide phosphodiesterase as an accessory ribonucleolytic activity during phosphate starvation of cultured tomato cells. Plant Physiol 122:543–552

Abel S, Ticconi CA, Delatorre CA (2002) Phosphate sensing in higher plants. Physiol Plant 115:1–8

Abelson PH (1999) A potential phosphate crisis. Science 283:2015–2020

Al-Ghazi Y, Muller B, Pinloche S, Tranbarger TG, Nacry P, Rossignol M, Tardieu F, Doumas P (2003) Temporal responses of *Arabidopsis* root architecture to phosphate starvation: evidence for the involvement of auxin signalling. Plant Cell Environ 26:1053–1066

Amtmann A, Hammond JP, Armengaud P, White PJ (2006) Nutrient sensing and signalling in plants: potassium and phosphorus. Adv Bot Res 43:209–257

Andersson MX, Larsson KE, Tjellstrom H, Liljenberg C, Sandelius AS (2005) The plasma membrane and the tonoplast as major targets for phospholipid-to-glycolipid replacement and stimulation of phospholipases in the plasma membrane. J Biol Chem 280:27578–27586

Andersson MX, Stridh MH, Larsson KE, Liljenberg C, Sandelius AS (2003) Phosphate-deficient oat replaces a major portion of the plasma membrane phospholipids with the galactolipid digalactosyldiacylglycerol. FEBS Lett 537:128–132

Aung K, Lin S-I, Wu C-C, Huang Y-T, C-l Su, Chiou T-J (2006) pho2, a phosphate over-accumulator, is caused by a nonsense mutation in a microRNA399 target gene. Plant Physiol 141:1000–1011

Bar-Yosef O (1991) Root excretions and their environmental effects. Influence on availability of phosphorus. In: Waisel Y, Eshel A (eds) Plant roots: the hidden half. Marcel Dekker, New York, pp 529–557

Barea JM, Ferrol N, Azcón-Aguilar C, Azcón R (2008) Mycorrhizal symbioses. In: White PJ, Hammond JP (eds) The ecophysiology of plant-phosphorus interactions, vol 7, Springer, Netherlands, pp 143–163

Bari R, Pant B, Stitt M, Scheible W-R (2006) PHO2, microRNA399, and PHR1 define a phosphate-signaling pathway in plants. Plant Physiol 141:988–999

Bariola PA, Howard C, Taylor C, Verburg M, Jaglan V, Green P (1994) The *Arabidopsis* ribonuclease gene RNS1 is tightly controlled in response to phosphate limitation. Plant J 6:673–685

Bates TR, Lynch JP (2001) Root hairs confer a competitive advantage under low phosphorus availability. Plant Soil 236:243–250

Benning C, Ohta H (2005) Three enzyme systems for galactoglycerolipid biosynthesis are coordinately regulated in plants. J Biol Chem 280:2397–2400

Benton JJ (1998) Plant nutrition. CRC press, New York

Bieleski RL (1973) Phosphate pools, phosphate transport, and phosphate availability. Ann Rev Plant Physiol 24:225–252

Borch K, Bouma TJ, Lynch JP, Brown KM (1999) Ethylene: a regulator of root architectural responses to soil phosphorus availability. Plant Cell Environ 22:425–431

Bould C, Hewitt EJ, Needham P (1986) Diagnosis of mineral disorders in plants, 1st edn. Chemical Publishing, London

Bozzo GG, Dunn EL, Plaxton WC (2006) Differential synthesis of phosphate-starvation inducible purple acid phosphatase isozymes in tomato (*Lycopersicon esculentum*) suspension cells and seedlings. Plant Cell Environ 29:303–313

Cakmak I (2002) Plant nutrition research: priorities to meet human needs for food in sustainable ways. Plant Soil 247:3–24

Cruz-Ramirez A, Oropeza-Aburto A, Razo-Hernandez F, Ramairez-Chavez E, Herrera-Estrella L (2006) Phospholipase DZ2 plays an important role in extraplastidic galactolipid biosynthesis and phosphate recycling in *Arabidopsis* roots. Plant Cell 103:6765–6770

Chen Y, Wang Y, Wu W (2008) Membrane transporters for nitrogen, phosphate and potassium uptake in plants. J Integr Plant Biol 50:835–848

Chen Z-H, Nimmo GA, Jenkins GI, Nimmo HG (2007) BHLH32 modulates several biochemical and morphological processes that respond to Pi starvation in *Arabidopsis*. Biochem J 405:191–198

Chiou T-J, Aung K, Lin S-I, Wu C-C, Chiang S-F, C-l Su (2006) Regulation of phosphate homeostasis by microRNA in *Arabidopsis*. Plant Cell 18:412–421

Chiou TJ (2007) The role of microRNAs in sensing nutrient stress. Plant Cell Environ 30:323–332

Daram P, Brunner S, Rausch C, Steiner C, Amrhein N, Bucher M (1999) Pht2;1 encodes a low-affinity phosphate transporter from *Arabidopsis*. Plant Cell 11:2153–2166

Delhaize E, Randall PJ (1995) Characterization of a phosphate-accumulator mutant of *Arabidopsis thaliana*. Plant Physiol 107:207–213

Devaiah BN, Karthikeyan AS, Raghothama KG (2007a) WRKY75 transcription factor is a modulator of phosphate acquisition and root development in *Arabidopsis*. Plant Physiol 143:1789–1801

Devaiah BN, Nagarajan VK, Raghothama KG (2007b) Phosphate homeostasis and root development in *Arabidopsis* are synchronized by the zinc finger transcription factor ZAT6. Plant Physiol 145:147–159

Dinkelaker B, Hengeler C, Marschner H (2005) Distribution and function of proteoid roots and other root clusters. Botanica Acta 108:183–200

Doerner P (2008) Phosphate starvation signaling: a threesome controls systemic Pi homeostasis. Curr Opin Plant Biol 11:536–540

Dolan L (2001) Plant development: the benefits of a change of scene 11:R702–R704

Dörmann P, Benning C (2002) Galactolipids rule in seed plants. Trends Plant Sci 7:112–118

Duan K, Yi K, Dang L, Huang H, Wu W, Wu P (2008) Characterization of a sub-family of *Arabidopsis* genes with the SPX domain reveals their diverse functions in plant tolerance to phosphorus starvation. Plant J 54:965–975

Essigmann B, Güler S, Narang RA, Linke D, Benning C (1998) Phosphate availability affects the thylakoid lipid composition and the expression of SQD1, a gene required for sulfolipid biosynthesis in *Arabidopsis thaliana*. Proc Natl Acad Sci USA 95:1950–1955

Franco-Zorrilla JM, Gonzalez E, Bustos R, Linhares F, Leyva A, Paz-Ares J (2004) The transcriptional control of plant responses to phosphate limitation. J Exp Bot 55:285–293

Franco-Zorrilla JM, Martin AC, Leyva A, Paz-Ares J (2005) Interaction between phosphate-starvation, sugar, and cytokinin signaling in *Arabidopsis* and the roles of cytokinin receptors CRE1/AHK4 and AHK3. Plant Physiol 138:847–857

Franco-Zorrilla JM, Martin AC, Solano R, Rubio V, Leyva A, Paz-Ares J (2002) Mutations at CRE1 impair cytokinin-induced repression of phosphate starvation responses in *Arabidopsis*. Plant J 32:353–360

Franco-Zorrilla JM, Valli A, Todesco M, Mateos I, Puga MI, Rubio-Somoza I, Leyva A, Weigel D, Garcia JA, Paz-Ares J (2007) Target mimicry provides a new mechanism for regulation of microRNA activity. Nat Genet 39:1033–1037

Fujii H, Chiou TJ, Lin S, Aung K, Zhu JK (2005) A miRNA involved in phosphate-starvation response in *Arabidopsis*. Curr Biol 15:2038–2043

Gahoonia TS, Asmar F, Giese H, Nielsen GG, Nielsen NE (2000) Root released organic acids and phosphorus uptake of two barley cultivars in laboratory and field experiments. Eur J Agron 12:281–289

Gardner WK, Parbery DG, Barber DA (1982) The acquisition of phosphorus by *Lupinus albus* L.I. Some characteristics of the soil/root interface. Plant Soil 68:19–32

Gaude N, Nakamura Y, Scheible W, Ohta H, Dörmann P (2008) Phospholipase C5 (NPC5) is involved in galactolipid accumulation during phosphate limitation in leaves of *Arabidopsis*. Plant J 56:28–39

Gilbert GA, Knight JD, Vance CP, Allan DL (2000) Proteoid root development of phosphorus deficient lupin is mimicked by auxin and phosphonate. Ann Bot 85:921–928

Goldstein AH (1992) Phosphate starvation indicuble enzymes and proteins in higher plants. In: Wray JL (ed) Inducible plant proteins their biochemistry and molecular biology. Cambridge University Press, Cambridge, UK, pp 25–44

Gottwald JR, Krysan PJ, Young JC, Evert RF, Sussman MR (2000) Genetic evidence for the in planta role of phloem-specific plasma membrane sucrose transporters. Proc Natl Acad Sci USA 97:13979–13984

Guo B, Jin Y, Wussler C, Blancaflor EB, Motes CM, Versaw WK (2008) Functional analysis of the *Arabidopsis* PHT4 family of intracellular phosphate transporters. New Phytol 177:889–898

Hamburger D, Rezzonico E, MacDonald-Comber P, Etétot J, Somerville C, Poirier Y (2002) Identification and characterization of *Arabidopsis* PHO1 gene involved in phosphate loading to the xylem. Plant Cell 14:889–902

Hammond JP, Bennett MJ, Bowen HC, Broadley MR, Eastwood DC, May ST, Rahn C, Swarup R, Woolaway KE, White PJ (2003) Changes in gene expression in *Arabidopsis* shoots during phosphate starvation and the potential for developing smart plants. Plant Physiol 132:578–596

Hammond JP, Broadley MR, Craigon DJ, Higgins J, Emmerson Z, Townsend H, White PJ, May ST (2005) Using genomic DNA-based probe-selection to improve the sensitivity of high-density oligonucleotide arrays when applied to heterologous species. Plant Methods 1:10

Hammond JP, White PJ (2008) Sucrose transport in the phloem: integrating root responses to phosphorus starvation. J Exp Bot 59:93–109

Härtel H, Dörmann P, Benning C (2000) DGD1-independent biosynthesis of extraplastidic galactolipids after phosphate deprivation in *Arabidopsis*. Proc Natl Acad Sci USA 97:10649–10654

Hinsinger P (2001) Bioavailability of soil inorganic P in the rhizosphere as affected by root-induced chemical changes: a review. Plant Soil 237:173–195

Holford ICR (1997) Soil phosphorus: its measurement and its uptake by plants. Aust J Soil Res 35:227–239

Jain A, Poling MD, Karthikeyan AS, Blakeslee JJ, Peer WA, Titapiwatanakun B, Murphy AS, Raghothama KG (2007) Differential effects of sucrose and auxin on localized phosphate deficiency-induced modulation of different traits of root system architecture in *Arabidopsis*. Plant Physiol 144:232–247

Jeschke WD, Peuke AD, Pate JS, Hartung W (1997) Transport, synthesis and catabolism of abscisic acid (ABA) in intact plants of castor bean (*Ricinus communis* L.) under phosphate deficiency and moderate salinity. J Exp Bot 48:1737–1747

Jones DL (1998) Organic acids in the rhizosphere – a critical review. Plant Soil 205:25–44

Jones DL, Darrah PR (1994) Role of root derived organic-acids in the mobilization of nutrients from the rhizosphere. Plant Soil 166:247–257

Jouhet J, Maréchal E, Baldan B, Bligny R, Joyard J, Block MA (2004) Phosphate deprivation induces transfer of DGDG galactolipid from chloroplast to mitochondria. J Cell Biol 167:863–874

Jungk A (2001) Root hairs and the acquisition of plant nutrients from soil. J Plant Nutr Soil Sci 164:121–129

Kai M, Takazumi K, Adachi H, Wasaki J, Shinano T, Osaki M (2002) Cloning and characterization of four phosphate transporter cDNAs in tobacco. Plant Sci 163:837–846

Karthikeyan AS, Varadarajan DK, Jain A, Held AM, Carpita NC, Raghothama KG (2006) Phosphate starvation responses are mediated by sugar signaling in *Arabidopsis*. Planta 225:907–918

Karthikeyan AS, Varadarajan DK, Mukatira UT, Panio DUM, Damsz B, Raghothama KG (2002) Regulated expression of *Arabidopsis* transporters. Plant Physiol 130:221–233

Keerthisinghe G, Hocking PJ, Ryan PR, Delhaize E (1998) Effect of phosphorus supply on the formation and function of proteoid roots of white lupin (*Lupinus albus* L.). Plant Cell Environ 21:467–478

Kirk GJ, Santos E, Findenegg GR (1999) Phosphate solubilization by organic anion secretion from rice (Oryza sativa L.) growing in aerobic soil. Plant Soil 211:11–18

Kirkby EA, Johnston AE (2008) Soil and fertilizer phosphorus in relation to crop nutrition. In: White PJH (ed) The ecophysiology of plant-phosphorus interactions. Springer, 223, p 177

Kobayashi K, Awai K, Nakamura M, Nagatani A, Masuda T, Ohta H (2009) Type-B monogalactosyldiacylglycerol synthases are involved in phosphate starvation-induced lipid remodeling, and are crucial for low-phosphate adaptation. Plant J 57:322–331

Kobayashi K, Masuda T, Takamiya K, Ohta H (2006) Membrane lipid alteration during phosphate starvation is regulated by phosphate signaling and auxin/cytokinin cross-talk. Plant J 47:238–248

Köck M, Loffler A, Abel S, Glund K (1995) cDNA structure and regulatory properties of a family of starvation-induced ribonucleases from tomato. Plant Mol Biol 27:477–485

Köck M, Theierl K, Stenzel I, Glund K (1998) Extracellular administration of phosphate sequestering metabolites induces ribonucleases in cultured tomato cells. Planta 204:404–407

Lambers HY, Shane MW, Cramer MD, Pearse SJ, Veneklass EJ (2006) Root structure and functioning for efficient acquisition of phosphorus: matching morphological and physiological traits. Ann Bot 98:693–713

Lan M, Comerford NB, Fox TR (1995) Organic anions effect on phosphorus release from spodic horizons. Soil Sci Soc Am J 59:1745–1749

Lejay L, Gansel X, Cerezo M, Tillard P, Muller C, Krapp A, von Wiren N, Daniel-Vedele F, Gojon A (2003) Regulation of root Ion transporters by photosynthesis: functional importance and relation with hexokinase. Plant Cell 15:2218–2232

Lenburg M, O'Shea E (1996) Signaling phosphate starvation. Trends Biochem Sci 21:383–387

Li MY, Welti R, Wang XM (2006) Quantitative profiling of *Arabidopsis* polar glycerolipids in response to phosphorus starvation. Roles of phospholipases Df1 and Df2 in phosphatidylcholine hydrolysis and digalactosyldiacylglycerol accumulation in phosphorus-starved plants. Plant Physiol 142:750–761

Liao H, Wan H, Shaff J, Wang X, Yan X, Kochian LV (2006) Phosphorus and aluminum interactions in soybean in relation to aluminum tolerance. Exudation of specific organic acids from different regions of the intact root system. Plant Physiol 141:674–684

Lin S-I, Chiang S-F, Lin W-Y, Chen J-W, Tseng C-Y, Wu P-C, Chiou T-J (2008) Regulatory network of microRNA399 and PHO2 by systemic signaling. Plant Physiol 147:732–746

Linkohr BI, Williamson LC, Fitter AH, Leyser O (2002) Nitrate and phosphate availability and distribution have different effects on root system architecture of *Arabidopsis*. Plant J 29:751–760

López-Bucio J, Cruz-Ramírez A, Herrera-Estrella L (2003) The role of nutrient availability in regulating root architecture. Curr Opin Plant Biol 6:280–287

López-Bucio J, de la Vega O, Guevara-Garcia A, Herrera-Estrella L (2000) Enhanced phosphorus uptake in transgenic tobacco plants that overproduce citrate. Nat Biotechnol 18:450–453

Lopez-Bucio J, Hernandez-Abreu E, Sanchez-Calderon L, Nieto-Jacobo MF, Simpson J, Herrera-Estrella L (2002) Phosphate availability alters architecture and causes changes in hormone sensitivity in the *Arabidopsis* root system. Plant Physiol 129:244–256

López-Bucio J, Hernández-Abreu E, Sánchez-Calderón L, Nieto-Jacobo MF, Simpson J, Herrera-Estrella L (2002) Phosphate availability alters architecture and causes changes in hormone sensitivity in the *Arabidopsis* root system. Plant Physiol 129:244–256

López-Bucio J, Hernández-Abreu E, Sánchez-Calderón L, Pérez-Torres A, Rampey RA, Bartel B, Herrera-Estrella L (2005) An auxin transport independent pathway is involved in phosphate stress-induced root architectural alterations in *Arabidopsis*. Identification of BIG as a mediator of auxin pericycle cell activation. Plant Physiol 137:681–691

Lynch J, van Beem J (1993) Growth and architecture of seedling roots of common bean genotypes. Crop Sci 33:1253–1257

Lloyd JC, Zakhleniuk OV (2004) Responses of primary and secondary metabolism to sugar accumulation revealed by microarray expression analysis of the *Arabidopsis* mutant, pho3. J Exp Bot 55:1221–1230

Ma Z, Baskin TI, Brown KM, Lynch JP (2003) Regulation of root elongation under phosphorus stress involves changes in ethylene responsiveness. Plant Physiol 131:1381–1390

Ma Z, Bielenberg DG, Brown KM, Lynch JP (2001) Regulation of root hair density by phosphorus availability in *Arabidopsis thaliana*. Plant Cell Environ 24:459–467

Manske G, Ortiz-Monasterio J, Van Grinkel M, González R, Rajaram S, Molina E, Vlek P (2000) Traits associated with improved P-uptake efficiency in CIMMYT's semidwarf spring bread wheat grown on an acid Andisol in Mexico. Plant Soil 221:189–204

Martín AC, del Pozo J, Iglesias J, Rubio V, Solano R, de la Peña A, Leyva A, Paz-Ares J (2000) Influence of cytokinins on the expression of phosphate starvation responsive genes in *Arabidopsis*. Plant J 24:559–567

Misson J, Raghothama KG, Jain A, Jouhet J, Block MA, Bligny R, Ortet P, Creff A, Somerville S, Rolland N, Doumas P, Nacry P, Herrerra-Estrella L, Nussaume L, Thibaud M-C (2005) A genome-wide transcriptional analysis using *Arabidopsis thaliana* Affymetrix gene chips determined plant responses to phosphate deprivation. Proc Natl Acad Sci USA 102:11934–11939

Miura K, Rus A, Sharkhuu A, Yokoi S, Karthikeyan AS, Raghothama KG, Baek D, Koo YD, Jin JB, Bressan RA, Yun D-J, Hasegawa PM (2005) The *Arabidopsis* SUMO E3 ligase SIZ1 controls phosphate deficiency responses. Proc Natl Acad Sci USA 102:7760–7765

Morcuende R, Bari R, Gibon Y, Zheng W, Pant BD, Blasing O, Usadel B, Czechowski T, Udvardi MK, Stitt M, Scheible W (2007) Genome-wide reprogramming of metabolism and regulatory networks of Arabidopsis in response to phosphorus. Plant Cell Environ 30:85–112

Mudge SR, Rae AL, Diatloff E, Smith FW (2002) Expression analysis suggests novel roles for members of the Pht1 family of phosphate transporters in *Arabidopsis*. Plant J 31:341–353

Müller R, Morant M, Jarmer H, Nilsson L, Nielsen TH (2007) Genome-wide analysis of the *Arabidopsis* leaf transcriptome reveals interaction of phosphate and sugar metabolism. Plant Physiol 143:156–171

Nacry P, Canivenc G, Muller B, Azmi A, Van Onckelen H, Rossignol M, Doumas P (2005) A role for auxin redistribution in the responses of the root system architecture to phosphate starvation in *Arabidopsis*. Plant Physiol 138:2061–2074

Nagy R, Vasconcelos MJ, Zhao S, McElver J, Bruce W, Amrhein N (2006) Differential regulation of five Pht1 phosphate transporters from maize (Zea mays L.). Plant Biol 8:186–197

Nakanishi H, Okumura N, Umehara Y, Nishizawa N-K, Chino M, Mori S (1993) Expression of a gene specific for iron deficiency (Ids3) in the roots of *Hordeum vulgare*. Plant Cell Physiol 34:401–410

Nielsen TH, Krapp A, Röper-Schwarz U, Stitt M (1998) The sugar-mediated regulation of genes encoding the small subunit of Rubisco and the regulatory subunit of ADP glucose pyrophosphorylase is modified by phosphate and nitrogen. Plant Cell Environ 21:443–454

Oberson A, Friesen DK, Rao IM, Bühler S, Frossard E (2001) Phosphorus transformations in an oxisol under contrasting land-use systems: the role of the soil microbial biomass. Plant Soil 237:197–210

Otani T, Ae N, Tanaka H (1996) Uptake mechanisms of crops grown in soils with low P status. II. Significance of organic acids in root exudates of pigeon pea. Soil Sci Plant Nutr 42:553–560

Paciorek T, Friml J (2006) Auxin signaling. J Cell Sci 119:1199–1202

Pant BD, Buhtz A, Kehr J, Wolf-Rüdiger S (2008) MicroRNA399 is a long-distance signal for the regulation of plant phosphate homeostasis. Plant J 53:731–738

Paszkowski U, Kroken S, Roux C, Briggs SP (2002) Rice phosphate transporters include an evolutionarily divergent gene specifically activated in arbuscular mycorrhizal symbiosis. Proc Natl Acad Sci USA 99:13324–13329

Pérez-Torres C, López-Bucio J, Cruz-Ramírez A, Ibarra-Laclette E, Dharmasiri S, Estelle M, Herrera-Estrella L (2008) Phosphate availability alters lateral root development in *Arabidopsis* by modulating auxin sensitivity via a mechanism involving the TIR1 auxin receptor. Plant Cell 20:3258–3272

Pierik R, Sasidharan R, Voesenek L (2007) Growth control by ethylene: adjusting phenotypes to the environment. J Plant Growth Regul 26:188–200

Plaxton WC (2004) Plant response to stress: biochemical adaptations to phosphate deficiency. Encyclopedia of plant and crop science. Taylor & Francis, UK, pp 976–980

Poirier Y, Thoma S, Somerville C, Schiefelbein J (1991) A mutant of *Arabidopsis* deficient in xylem loading of phosphate. Plant Physiol 97:1087–1093

Poleg YAR, Kang S, Hall JG, Metzenberg RL (1996) NUC-2, a component of the phosphate-regulated signal transduction pathway in *Neurospora crassa*, is an ankyrin repeat protein. Mol Gen Genet 252:709–716

Raghothama KG (1999) Phosphate acquisition. Annu Rev Plant Physiol Plant Mol Biol 50:665–693

Raghothama KG, Karthikeyan AS (2005) Phosphate acquisition. Plant Soil 274:37–49

Rausch C, Bucher M (2002) Molecular mechanisms of phosphate transport in plants. Planta 216:23–37

Rolland F, Sheen J (2005) Sugar sensing and signalling networks in plants. Biochem Soc Trans 33:269–271

Rubio V, Linhares F, Solano R, Martín AC, Iglesias J, Leyva A, Javier P-A (2001) A conserved MYB transcription factor involved in phosphate starvation signaling both in vascular plants and in unicellular algae. Genes Dev 15:2122–2133

Ruzicka K, Ljung K, Vanneste S, Podhorska R, Beeckman T, Friml J, Benkova E (2007) Ethylene regulates root growth through effects on auxin biosynthesis and transport-dependent auxin distribution. Plant Cell 19:2197–2212

Sadka A, DeWald DB, May GD, Park WD, Mullet JE (1994) Phosphate modulates transcription of soybean VspB and other sugar-inducible genes. Plant Cell 6:737–749

Sánchez-Calderón L, López-Bucio J, Chacón-López A, Cruz-Ramírez A, Nieto-Jacobo F, Dubrovsky JG, Herrera-Estrella L (2005) Phosphate starvation induces a determinate developmental program in the roots of *Arabidopsis thaliana*. Plant Cell Physiol 46:174–184

Sanchez-Calderon L, Lopez-Bucio J, Chacon-Lopez A, Gutierrez-Ortega A, Hernandez-Abreu E, Herrera-Estrella L (2006) Characterization of low phosphorus insensitive mutants reveals a crosstalk between low phosphorus-induced determinate root development and the activation of genes involved in the adaptation of *Arabidopsis* to phosphorus deficiency. Plant Physiol 140:879–889

Sanda S, Leustek T, Theisen MJ, Garavito RM, Benning C (2001) Recombinant *Arabidopsis* SQD1 converts UDP-glucose and sulfite to the sulfolipid head group precursor UDP-sulfoquinovose in vitro. J Biol Chem 276:3941–3946

Schachtman DP, Reid RJ, Ayling SM (1998) Phosphorus uptake by plants: from soil to cell. Plant Physiol 116:447–453

Schmidt W, Schikora A (2001) Different pathways are involved in phosphate and iron stress-induced alterations of root epidermal cell development. Plant Physiol 125:2078–2084

Schunmann PH, Richardson AE, Smith FW, Delhaize E (2004) Characterization of promoter expression patterns derived from the Pht1 phosphate transporter genes of barley (*Hordeum vulgare* L.). J Exp Bot 55:855–865

Shin H, Shin H, Chen R, Harrison M (2006) Loss of At4 function impacts phosphate distribution between the roots and the shoots during phosphate starvation. Plant J 45:712–726

Stals H, Inzé D (2001) When plant cells decide to divide. Trends Plant Sci 6:359–364

Stepanova AN, Yun J, Likhacheva AV, Alonso JM (2007) Multilevel interactions between ethylene and auxin in *Arabidopsis* roots. Plant Cell tpc.107.052068

Svistoonoff S, Creff A, Reymond M, Sigoillot-Claude C, Ricaud L, Blanchet A, Nussaume L, Desnos T (2007) Root tip contact with low-phosphate media reprograms plant root architecture. Nat Genet 39:792–796

Swarup R, Perry P, Hagenbeek D, Van Der Straeten D, Beemster GTS, Sandberg G, Bhalerao R, Ljung K, Bennett MJ (2007) Ethylene upregulates auxin biosynthesis in *Arabidopsis* seedlings to enhance inhibition of root cell elongation. Plant Cell 19:2186–2196

Ticconi CA, Abel S (2004) Short on phosphate: plant surveillance and countermeasures. Trends Plant Sci 9:548–555

Ticconi CA, Delatorre CA, Lahner B, Salt DE, Abel S (2004) Arabidopsis pdr2 reveals a phosphate-sensitive checkpoint in root development. Plant J 37:801–814

Tiessen H (2008) Phosphorus in the global environment. In: White PJ, Hammond JP (eds) The ecophysiology of plant-phosphorus interactions, vol 7, Springer, Netherlands, pp 1–7

Ulker B, Somssich IE (2004) WRKY transcription factors: From DNA binding towards biological function. Curr Opin Plant Biol 7:491–498

Valdés-López O, Hernández G (2008) Transcriptional regulation and signaling in phosphorus starvation: what about legumes? J Integr Plant Biol 10:1213–1222

Vance CP (2008) Plants without arbuscular mycorrhizae. In: White PJ, Hammond JP (eds) The ecophysiology of plant-phosphorus interactions, vol 7, Springer, Netherlands, pp 117–142

Vance CP, Uhde-Stone C, Allan DL (2003) Phosphorus acquisition and use: critical adaptations by plants for securing a nonrenewable resource. New Phytol 157:423–447

Versaw WK, Harrison MJ (2002) A chloroplast phosphate transporter, PHT2;1 influences allocation of phosphate within the plant and phosphate-starvation responses. Plant Cell 14:1751–1766

Wang C, Ying S, Huang H, Li K, Wu P, Shou H (2009) Involvement of OsSPX1 in phosphate homeostasis in rice. Plant J 57:895–904

Wang X, Yi K, Tao Y, Wang F, Wu Z, Jiang D, Chen X, Zhu L, Wu P (2006) Cytokinin represses phosphate-starvation response through increasing of intracellular phosphate level. Plant Cell Environ 29:1924–1935

Wanner BL (1996) Signal transduction in the control of phosphate-regulated genes of *Escherichia coli*. Kidney Int 49:964–967

White PJ, Broadley MR, Greenwood DJ, Hammond JP (2005) Genetic modifications to improve phosphorus acquisition by roots. In Proceedings 568. International Fertiliser Society, York, UK

White PJ, Hammond JP (2008) Phosphorous nutrition of terrestrial plants. In: White PJ, Hammond JP (eds) The ecophysiology of plant-phosphorus interactions, vol 7, Springer, Netherlands, pp 51–81

Williamson LC, Ribrioux SPCP, Fitter AH, Leyser HMO (2001) Phosphate availability regulates root system architecture in *Arabidopsis*. Plant Physiol 126:875–882

Wu P, Ma L, Hou X, Wang M, Wu Y, Liu F, Deng XW (2003) Phosphate starvation triggers distinct alterations of genome expression in *Arabidopsis* roots and leaves. Plant Physiol 132:1260–1271

Yu B, Xu C, Benning C (2002) *Arabidopsis* disrupted in SQD2 encoding sulfolipid synthase is impaired in phosphate-limited growth. Proc Natl Acad Sci USA 99:5732–5737

Zakhleniuk OV, Raines CA, Lloyd JC (2001) pho3: a phosphorus-deficient mutant of *Arabidopsis thaliana* (L.) Heynh. Planta 212:1432–2048

Zhang F, Ma J, Cao YP (1997) Phosphorus deficiency enhances root exudation of low molecular weight organic acids and utilization of sparingly soluble inorganic phosphates by radish (*Raphanus sativus* L.) and rape (*Brasicca napus* L.) plants. Plant Soil 196:261–264

Zhang YJ, Lynch JP, Brown KM (2003) Ethylene and phosphorus availability have interacting yet distinct effects on root hair development. J Exp Bot 54:2351–2361

Zhou J, Jiao FC, Wu ZC, Li Y, Wang X, He X, Zhong W, Wu P (2008) OsPHR2 Is involved in phosphate-starvation signaling and excessive phosphate accumulation in shoots of plants. Plant Physiol 146:1673–1686

Zhu J, Lynch JP (2004) The contribution of lateral rooting to phosphorus acquisition efficiency in maize (*Zea mays*) seedlings. Funct Plant Biol 31:949–958

Zimmermann P, Regierer B, Kossmann J, Frossard E, Amrhein N, Bucher M (2004) Differential synthesis of three purple acid phosphatases from potato. Plant Biol 6:519–528

Potassium

Philip J. White and Alison J. Karley

Abstract Potassium (K) is the most abundant inorganic cation in plants. It is required for the activation of many enzymes, as a cellular osmoticum for rapidly expanding cells, and as a counter cation for anion accumulation and electrogenic transport processes. This chapter describes (1) the symptoms of potassium deficiency and the acclimatory responses of plants to potassium starvation, (2) the mechanisms by which roots acquire K^+ from the soil and K^+ is transported between tissues, and (3) the molecular biology of the transport proteins that catalyse K^+ influx and efflux across the plasma membrane and tonoplast of plant cells to effect K^+ uptake and redistribution within the plant, cell expansion and shrinking, and cytoplasmic K^+ homeostasis.

1 Potassium is an Essential Mineral Element

1.1 Physiological Functions of Potassium

Potassium (K) is the most abundant inorganic cation in plants, comprising up to 10% of a plant's dry weight (Broadley et al. 2004; Watanabe et al. 2007). It is concentrated in growing tissues and reproductive organs, reflecting the vital functions of K^+ in cell metabolism and extension growth. Potassium is required for the activation of many enzymes including those of energy metabolism, protein synthesis and solute transport (Leigh and Wyn Jones 1984; Mengel et al. 2001; Amtmann et al. 2008; Britto and Kronzucker 2008). For optimal performance, K^+ concentrations in metabolically active compartments, such as the cytosol, the nucleus, the stroma of chloroplasts and the matrix of mitochondria, must be maintained at about

P.J. White (✉) and A.J. Karley
Scottish Crop Research Institute, Invergowrie, Dundee DD2 5DA, UK
e-mail: philip.white@scri.ac.uk; alison.karley@scri.ac.uk

R. Hell and R.-R. Mendel (eds.), *Cell Biology of Metals and Nutrients*, Plant Cell Monographs 17,
DOI 10.1007/978-3-642-10613-2_9,

100 to 150 mM (Fig. 1; Leigh and Wyn Jones 1984). Potassium is also required as a counter cation for the neutralisation of fixed negative charges, for the maintenance of trans-membrane voltage gradients, for cytoplasmic pH homeostasis, and for the transport of inorganic anions and metabolites both within and outside the cell (Leigh and Wyn Jones 1984; Mengel et al. 2001; Britto and Kronzucker 2008).

The uptake of K^+ by plant cells, and its accumulation in vacuoles, is the primary driver for their osmotic expansion (Mengel et al. 2001). Rapid cell expansion relies on high mobility of the active osmoticum and, for this reason, only a few other inorganic ions can replace K^+ in this role (Amtmann et al. 2006). However, once cell expansion is over, K^+ can be removed from the vacuole and turgor maintained by less mobile osmotica, such as sugars, organic acids and compatible solutes (Amtmann et al. 2006). The lowest limit for vacuolar K^+ concentration appears to be 10–20 mM, which is thought to reflect a maximum trans-tonoplast voltage of about −40 to −60 mV (Fig. 1; Leigh and Wyn Jones 1984).

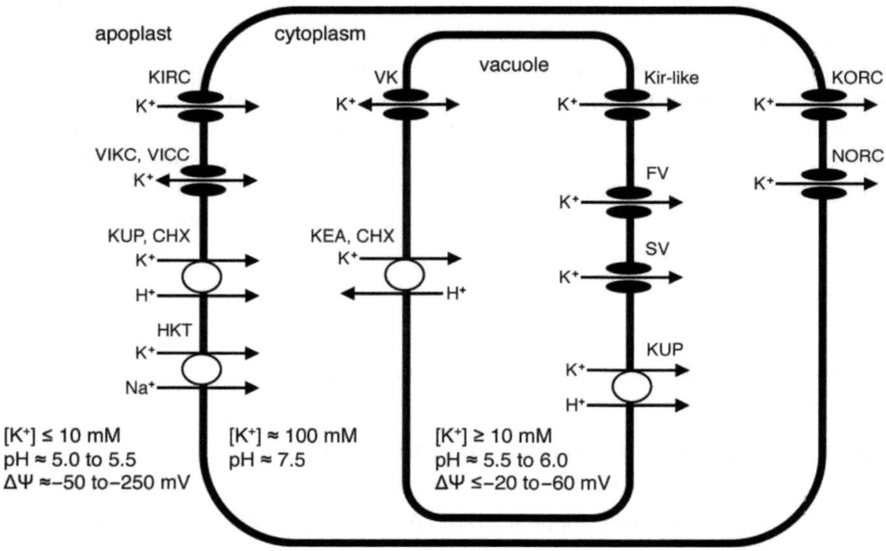

Fig. 1 Electrochemical gradients and subcellular locations of K transporters in a stereotypical cell of *Arabidopsis thaliana*. In the plasma membrane there are inward-rectified K channels (KIRCs) encoded by *AtAKT1*, *AtAKT5*, *AtSPIK*, *AtKAT1*, *AtKAT2*, and *AtKC1*, outward-rectified K channels (KORCs) encoded by *AtSKOR* and *AtGORK*, voltage-independent K^+-channels (VIKCs) encoded by *AtAKT2/3* and *AtTPK4*, voltage-independent cation channels (VICCs) encoded by members of the *AtCNGC* and *AtGLR* gene families, H^+/K^+-symporters (KUPs) encoded by members of the *AtKT/HAK/KUP* gene family, H^+/cation-symporters (CHX), such as *AtCHX13*, and the Na^+/K^+-symporter *AtHKT1*. In the tonoplast there are fast-activating vacuolar (FV) channels, whose genetic identity is unknown, slowly-activating vacuolar (SV) channels, encoded by *AtTPC1*, vacuolar K^+ (VK) channels, encoded by members of the *AtTPK* gene family, Kir-like channels, encoded by *AtKCO3*, H^+/K^+-syporters (KUPs) encoded by members of the *AtKT/HAK/KUP* gene family, and H^+/cation-antiporters (KEA, CHX) encoded by members of the *AtKEA*, *AtCHX* and *AtNHX* gene families

The accumulation of K^+ is essential for the growth of the root system, both for cell expansion in the elongation zone and for the elongation of root hair cells (Dolan and Davies 2004). It is also required for leaf expansion, for the elongation of pollen tubes towards fertile ovules (Mouline et al. 2002) and for the enlargement of fruits and tubers. The rapid accumulation and loss of K^+ by guard cells controls the opening and closing of stomata and, thereby, gas exchange and transpiration (Amtmann and Blatt 2009). The redistribution of K^+ between cells within tissues underpins the bending of roots and coleoptiles in response to gravity (Philippar et al. 1999), leaf movements in sensitive plants in response to shaking and touch, the closing of traps in carnivorous plants, and the diurnal and circadian movements of leaves in response to light signals or an endogenous "clock" (Moran 2007). These cellular phenomena are often attributed to plasma-membrane hyperpolarisation, apoplastic acidification, and an increase in the number and/or activity of transport proteins catalysing K^+ influx across the plant membrane and sequestration in the vacuole. Expanding cells are often characterised by high cytosolic Ca^{2+} concentrations ($[Ca^{2+}]_{cyt}$), and the direction of elongation growth is determined by an elevated $[Ca^{2+}]_{cyt}$ at the apex of the growing cell (White and Broadley 2003; Dolan and Davies 2004; Cheung and Wu 2008; Frietsch et al. 2008). The rapid loss of K^+ from guard cells during stomatal closure (Amtmann and Blatt 2009) and from the shrinking pulvinor cells during leaf movements (Moran 2007), is effected by plasma membrane depolarisation and the opening of K^+ channels that facilitate K^+ efflux from the vacuole and across the plasma membrane. The latter responses appear to be controlled by strictly coordinated temporal changes in both cytosolic pH and Ca^{2+} concentrations through cascades of protein phosphorylation (White 2000; Moran 2007; Amtmann and Blatt 2009).

1.2 Symptoms of Potassium Deficiency

The response of plant growth to increasing K^+ availability follows a hyperbolic relationship and plants can acquire sufficient K^+ for growth from solutions containing micromolar K^+ concentrations, provided the K^+ supply to the roots matches the minimal demand of the plant and NH_4^+ is absent from the rhizosphere (e.g. Asher and Ozanne 1967; Wild et al. 1974; Spear et al. 1978a; Siddiqi and Glass 1983a; White 1993, 1997b). The 'critical' tissue K^+ concentration, at which growth and development attain 90% of their maxima, approximates 5–20 μmol g^{-1} FW (Leigh and Wyn Jones 1984). Shoot tissues reach their critical tissue K^+ concentrations at a lower K^+ supply than root tissues (Fig. 2a, Asher and Ozanne 1967; Spear et al. 1978a; White 1993, 1997b). Tissue concentrations of readily-available cations, such as Na^+, Ca^{2+} and Mg^{2+}, are generally higher when K^+ is in short supply, and lower critical tissue K^+ concentrations are reported when other cations that can be used as cellular osmotica are available to the plant (Johnson 1973; Leigh and Wyn Jones 1984; Barraclough and Leigh 1993). When K^+ is readily available, tissue K^+

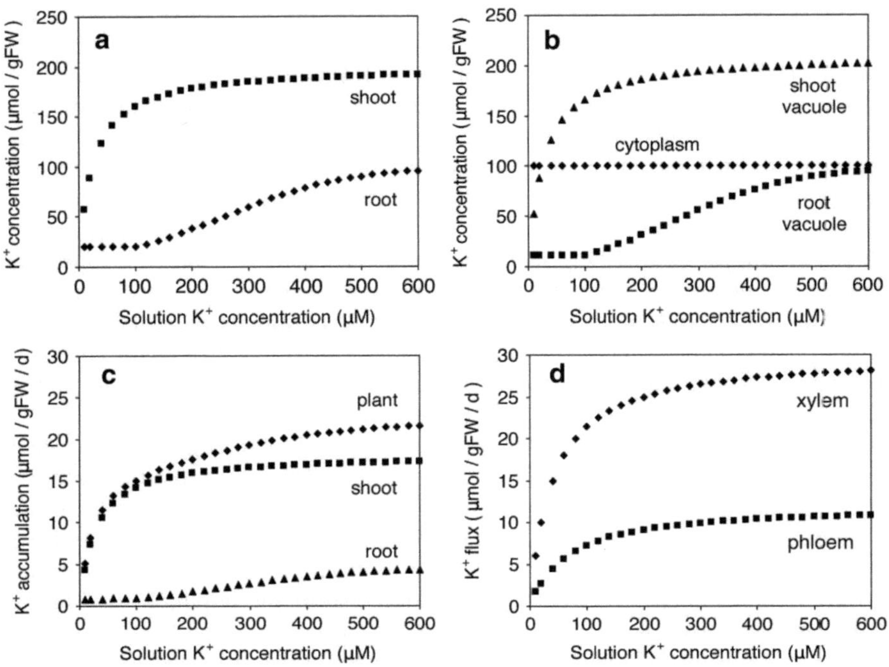

Fig. 2 Potassium nutrition of 14-day old rye plants growing hydroponically. (**a**) Relationships between solution K^+ concentration ($[K^+]_{ext}$) and root (*diamonds*) and shoot (*squares*) K^+ concentrations. (**b**) Predicted relationships between $[K^+]_{ext}$ and vacuolar K^+ concentrations in roots (*squares*) and shoots (*triangles*). These calculations assumed a cytoplasmic volume of 10% containing 100 mM K^+ and no apoplastic contribution (Leigh and Wyn Jones 1984). (**c**) Relationships between $[K^+]_{ext}$ and K^+ uptake (*diamonds*), root K^+ accumulation (*triangles*) and shoot K^+ accumulation (*squares*). (**d**) Relationships between $[K^+]_{ext}$ and K^+ fluxes in the xylem (*diamonds*) and phloem (*squares*). All relationships based on data from White (1993) and White (1997b)

concentrations often exceed those required for maximal growth (Leigh and Wyn Jones 1984).

Plants experiencing mild K-deficiency rarely show overt visible symptoms (Mengel et al. 2001; Fageria 2009). One reason for this is that K^+ is readily redistributed within the plant via the phloem from mature to developing tissues. Plants experiencing more severe K-deficiency exhibit symptoms consistent with the vital functions of this element (Johnson 1973; Bould et al. 1983; Amtmann et al. 2008). They exhibit a scorching along the margins of older leaves. They grow slowly, have short stature and poorly-developed root systems, and are more susceptible to frost damage, pests and diseases. Both leaves and roots of K-deficient plants are short-lived. Stems are weak, and seed and fruit are small and shrivelled. The physiological symptoms of K-deficiency include impaired phloem transport, particularly of sucrose, increased leaf carbohydrate concentrations, a reduction in chlorophyll concentrations and photosynthetic capacity, decreased water content, decreased turgor, impaired stomatal regulation and reduced transpiration (Cakmak et al. 1994; Mengel et al. 2001; Hermans et al. 2006; Amtmann et al. 2008).

The decline in photosynthesis observed in K-deficient plants appears to be a consequence of sucrose accumulation in leaves and its effects on gene expression (Hermans et al. 2006). Leaves of K-deficient plants accumulate sugars, including sucrose, but rarely starch (Bould et al. 1983; Hermans et al. 2006; Amtmann et al. 2008). Although these solutes can replace K^+ as a cellular osmoticum, this phenomenon is likely to be a consequence of impaired sucrose export from leaves of K-deficient plants, which can be attributed to a requirement for K^+ for loading sucrose into the phloem (Mengel et al. 2001; Deeken et al. 2002; Hermans et al. 2006). An inverse relationship between phloem sucrose concentration and plant K status has been observed across a wide range of nutritional treatments. In addition, concentrations of amide nitrogen, amino acids (lysine, arginine and tyrosine), and polyamines, such as putrescine and agmatine, often increase dramatically in K-deficient plants, and have been used to diagnose K-deficiency in crop plants (Bould et al. 1983).

1.3 Acclimatory Responses to Potassium Starvation

Plants have evolved various morphological and physiological adaptations to acquire K^+ and cope with low tissue K^+ concentrations when this element is in short supply. In contrast to N and P deficiencies, K-deficiency does not generally result in greater biomass partitioning to roots or in major alterations to root architecture (Hermans et al. 2006). However, the expression of genes encoding high-affinity K^+-influx systems increases when plants lack K^+ (e.g. Wang et al. 1998; Shin and Schachtman 2004; Gierth et al. 2005; Hampton et al. 2005; Qi et al. 2008). Increased K^+ uptake is likely to reduce rhizosphere K^+ concentrations and accelerate K^+ diffusion to the root surface and desorption from "exchangeable" binding sites in the soil. Physiological adaptations to low K supply include the replacement of vacuolar K with alternative osmotica, redistribution of K^+ from mature to developing tissues, and reducing plant growth to maintain appropriate tissue K^+ concentrations for cell function.

Amtmann et al. (2006) suggested that fluctuations in apoplastic K^+ concentration and the membrane potential of root cells are most likely to initiate immediate acclimatory responses to reduced K^+ phytoavailability, since cytoplasmic K concentrations ($[K^+]_{cyt}$) are relatively unaffected by K^+ supply (Walker et al. 1996, 1998), and that K^+ channels are the immediate targets for regulating K^+ fluxes (Fig. 3a). The voltage-dependence of both inward rectifying K^+ channels (KIRCs) and outward-rectifying K^+ channels (KORCs) in the plasma membrane of root cells respond to the K^+ gradient between apoplast and cytoplasm, such that they mediate only K^+ influx and K^+ efflux, respectively (White 1997a; Amtmann et al. 2006; Amtmann and Blatt 2009). The opening of these channels determines the direction of K^+ uptake across the plasma membrane, which can be promoted if an appropriate hyperpolarisation of the plasma membrane can be maintained by the activity of the

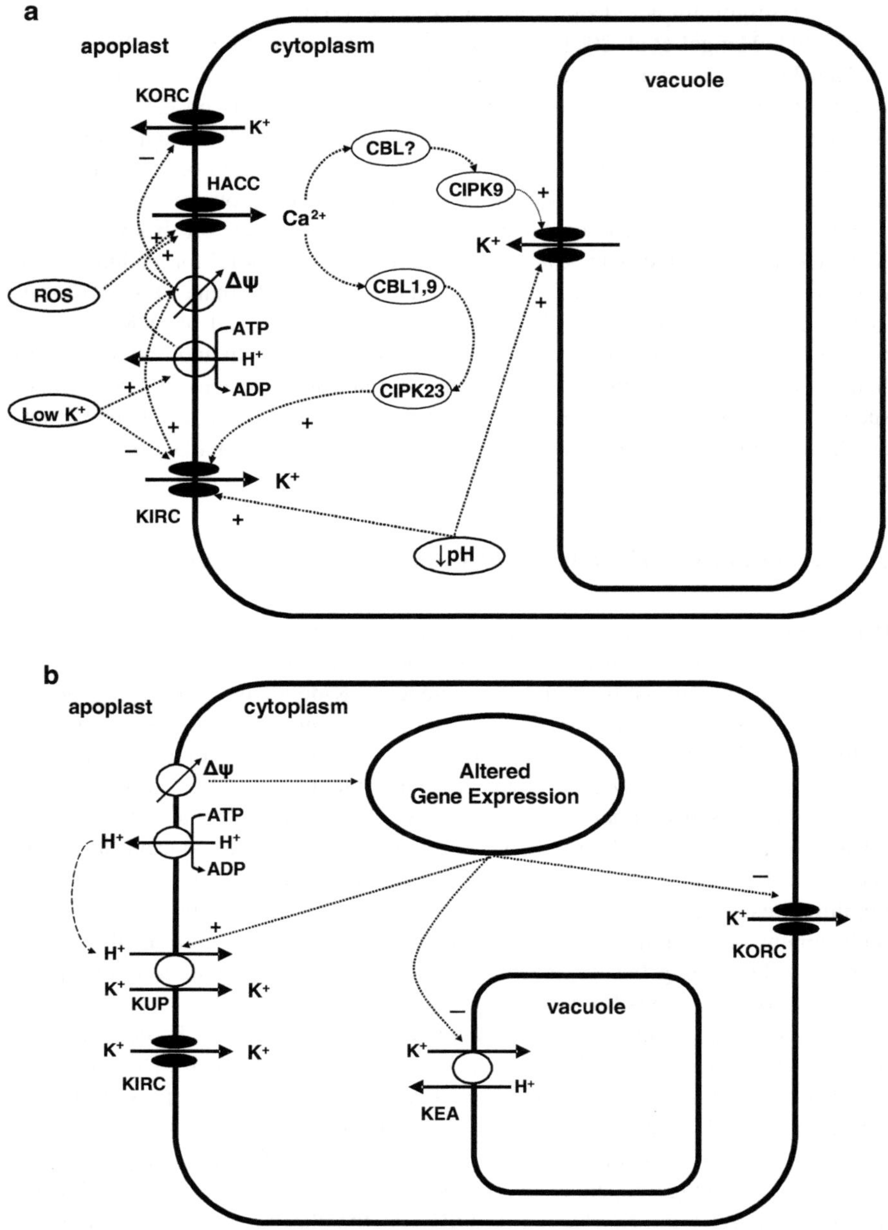

Fig. 3 Immediate acclimatory responses in root cells to reduced K^+ supply. (**a**) *Initial events.* Low external K^+ causes hyperpolarisation of the plasma membrane through reduced K^+ influx and increased activity of the plasma membrane H^+-ATPase. Hyperpolarisation of the plasma membrane results in a greater inward K^+ electrochemical gradient, reduced opening of outward-rectified K^+ channels (KORCs), and increased opening of inward-rectified K^+ channels (KIRCs).

plasma membrane H^+-ATPase (Amtmann et al. 2006; Amtmann and Blatt 2009). The activities of KIRCs and KORCs are regulated by apoplastic and cytosolic pH, and by $[Ca^{2+}]_{cyt}$, both directly and indirectly through posttranslational modification (Amtmann et al. 2006; Amtmann and Blatt 2009). In *Arabidopsis thaliana*, coupling between a calcineurin B-like protein (CBL)-interacting protein kinase (CIPK23) and two upstream Ca^{2+}-binding proteins (CBL1 and CBL9) in tandem with a 2C-type protein phosphatase regulates the activity of AtAKT1, the major K^+ channel involved in nutritional K^+ uptake by roots (Amtmann et al. 2006; Amtmann and Blatt 2009). It is thought that low rhizosphere K^+ concentrations cause plasma membrane hyperpolarisation and initiate the production of ROS through the activity of the NADPH oxidase, AtrbohC. These events increase Ca^{2+} influx through hyperpolarisation-activated Ca^{2+} channels in the plasma membrane and $[Ca^{2+}]_{cyt}$, which results in the phosphorylation and opening of AtAKT1 through the CBL/CIPK23 cascade. In parallel, a CBL/CIPK9 cascade initiated by increased $[Ca^{2+}]_{cyt}$ and subject to transcriptional regulation is thought to open K^+ channels in the tonoplast. A decrease in cytosolic pH, which is associated with decreasing cytosolic K^+ concentrations in particular root cells (Walker et al. 1996, 1998), would also increase K^+ influx through KIRCs and K^+ release from the vacuole. The magnitude and direction of K^+ fluxes across both plasma membrane and tonoplast can also be influenced by heteromerisation of subunits, and interactions with beta-subunits, G-proteins, and 14-3-3 proteins (Zimmermann and Chérel 2005; Gambale and Uozumi 2006; Amtmann et al. 2006; Lebaudy et al. 2007, 2008; Amtmann and Blatt 2009).

In addition to the immediate effects of hyperpolarisation of root cells on the activity of plasma membrane KIRCs and KORCs, it has recently been observed that the expression of genes encoding high-affinity H^+/K^+ symporters, such as *AtHAK5* in *A. thaliana*, is increased in response to prolonged hyperpolarisation of root cells (Fig. 3b; Nieves-Cordones et al. 2008). This serves to increase K^+ uptake by roots over a period of hours to days. The initial signal transduction pathway effecting this

Fig. 3 (continued) Thus, K^+ influx is accelerated. Increased opening of KIRCs is effected both by the more negative membrane potential and by protein phosphorylation coordinated through a cytoplasmic signal transduction cascade involving CBL1, CBL9, and CIPK23, which is initiated by Ca^{2+} influx through hyperpolarisation-activated calcium channels (HACCs). Increased opening of HACCs is promoted by the production of reactive oxygen species (ROS) through the activity of a NADPH oxidase, AtrbohC. Increased cytosolic Ca^{2+} also results in increased opening of vacuolar K^+ channels through a cytoplasmic signal transduction cascade involving CBLs and CIPK9. A decrease in cytosolic pH, which is associated with decreasing cytosolic K^+ concentrations in particular root cells, increases the opening of KIRCs and vacuolar K^+ channels. (**b**) *Altered gene expression*. Prolonged hyperpolarisation of the plasma membrane increases the expression of genes encoding high-affinity H^+/K^+ symporters, such as AtHAK5 in *Arabidopsis thaliana*, thereby increasing root K^+ influx capacity. Also in *A. thaliana*, the expression of genes encoding a putative vacuolar K^+/H^+ antiporter (AtKEA5), the KORC AtSKOR, which is thought to load K^+ into the xylem, and AtAKT2, a KIRC responsible for the recirculation of K^+ from the shoot to the root, are all reduced by low K^+ phytoavailability

could also include $[Ca^{2+}]_{cyt}$ signals. In *A. thaliana*, the expression of genes encoding a putative vacuolar K^+/H^+ antiporter (AtKEA5), the KORC AtSKOR, which is thought to load K^+ into the xylem, and AtAKT2, a KIRC responsible for the recirculation of K^+ from the shoot to the root, are all reduced by low K^+ phytoavailability (Zimmermann and Chérel 2005; Schachtman and Shin 2007). These are thought to maintain the $[K^+]_{cyt}$ of root cells and to restrict long-distance K^+ transport, which might act as a signal of plant K^+ status (Drew et al. 1990; White 1997b; Amtmann et al. 2006; Schachtman and Shin 2007).

Subsequent morphological and physiological adaptations in response to prolonged K^+ starvation could be triggered by an increase in ethylene production, which, together with an increase in ROS, stimulates the initiation and elongation of root hairs (Dolan and Davies 2004, White et al. 2005), and jasmonic acid biosynthesis, which is thought to be responsible for the characteristic accumulation of polyamines observed in K^+-deficient plants and for an increased systemic resistance to particular pests and pathogens (Amtmann et al. 2008). Other K-deficiency symptoms appear to arise as secondary consequences of impaired energy metabolism, redistribution of solutes within the plant and/or reduced growth.

2 The Acquisition and Cellular Distribution of Potassium

2.1 Potassium Acquisition by Plant Roots

The solution K^+ concentration in most soils lies between 0.1 and 1 mM. This represents only 0.1%–0.2% of the total soil K, of which 1%–2% is “exchangeable” K^+, 1%–10% is “non-exchangeable” K^+ associated with clay lattices, and 90%–98% is present as K-minerals (Mengel et al. 2001; Rengel and Damon 2008; Fageria 2009). A large fraction of the total soil K available to plants resides in the topsoil. Diffusion through and mass flow of the soil solution contribute most to the delivery of K^+ to the root surface (Jungk and Claassen 1997). In circumstances when the roots' capacity for K^+ uptake exceeds the rate at which K^+ can be delivered to the rhizosphere, K^+ acquisition by plants is determined by the K^+ concentration gradient between the rhizosphere and the soil solution and by the flow of water to the root. Thus, factors influencing K^+ acquisition by plants include (1) the rate of K^+ uptake across the plasma membrane of root cells, which reduces the K^+ concentration in the rhizosphere solution, (2) the release of non-exchangeable K^+ by root exudates, which increases K^+ concentration and availability in the soil solution, (3) the proliferation of roots into the soil volume, which increases the area for K^+ uptake and also reduces the distance required for K^+ diffusion and water flow, and (4) the transpiration rate of the plant, which drives mass flow of the soil solution to the root (Jungk and Claassen 1997; Rengel and Damon 2008).

Potassium uptake by roots and accumulation by plants are determined by the K^+ uptake capacity of the roots, the K^+ concentration at the root surface, and the replenishment of rhizosphere K^+. The relationship between root K^+ uptake (plant accumulation) and K^+ concentration in the rhizosphere generally follows the sum of a hyperbolic and a linear function (Fig. 2c; Kochian and Lucas 1988; Leigh and Wyn Jones 1984; Mengel et al. 2001; Britto and Kronzucker 2008). When grown under identical conditions, plant species differ in (a) the relationship between K^+ influx (and accumulation) and rhizosphere K^+ concentration (e.g. Asher and Ozanne 1967; Wild et al. 1974; Spear et al. 1978a, b; Pettersson and Jensén 1983; Steingrobe and Claassen 2000; Jungk 2001; El Dessougi et al. 2002) and (b) the selectivity of monovalent cation accumulation (e.g. Broadley et al. 2004; White et al. 2004). These observations suggest that the complement of proteins catalysing K^+ uptake by root cells differs between plant species. Similarly, genotypes of crop species differ in the relationship between rhizosphere K^+ concentration and K^+ uptake by roots (e.g. Glass and Perley 1980; Siddiqi and Glass 1983a, b; Siddiqi et al. 1987; Chen and Gabelman 1995, 2000; Trehan 2005; Zhang et al. 2007; Rengel and Damon 2008).

2.2 *Thermodynamic Consideration of K^+ Uptake and Distribution in Root Cells*

The hypothesis that K^+ influx to root cells is mediated by distinct "high-affinity" and "low-affinity" transporters, operating at low (<1 mM) and high (>1 mM) rhizosphere K^+ concentrations, respectively, has been modified little since its conception over 50 years ago (Epstein and Bloom 2005; Britto and Kronzucker 2008). However, it is not the "affinity" for K^+ that differentiates K^+ transport mechanisms in the root plasma membrane, but their coupling to pH and voltage gradients (Fig. 1, Gierth and Mäser 2007; Britto and Kronzucker 2008; Karley and White 2009). Electrophysiological studies indicate that K^+ influx across the plasma membrane of root cells occurs against its electrochemical gradient at rhizosphere concentrations less than about 1 mM K^+ (Maathuis and Sanders 1993; Walker et al. 1996). This can be catalysed by H^+/K^+ symporters in the plasma membrane, energized by the pH and voltage gradients generated at the plasma membrane H^+-ATPase, which are capable of accumulating K^+ from rhizosphere solutions containing less than 100 nM K^+ (Fig. 1). At rhizosphere K^+ concentrations above 1 mM, which are common in well-fertilised agricultural soils, K^+ influx to root cells can be energised by the voltage gradient alone and facilitated by K^+ channels. In roots of K-starved plants, K^+ appears to be close to thermodynamic equilibrium across the tonoplast, suggesting that K^+ channels dominate K^+ fluxes across this membrane under these conditions (Maathuis and Sanders 1993; Walker et al. 1996). In K-replete plants, however, which have substantially higher vacuolar K^+ concentrations than K-starved plants, K^+ must be actively transported into the vacuoles of root cells. This is thought to be catalysed by K^+/H^+-antiporters energised by the H^+ gradient generated by the vacuolar H^+-ATPase and/or H^+-PPiase. Potassium efflux from the vacuole can be mediated by K^+-channels in root cells of both K-starved and K-replete plants.

2.3 Cellular K^+ Homeostasis

Cytosolic K^+ concentrations around 100 mM are generally maintained in plant cells to ensure optimal function (Jeschke 1984; Leigh and Wyn Jones 1984, 1986; Memon et al. 1985; Drew et al. 1990; White et al. 1991; Britto and Kronzucker 2008). This is effected by the redistribution of K^+ between vacuolar and cytosolic compartments within the cell and by the redistribution of K^+ from mature and/or senescing tissues to developing tissues within the plant.

In K^+ replete plants, $[K^+]_{cyt}$ is similar in both root and leaf cells, but vacuolar K^+ concentrations in leaf cells are about double those in root cells (Fig. 2b; Cuin et al. 2003). When plants lack K^+, vacuolar K^+ is redistributed to the cytoplasm (Memon et al. 1985; Huang and van Steveninck 1989; Walker et al. 1996, 1998; Cuin et al. 2003). Root tissues require a higher K^+ supply than shoot tissues to achieve their critical K^+ concentration (Fig. 2a), which suggests that root cells might be able to tolerate lower vacuolar K^+ concentrations than shoot cells (White 1993, 1997b). Differences between barley varieties in their ability to mobilise K^+ from the vacuole to the cytoplasm of root cells at low K^+ supply appear to correlate with their sensitivity of growth to K-starvation (Memon et al. 1985). In addition, different cell types within the root and shoot display distinct responses to K-starvation in the redistribution of K^+ between vacuolar and cytosolic compartments. In barley roots, the $[K^+]_{cyt}$ of epidermal cells declines when the vacuolar K^+ activity falls below about 25 mM, but the $[K^+]_{cyt}$ of cortical cells remains constant irrespective of K^+ status (Walker et al. 1996, 1998). The $[K^+]_{cyt}$ in expanding cells of the seminal and nodal roots of barley also declines during K-starvation (Walker et al. 1998), despite higher vacuolar K^+ concentrations being present in cells closer to the root meristem during K^+ starvation, presumably to drive cell expansion and to buffer essential meristematic activities against the vagaries of K^+ supply (Huang and van Steveninck 1989). By contrast, in barley plants under salt stress, $[K^+]_{cyt}$ in leaf epidermal cells can be as low as 15 mM, despite vacuolar K^+ activities of 50 mM, whereas $[K^+]_{cyt}$ in leaf mesophyll cells is maintained at approximately 70 mM, presumably to minimise any detrimental effects on photosynthesis (Cuin et al. 2003). Cellular K^+ concentrations in leaf mesophyll cells also exceed those in epidermal cells in K-starved plants (James et al. 2006).

3 Potassium Transport Within the Plant

At submillimolar rhizosphere K^+ concentrations, K^+ influx to root cells appears to be catalysed by H^+/K^+-symporters, whereas at rhizosphere K^+ concentrations greater than about 1 mM, K^+ influx can be mediated by K^+ channels (Sect. 2.2).

It is noteworthy, however, that unidirectional K^+ influx and K^+ efflux across the plasma membrane of root cells are far greater than the rate of K^+ uptake (accumulation) by the plant (Jeschke 1983; White et al. 1991; Britto and Kronzucker 2008). This is thought to reflect (1) the role of K^+ in charge-balancing fluxes of other ions important for plant nutrition and/or cell signalling and (2) an absolute requirement for $[K^+]_{cyt}$ homeostasis. Rapid K^+ efflux from root cells is effected by depolarisation of the plasma membrane and is mediated by the opening of KORCs (White 1997a; Moran 2007; Amtmann and Blatt 2009). Voltage-insensitive cation channels (VICCs) are also present in the plasma membrane of root cells (White 1997a; Hampton et al. 2005; Demidchik and Maathuis 2007). These channels, which can catalyse both K^+ influx and K^+ efflux from root cells, do not appear to contribute to nutritional K^+ uptake, but are thought to balance electrically other transport processes and, since they also catalyse Ca^{2+} influx, to contribute to cytosolic Ca^{2+} homeostasis and signalling (White and Broadley 2003; Hampton et al. 2005; Demidchik and Maathuis 2007).

The capacity for influx of K^+ and Rb^+, which is often used as a tracer for K^+, to roots increase dramatically with decreasing root K^+ concentration in K-starved plants (e.g. Glass 1976; Pettersson and Jensén 1979, 1983; Wrona and Epstein 1985; White et al. 1987; White 1997b; Shin and Schachtman 2004). Recently, this has been attributed to increased expression of genes encoding high-affinity H^+/K^+ symporters, such as *AtHAK5* in *A. thaliana* and its homologs in tomato (*LeHAK5*), pepper (*CaHAK1*), barley (*HvHAK1*), rice (*OsHAK1*), and other plant species (Hampton et al. 2005; Gierth and Mäser 2007; Nieves-Cordones et al. 2008; Qi et al. 2008). There is some evidence that these transcriptional responses to K-starvation are initiated by the prolonged hyperpolarisation of root cells when rhizosphere K^+ concentrations are low (Nieves-Cordones et al. 2008). However, older experiments on plants whose root systems were divided between solutions with high and low K^+ concentrations demonstrated that roots supplied with high K^+-concentrations have enhanced K^+ uptake in K-starved plants, suggesting that, in addition to cell membrane potential, plant K-status controls K^+ uptake through a systemic signal (Drew et al. 1990). Phloem K^+ concentration and/or K^+ flux have been postulated to be this signal.

Following uptake by epidermal and cortical cells, K^+ is transported symplastically across the root through plasmodesmata to the stelar parenchyma cells, where it is loaded into the xylem (Kochian and Lucas 1988). All regions of the root contribute to loading the xylem with K^+, and it can be estimated that over 90% of the K^+ entering the xylem is delivered through a symplastic route (P.J. White, unpublished calculations). This is consistent with the recent observation that increased suberisation of the root endodermis in the *enhanced suberin 1* (*esb1*) mutant of *A. thaliana* has little effect on shoot K concentration (Baxter et al. 2009). The voltage across the symplast/xylem boundary is about −80 mV (De Boer and Volkov 2003), which allows KORCs to load the xylem with K^+ concentrations up to about 4 mM. Loading the xylem with higher K^+ concentrations through KORCs requires a substantial depolarisation of the stelar parenchyma cells. The xylem K^+ concentration ranges from about 2 to 25 mM, depending upon a variety of factors

(Marschner et al. 1997). Xylem K^+ concentration increases with increasing K^+ concentration in the rhizosphere (Fig. 2d; e.g. Armstrong and Kirby 1979; White 1997b; Peuke et al. 2002). The presence of high concentrations of Na^+, Ca^{2+}, or NH_4^+ in the rhizosphere reduces K^+ uptake and xylem K^+ concentrations (e.g. Munns 1985; Jeschke et al. 1992; Lu et al. 2005). Xylem sap K^+ concentrations are also reduced in P-starved plants (Jeschke et al. 1997). Potassium uptake, xylem K^+ concentration, and K^+ flux to the shoot are all affected by transpiration and also exhibit diurnal cycles driven by illumination (e.g. Armstrong and Kirby 1979; Jeschke 1984; Schurr and Schulze 1995; Macduff and Dhanoa 1996; Macduff et al. 1997; Herdel et al. 2001; Peuke et al. 2001; Malone et al. 2002; Siebrecht et al. 2003; Goodger et al. 2005). Although greater water flow through the xylem reduces sap K^+ concentration, it generally increases K^+ flux to the shoot and K^+ uptake by roots. Both xylem K^+ concentrations and K^+ fluxes to the shoot are reduced in plants during drought, which is thought to be a consequence of (1) reduced expression of genes encoding K^+ channels that load K^+ into the xylem, such as *AtSKOR* in *A. thaliana*, and (2) reduced transpirational water losses through stomatal closure (Gaymard et al. 1998; De Boer and Volkov 2003; Goodger et al. 2005).

The delivery of K^+ within the shoot via the xylem is largely determined by transpirational water flows. The larger vessels are designed for the rapid onward movement of sap, whereas the smaller vessels are important for solute transfer between the xylem and the surrounding tissues. The apoplastic K^+ concentration at the point of xylem unloading approximates 5–20 mM, which allows K^+ to enter the shoot symplast through KIRCs and VICCs in the plasma membrane of the bundle sheath cells of the smaller veins (Keunecke et al. 2001). With the obvious exception of guard cells, which adjust their K^+ concentration to regulate stomatal aperture, and their neighbouring epidermal cells, most cells in leaves of K^+-replete plants appear to have similar K^+ concentrations (Leigh and Storey 1993; Fricke et al. 1994). Potassium is redistributed from mature leaves to developing tissues via the phloem. The K^+ concentration in phloem sap ranges from about 10 to 150 mM, depending upon a variety of environmental factors including K availability (Marschner et al. 1997). The resting potential of the sieve element plasma membrane lies between -150 mV and -50 mV, depending upon plant species, and contains a weakly inwardly-rectifying KIRC with electrophysiological properties resembling AtAKT2/3 of *A. thaliana*, which facilitates K^+ influx to the phloem (Deeken et al. 2002; Hafke et al. 2007). Since phloem K^+ concentrations are lower in plants with lower K-status (Fig. 2d; Mengel and Haeder 1977; Drew et al. 1990; Cakmak et al. 1994; Peuke et al. 2002; Gould et al. 2004), it is thought that phloem K^+ concentration and/or the K^+ flux from the shoot to the root might regulate K^+ uptake by roots in response to shoot K status (Drew et al. 1990; White 1997b; Amtmann et al. 2006). To effect charge balance, concentrations of other cations, such as Na^+, are often increased in the phloem sap of K-starved plants (e.g. Peuke et al. 2002).

In addition to being a putative signal for plant K status, K^+ recirculation within the plant via the phloem serves a number of other functions, such as (1) maintaining cation-anion balance within the plant, especially when nitrate assimilation occurs in the shoot, (2) enabling the loading of sugars, organic acids, and amino acids into the phloem, (3) contributing to the driving force for mass flow of solution, (4) redistributing K^+ from senescing to developing tissues, (5) meeting the K demand of elongating cells in plants subject to variable rhizosphere K availability, and (6) maintaining high K/Na quotients in sensitive meristematic tissues (Marschner et al. 1997). It has been estimated that up to 90% of the K^+ delivered to the shoot via the xylem is exported back to the root via the phloem (Armstrong and Kirby 1979; Jeschke and Pate 1991; Jeschke et al. 1992, 1995, 1997; Peuke and Jeschke 1993; Marschner et al. 1997; White 1997b; Peuke et al. 2002; Lu et al. 2005).

4 The Molecular Biology of K^+ Transporters

Potassium influx to plant cells can be mediated by cation/H^+ symporters, K^+/Na^+ symporters, and/or K^+-permeable cation channels, such as KIRCs, voltage-independent K^+ channels (VIKCs), and VICCs, depending upon the K^+ electrochemical gradient (Fig. 1, Sect. 2.2). Potassium efflux from plant cells occurs through KORCs and non-specific outward-rectifying cation channels (NORCs). Potassium sequestration in vacuoles can be mediated by cation/H^+ antiporters and/or vacuolar K^+ (VK) channels, while K^+ efflux from vacuoles occurs through fast-activating vacuolar (FV) channels, slowly-activating vacuolar (SV) channels, VK channels, and/or cation/H^+ symporters. Since orthologues of *A. thaliana* genes encoding all these K^+ transporters have been found in every angiosperm species studied to date, this section will summarise the molecular biology of K^+ transporters in plant cells with specific reference to *A. thaliana* (Zimmermann and Chérel 2005; Gambale and Uozumi 2006; Lebaudy et al. 2007; Gupta et al. 2008).

Members of two gene families, the K^+ uptake permeases (*KT/HAK/KUP*) and the cation-H^+ exchangers (*CHXs*) encode plasma membrane K^+/H^+ symporters (Table 1; Gierth and Mäser 2007; Britto and Kronzucker 2008; Zhao et al. 2008). In *A. thaliana*, AtHAK5, AtKUP1, AtKUP2, AtKUP11 and AtCHX13 have been found in plasma membranes of various cell types and are thought to catalyse K^+ influx to cells at low apoplastic K^+ concentrations (Gierth and Mäser 2007; Qi et al. 2008; Rubio et al. 2008; Zhao et al. 2008). Most genes encoding AtKUPs are expressed in roots, with *AtHAK5* and, occasionally, *AtKUP3* being induced by K-starvation (Shin and Schachtman 2004; Hampton et al. 2005; Gierth et al. 2005; Zimmermann and Chérel 2005; Qi et al. 2008; Rubio et al. 2008). The expression of *AtCHX13* is also increased in roots of K-starved plants (Zhao et al. 2008). It is thought that AtHAK5 dominates nutritional K^+ influx to roots of K-starved *A. thaliana* plants (Gierth et al. 2005; Gierth and Mäser 2007; Qi et al. 2008; Rubio et al. 2008). The KUPs are characteristically inhibited, and transcription of their genes reduced, by NH_4^+, which can serve as a useful pharmacological tool to dissect the physiological roles of these transporters (Bañuelos et al. 2002; Martínez-Cordero et al. 2005; Fulgenzi et al. 2008; Nieves-Cordones et al. 2008;

Table 1 Transport proteins implicated in the uptake and distribution of K^+ between plant organs, cell types, and sub-cellular compartments. See text for details

Gene family	Member	Locus	Membrane	Organ	Proposed mechanism	Putative function (s)
KT/HAK/KUP	AtKUP1	At2g30070	PM	R,St,L,F,Si	K^+/H^+ symport	K^+ influx
	AtKUP2	At2g40540	PM	R,St,L,F,Si	K^+/H^+ symport	K^+ influx
	AtKUP3	At3g02050		R,L,F,Si		
	AtKUP4	At4g23640	TP	R,L,F,Si	K^+/H^+ symport	
	AtKUP5	At4g33530	TP	R,L,F,Si	K^+/H^+ symport	
	AtKUP6	At1g70300		R,L,F,Si	K^+/H^+ symport	
	AtKUP7	At5g09400	TP	R,L,F,Si	K^+/H^+ symport	
	AtKUP8	At5g12880		R,L,F,Si		
	AtKUP9	At4g19960		R,L,F,Si		
	AtKUP10	At1g31120		R,L	K^+/H^+ symport	
	AtKUP11	At2g35060	PM	L	K^+/H^+ symport	
	AtKUP12	At1g60160	CP	R,L,F,Si		
	AtHAK5	At4g13420	PM, ER	R,L,F,Si	K^+/H^+ symport	nutritional K^+ uptake
HKT/Trk	AtHKT1	At4g10310	PM	R,S	$(K^+)/Na^+$ co-transport	phloem Na^+ recirculation
CPA1	AtNHX1	At5g27150	TP	R,L,St,F,P,Si	Cation/H^+ exchange	vacuolar K^+ and Na^+ accumulation
	AtNHX2	At3g05030	TP	R,L,F	Cation/H^+ exchange	vacuolar Na^+ sequestration
	AtNHX3	At5g55470	TP	F,P	Cation/H^+ exchange	vacuolar Na^+ sequestration
	AtNHX4	At3g06370	TP	R	Cation/H^+ exchange	vacuolar Na^+ sequestration
	AtNHX5	At1g54370	TP,EN	R,L,St,F	K^+/H^+ exchange	vacuolar Na^+ sequestration endosomal pH regulation
	AtNHX6	At1g79610	EN			endosomal pH regulation
	AtSOS1 (AtNHX7)	At2g01980	PM		Na^+/H^+ exchange	Na^+ efflux salt tolerance
	AtNHX8	At1g14660	PM	R,H,L	Li^+/H^+ exchange	Li^+ efflux
CPA2	AtCHX1	At1g16380		P		
	AtCHX2	At1g79400		L,P		
	AtCHX3	At5g22900		P		
	AtCHX4	At3g44900		L,P,Sp		
	AtCHX5	At1g08150		L,P,Sp		
	AtCHX6A	At1g08140		R,P		

	AtCHX7	At2g28170		P		
	AtCHX8	At2g28180		L,P		
	AtCHX9	At5g22910		L,P		
	AtCHX10	At3g44930		P		
	AtCHX11	At3g44920		P		
	AtCHX12	At3g44910		L,P		
	AtCHX13	At2g30240	PM	R,F,Si,P	K^+/H^+ symport	K^+ influx
	AtCHX14	At1g06970		R,L,F,P,Sp		
	AtCHX15	At2g13620		L,P	K^+/H^+ symport (?)	
	AtCHX16	At1g64170		R,L,Sp		
	AtCHX17	At4g23700	EN	R,L,F,P,Sp	K^+/H^+ exchange	K^+ homeostasis cytosolic pH regulation
	AtCHX18	At5g41610		R,L,P,Sp		
	AtCHX19	At3g17630		R,L,P,Sp		
	AtCHX20	At3g53720	EN	R,L,F,P,Sp	K^+/H^+ exchange	K^+ homeostasis cytosolic pH regulation
	AtCHX21	At2g31910	PM	R,L,P,Sp	Na^+/H^+ exchange	xylem Na^+ fluxes
	AtCHX22	At2g37910		P		
	AtCHX23	At1g05580	CP	P	K^+/H^+ exchange	K^+ homeostasis cytosolic pH regulation
	AtCHX24	At5g37060		P		
	AtCHX25	At5g58460		P		
	AtCHX26	At5g01680		L,P,Sp		
	AtCHX27	At5g01690		L,P,Sp		
	AtCHX28	At3g52080		R,P,Sp		
CPA2	AtKEA1	At1g01790	TP (?)	R,L,St,F,Si	K^+/H^+ antiport	vacuolar K^+ accumulation
(KEA)	AtKEA2	At4g00630		R,L,F,Si		
	AtKEA3	At4g04850				
	AtKEA4	At2g19600		R,L,F,Si		
	AtKEA5	At5g51710		R,L,F,Si		
	AtKEA6	At5g11800		R,L,F,Si		
Shaker-type	AtAKT1	At2g26650	PM	R,L,P	K^+ channel (KIRC)	nutritional K^+ uptake
	AtAKT2/3	At4g22200	PM	R,St,L,F,Si	K^+ channel (KIRC)	phloem K^+ loading
	AtSPIK	At2g25600	PM	F,P	K^+ channel (KIRC)	osmotic K^+ influx
	AtAKT5	At4g32500	PM	F,P,Si	K^+ channel (KIRC)	

(*continued*)

Table 1 (continued)

Gene family	Member	Locus	Membrane	Organ	Proposed mechanism	Putative function (s)
	AtKAT1	At5g46240	PM	R,St,L,Si	K^+ channel (KIRC)	osmotic K^+ influx
	AtKAT2	At4g18290	PM	St,L,F	K^+ channel (KIRC)	osmotic K^+ influx
	AtKC1	At4g32650	PM	R,L	K^+ channel subunit	regulation of AtAKT1
	AtSKOR	At3g02850	PM	R,F,P	K^+ channel (KORC)	xylem K^+ loading
	AtGORK	At5g37500	PM	R,St,L,F,P,Si	K^+ channel (KORC)	K^+ efflux for charge compensation
CNGC	AtCNGC1	At5g53130	PM	R,S,F,Si	Cation channel	K^+ influx
	AtCNGC2	At5g15410	PM	R,St,L,F,Si	Cation channel	Ca^{2+} signalling
	AtCNGC3	At2g46430	PM	R,L,F,Si	Cation channel	cation uptake
	AtCNGC4	At5g54250		L,F	Cation channel	Ca^{2+} signalling
	AtCNGC5	At5g57940	PM	R,L,F,Si		
	AtCNGC6	At2g23980		R,L,F,Si		
	AtCNGC7	At1g15990		P		
	AtCNGC8	At1g19780		R,L,F,P,Si		
	AtCNGC9	At4g30560		R,S,F,Si		
	AtCNGC10	At1g01340	PM	R,L,F,Si	Cation channel	K^+ influx
	AtCNGC11	At2g46440	PM	L,F,Si	Cation channel	Ca^{2+} signalling
	AtCNGC12	At2g46450	PM	R,L,F,Si	Cation channel	Ca^{2+} signalling
	AtCNGC13	At4g01010		R,S		
	AtCNGC14	At2g24610		R		
	AtCNGC15	At2g28260		R		
	AtCNGC16	At3g48010		P		
	AtCNGC17	At4g30360		R,L,F		
	AtCNGC18	At5g14870	PM	R,P	Cation channel	Ca^{2+} signalling
	AtCNGC19	At3g17690		R,Si		
	AtCNGC20	At3g17700		S		
GLR	AtGLR1.1	At3g04110		R,St,L,F,Si	Cation channel	Ca^{2+} signalling
	AtGLR1.2	At5g48400		R,St,L,F,Si		
	AtGLR1.3	At5g48410		R,St,L,F,Si		
	AtGLR1.4	At3g07520		R,St,L,F,Si	Cation channel	
	AtGLR2.1	At5g27100		R,St,L		
	AtGLR2.2	At2g24720		R,S		

	AtGLR2.3	At2g24710		R,S		
	AtGLR2.4	At4g31710		R,Si		
	AtGLR2.5	At5g11210		R,St,L,F,Si		
	AtGLR2.6	At5g11180		R		
	AtGLR2.7	At2g29120		R,St,L,Si		
	AtGLR2.8	At2g29110		R,St,L,Si		
	AtGLR2.9	At2g29100		R,St,L		
	AtGLR3.1	At2g17260		R,St,L,F,Si		
	AtGLR3.2	At4g35290		R,St,L,F,Si		Ca^{2+} signalling
	AtGLR3.3	At1g42540	PM (?)	R,St,L,F,Si	Cation channel (?)	Ca^{2+} signalling
	AtGLR3.4	At1g05200	PM	R,St,L,F,Si	Cation channel	Ca^{2+} signalling
	AtGLR3.5	At2g32390		R,St,L,F,Si		
	AtGLR3.6	At3g51480		R,St,L,F,Si		
	AtGLR3.7	At2g32400		R,St,L,F,Si	Cation channel	Ca^{2+} signalling
TPK/KCO	AtTPK1	At5g55630	TP	R,L,F,P,Si	K^+ channel (VK)	cellular K^+ homeostasis
	AtTPK2	At5g46370	TP	R,F,P		
	AtTPK3 (=AtKCO6)	At4g18160	TP	R,L,F,P,Si		
	AtTPK4	At1g02510	PM	R,H,F,P	K^+ channel	K^+ fluxes for charge compensation
	AtTPK5	At4g01840	TP	R,St,L,F,Si		
Kir-like	AtKCO3	At5g46360	TP	R,St,L,F	K^+ channel	vacuolar K^+ release
TPC1	AtTPC1	At4g03560	TP		Cation channel (SV)	vacuolar Ca^{2+} release

PM plasma membrane, *TP* tonoplast, *CP* chloroplast envelope, *EN* endosomal membranes, *ER* endoplasmic reticulum, *R* root, *S* shoot, *L* leaf, *St* stem, *F* flower, *P* pollen, *Si* silique, *Sp* sporophyte

Qi et al. 2008; Rubio et al. 2008). In addition to K^+/H^+ symporters, K^+/Na^+ co-transporters encoded by members of the *HKT/Trk* gene family are also found in the plasma membranes of plant cells (Gierth and Mäser 2007). Although, AtHKT1 does not appear to catalyse K^+ transport in *A. thaliana*, homologs in other plant species, including wheat, rice, eucalyptus (*Eucalyptus camaldulensis*), and ice plant (*Mesembryanthemum crystallinum*), do contribute to K^+ influx to plant cells (Gierth and Mäser 2007).

In *A. thaliana*, plasma membrane KIRCs are encoded by several members of the voltage-gated Shaker-type channel family and VIKCs are encoded by *AtAKT2/3* and by one member (*AtTPK4=AtKCO4*) of the tandem pore K^+ (*TPK/KCO*) channel family (Table 1). The main K^+ channel involved in K^+ nutrition of *A. thaliana* is AtAKT1 (Hirsch et al. 1998; Broadley et al. 2001; Gierth et al. 2005; Rubio et al. 2008) and AtKC1 appears to be a regulatory subunit for AtAKT1 in root hairs (Reintanz et al. 2002; Pilot et al. 2003). *AtAKT2/3* is expressed in the phloem and xylem parenchyma and has been implicated in both loading and unloading of the phloem (Deeken et al. 2002). AtKAT1 is primarily responsible for K^+ influx to guard cells and AtKAT2 contributes both to K^+ influx to guard cells and phloem K^+ loading (Zimmermann and Chérel 2005; Lebaudy et al. 2007; Amtmann and Blatt 2009). AtSPIK and AtTPK4 are primarily responsible for the K^+-influx that enables the elongation of pollen tubes (Becker et al. 2004). The VICCs are thought to be encoded by members of the cyclic nucleotide gated channel (*CNGC*) and glutamate receptor (*GLR*) gene families (White and Broadley 2003; Hampton et al. 2005; Demidchik and Maathuis 2007). Many genes encoding CNGCs and GLRs are expressed throughout the plant (Table 1; Chiu et al. 2002; Talke et al. 2003; Hampton et al. 2005; Christopher et al. 2007; Kaplan et al. 2007; Urquhart et al. 2007; Frietsch et al. 2008; Roy et al. 2008), where they are implicated in cytosolic Ca^{2+} homeostasis and signalling (White and Broadley 2003; Demidchik and Maathuis 2007; Stephens et al. 2008; Tapken and Hollmann 2008). Recently, the *A. thaliana* annexin AnxAt1 has also been found to form K^+-permeable channels in artificial lipid bilayers, with channel formation increasing in response to reduced cytosolic pH (Gorecka et al. 2007). It has been proposed that annexins mediate Ca^{2+} influx to plant cells, but they could also mediate K^+ influx (White et al. 2002; White and Broadley 2003; Mortimer et al. 2008, Laohavisit et al. 2009).

Potassium efflux from plant cells, whether to the apoplast or to the xylem, appears to be mediated by both KORCs and NORCs (Fig. 2). These are also encoded by members of the voltage-gated Shaker-type channel family (Table 1). The KORC AtGORK is present in cells throughout the *A. thaliana* plant, where it is thought to be involved in electrical charge compensation, and also dominates K^+ efflux from guard cells during stomatal closure (Ivashikina et al. 2001; Reintanz et al. 2002; Fizames et al. 2004; Lebaudy et al. 2007). The KORC AtSKOR, which is present in the root pericycle and stelar parenchyma, is thought to mediate K^+ loading of the xylem (Gaymard et al. 1998; De Boer and Volkov 2003; Johansson et al. 2006). Genes encoding NORCs are currently unknown.

Several members of the monovalent cation/proton-antiporter (CPA) family, which in *A. thaliana* comprises eight Na^+/H^+-eXchangers (AtNHXs), 28 AtCHXs, six AtKEAs and two AtNHDs resembling NhaD, can mediate K^+ influx to the vacuoles and endosomes of plant cells (Table 1: Sze et al. 2004; Pardo et al. 2006; Gierth and Mäser 2007). These include the cation/H^+-antiporters AtNHX1, AtNHX5, AtCHX17, and AtCHX20, which have all been implicated in cellular K^+ homeostasis and the regulation of cytosolic pH (Cellier et al. 2004; Gierth and Mäser 2007; Padmanaban et al. 2007; Morris et al. 2008), plus AtNHX2, AtNHX3 and AtNHX4 (Pardo et al. 2006; Gierth and Mäser 2007; Jaquinod et al. 2007). Some *AtCHX* genes are expressed exclusively during microgametogenesis or in sporophytic tissue in *A. thaliana*, suggesting that they are specifically involved in maintaining K^+ homeostasis during pollen development and germination (Sze et al. 2004), but most *AtCHX*s are expressed in several tissues (Sze et al. 2004; Padmanaban et al. 2007). Recently, two members of the calcium cation exchanger (CCX) family of transporters (AtCCX3 and AtCCX4) have also been suggested to function as K^+/H^+ exchangers and catalyse K^+ influx to the vacuole (Morris et al. 2008). *AtCCX3* is expressed principally in flowers, whereas *AtCCX4* is expressed throughout the plant (Morris et al. 2008).

Potassium is released from the vacuole through K^+-permeable cation channels. These include (1) fast-activating vacuolar (FV) channels, whose genetic identities are currently unknown (Demidchik and Maathuis 2007), (2) slowly-activating (SV) channels, one of which appears to be encoded by *AtTPC1* in *A. thaliana* (Peiter et al. 2005; Ranf et al. 2008; Gradogna et al. 2009), (3) voltage-independent, Ca^{2+}-activated VK channels, one of which appears to be encoded by *AtTPK1* (=*AtKCO1*) in *A. thaliana* (Bihler et al. 2005; Gobert et al. 2007; Latz et al. 2007) and (4) K^+ channels encoded by other members of the *TPK*/*KCO* and Kir-like (*KCO3*) gene families (Table 1, Zimmermann and Chérel 2005; Voelker et al. 2006; Lebaudy et al. 2007). Several KUPs, such as AtKUP4, AtKUP5 and AtKUP7 have also been found in the tonoplast and might catalyse K^+ efflux from the vacuole (Jaquinod et al. 2007).

5 Summary

Potassium is the most abundant inorganic cation in plants. It is required for the activation of many enzymes in metabolically-active cellular compartments, as a vacuolar osmoticum for rapidly expanding cells, and as a counter cation for anion accumulation and electrogenic transport processes. Plants that lack K have lower water content, impaired stomatal regulation, reduced transpiration, impaired phloem transport, higher leaf carbohydrate concentrations, higher polyamine concentrations, lower leaf chlorophyll concentrations and reduced photosynthetic capacity. Visible symptoms of K-deficiency include scorching along the margins of older leaves, reduced growth, reduced fecundity, and a greater susceptibility to abiotic stresses, pests and diseases. Plants acclimate to low K supply by increasing their root K^+ uptake capacity, replacing vacuolar K^+ with alternative osmotica, redistributing K^+ from mature to developing tissues, and reducing plant growth to maintain appropriate tissue K^+ concentrations for cellular functions. Potassium is highly mobile within the plant and many genes encoding K^+ transport proteins responsible for distributing K^+ within cells and between tissues are known. It may be possible to use this knowledge of molecular biology to develop crops that utilize K-fertilisers more effectively, to improve both plant and animal nutrition (Karley and White 2009).

References

Amtmann A, Blatt MR (2009) Regulation of macronutrient transport. New Phytol 181:35–52

Amtmann A, Hammond JP, Armengaud P, White PJ (2006) Nutrient sensing and signalling in plants: potassium and phosphorus. Adv Bot Res 43:209–257

Amtmann A, Troufflard S, Armengaud P (2008) The effect of potassium nutrition on pest and disease resistance in plants. Physiol Plant 133:682–691

Armstrong MJ, Kirby EA (1979) Estimation of potassium recirculation in tomato plants by comparison of the rates of potassium and calcium accumulation in the tops with their fluxes in the xylem stream. Plant Physiol 63:1143–1148

Asher CJ, Ozanne PG (1967) Growth and potassium content of plants in solution cultures maintained at constant potassium concentrations. Soil Sci 103:155–161

Bañuelos MA, Garciadeblas B, Cubero B, Rodríguez-Navarro A (2002) Inventory and functional characterization of the HAK potassium transporters of rice. Plant Physiol 130:784–795

Barraclough PB, Leigh RA (1993) Critical plant K concentrations for growth and problems in the diagnosis of nutrient deficiencies by plant analysis. Plant Soil 155/156:219–222

Baxter I, Hosmani PS, Rus A, Lahner B, Borevitz JO, Muthukumar B, Mickelbart MV, Schreiber L, Franke RB, Salt DE (2009) Root suberin forms an extracellular barrier that affects water relations and mineral nutrition in *Arabidopsis*. PLoS Genet 5(5):e1000492

Becker D, Geiger D, Dunkel M, Roller A, Bertl A, Latz A, Carpaneto A, Dietrich P, Roelfsema MRG, Voelker C, Schmidt D, Mueller-Roeber B, Czempinski K, Hedrich R (2004) AtTPK4, an *Arabidopsis* tandem-pore K^+ channel, poised to control the pollen membrane voltage in a pH- and Ca^{2+}-dependent manner. Proc Natl Acad Sci U S A 101:15621–15626

Bihler H, Eing C, Hebeisen S, Roller A, Czempinski K, Bertl A (2005) TPK1 is a vacuolar ion channel different from the slow-vacuolar cation channel. Plant Physiol 139:417–424

Bould C, Hewitt EJ, Needham P (1983) Diagnosis of mineral deficiencies in plants, vol 1 principles. HMSO, London, UK

Britto DT, Kronzucker HJ (2008) Cellular mechanisms of potassium transport in plants. Physiol Plant 133:637–650

Broadley MR, Bowen HC, Cotterill HL, Hammond JP, Meacham MC, Mead A, White PJ (2004) Phylogenetic variation in the shoot mineral concentration of angiosperms. J Exp Bot 55:321–336

Broadley MR, Escobar-Gutiérrez AJ, Bowen HC, Willey NJ, White PJ (2001) Influx and accumulation of Cs^+ by the *akt1* mutant of *Arabidopsis thaliana* (L.) Heynh. lacking a dominant K^+ transport system. J Exp Bot 52:839–844

Cakmak I, Hengeler C, Marschner H (1994) Changes in phloem export of sucrose in leaves in response to phosphorous, potassium and magnesium deficiency in bean plants. J Exp Bot 45:1251–1257

Cellier F, Conejero G, Ricaud L, Luu DT, Lepetit M, Gosti F, Casse F (2004) Characterization of *AtCHX17*, a member of the cation/H^+ exchangers, CHX family, from *Arabidopsis thaliana* suggests a role in K^+ homeostasis. Plant J 39:834–846

Chen J, Gabelman WH (1995) Isolation of tomato strains varying in potassium acquisition using a sand-zeolite culture system. Plant Soil 176:65–70

Chen J, Gabelman WH (2000) Morphological and physiological characteristics of tomato roots associated with potassium-acquisition efficiency. Sci Hortic 83:213–225

Cheung AY, Wu HM (2008) Structural and signaling networks for the polar cell growth machinery in pollen tubes. Annu Rev Plant Biol 59:547–572

Chiu JC, Brenner ED, DeSalle R, Nitabach MA, Holmes TC, Coruzzi GM (2002) Phylogenetic and expression analysis of the glutamate-receptor-like gene family in *Arabidopsis thaliana*. Mol Biol Evol 19:1066–1082

Christopher DA, Borsics T, Yuen CYL, Ullmer W, Andème-Ondzighi C, Andres MA, Kang B-H, Staehelin A (2007) The cyclic nucleotide gated cation channel AtCNGC10 traffics from the ER via Golgi vesicles to the plasma membrane of *Arabidopsis* root and leaf cells. BMC Plant Biol 7:48

Cuin TA, Miller AJ, Laurie SA, Leigh RA (2003) Potassium activities in cell compartments of salt-grown barley leaves. J Exp Bot 54:657–661

De Boer AH, Volkov V (2003) Logistics of water and salt transport through the plant: structure and functioning of the xylem. Plant Cell Environ 26:87–101

Deeken R, Geiger D, Fromm J, Koroleva O, Ache P, Langenfeld-Heyser R, Sauer N, May ST, Hedrich R (2002) Loss of the AKT2/3 potassium channel affects sugar loading into the phloem of *Arabidopsis*. Planta 216:334–344

Demidchik V, Maathuis FJM (2007) Physiological roles of nonselective cation channels in plants: from salt stress to signaling and development. New Phytol 175:387–404

Dolan L, Davies J (2004) Cell expansion in roots. Curr Opin Plant Biol 7:33–39

Drew MC, Webb J, Saker LR (1990) Regulation of K^+ uptake and transport to the xylem in barley roots, K^+ distribution determined by electron probe X-ray microanalysis of frozen-hydrated cells. J Exp Bot 41:815–825

El Dessougi H, Claassen N, Steingrobe B (2002) Potassium efficiency mechanisms of wheat, barley, and sugar beet grown on a K fixing soil under controlled conditions. J Plant Nutr Soil Sci 165:732–737

Epstein E, Bloom AJ (2005) Mineral nutrition of higher plants: principles and perspectives, 2nd edn. Sinauer Associates, Sunderland, MA

Fageria NK (2009) The use of nutrients in crop plants. CRC Press, Boca Raton, Florida

Fizames C, Munos S, Cazettes C, Nacry P, Boucherez J, Gaymard F, Piquemal D, Delorme V, Commes T, Doumas P, Cooke R, Marti J, Sentenac H, Gojon A (2004) The *Arabidopsis* root transcriptome by serial analysis of gene expression. Gene identification using the genome sequence. Plant Physiol 134:67–80

Fricke W, Leigh RA, Tomos AD (1994) Concentrations of inorganic and organic solutes in extracts from individual epidermal, mesophyll and bundle-sheath cells of barley leaves. Planta 192:310–316

Frietsch S, Wang Y-F, Sladek C, Poulsen LR, Romanowsky SM, Schroeder JI, Harper JF (2008) A cyclic nucleotide-gated channel is essential for polarized tip growth of pollen. Proc Natl Acad Sci USA 104:14531–14536

Fulgenzi FR, Peralta ML, Mangano S, Danna CH, Vallejo AJ, Puigdomenech P, Santa-María GE (2008) The ionic environment controls the contribution of the barley HvHAK1 transporter to potassium acquisition. Plant Physiol 147:252–262

Gambale F, Uozumi N (2006) Properties of *Shaker*-type potassium channels in higher plants. J Membr Biol 210:1–19

Gaymard F, Pilot G, Lacombe B, Bouchez D, Bruneau D, Boucherez J, Michaux-Ferriere N, Thibaud J-B, Sentenac H (1998) Identification and disruption of a plant shaker-like outward channel involved in K^+ release into the xylem sap. Cell 94:647–655

Gierth M, Mäser P (2007) Potassium transporters in plants – involvement in K^+ acquisition, redistribution and homeostasis. FEBS Lett 581:2348–2356

Gierth M, Mäser P, Schroeder JI (2005) The potassium transporter *AtHAK5* functions in K^+ deprivation-induced high-affinity K^+ uptake and *AKT1* K^+ channel contribution to K^+ uptake kinetics in *Arabidopsis* roots. Plant Physiol 137:1105–1114

Glass ADM (1976) Regulation of potassium absorption in barley roots. An allosteric model. Plant Physiol 58:33–37

Glass ADM, Perley JE (1980) Varietal differences in potassium uptake by barley. Plant Physiol 65:160–164

Gobert A, Isayenkov S, Voelker C, Czempinski K, Maathuis FJM (2007) The two-pore channel *TPK1* gene encodes the vacuolar K^+ conductance and plays a role in K^+ homeostasis. Proc Natl Acad Sci U S A 104:10726–10731

Goodger JQD, Sharp RE, Marsh EL, Schachtman DP (2005) Relationships between xylem sap constituents and leaf conductance of well-watered and water-stressed maize across three xylem sap sampling techniques. J Exp Bot 56:2389–2400

Gorecka KM, Thouverey C, Buchet R, Pikula S (2007) Potential role of annexin AnnAt1 from *Arabidopsis thaliana* in pH-mediated cellular response to environmental stimuli. Plant Cell Physiol 48:792–803

Gould N, Thorpe MR, Minchin PEH, Pritchard J, White PJ (2004) Solute is imported to elongating root cells of barley as a pressure driven-flow of solution. Funct Plant Biol 31:391–397

Gradogna A, Scholz-Starke J, Gutla PVK, Carpaneto A (2009) Fluorescence combined with excised patch: measuring calcium currents in plant cation channels. Plant J 58:175–182

Gupta M, Qiu X, Wang L, Xie W, Zhang C, Xiong L, Lian X, Zhang Q (2008) KT/HAK/KUP potassium transporters gene family and their whole-life cycle expression profile in rice (*Oryza sativa*). Mol Genet Genomics 280:437–452

Hafke JB, Furch ACU, Reitz MU, van Bel AJE (2007) Functional sieve element protoplasts. Plant Physiol 145:703–711

Hampton CR, Broadley MR, White PJ (2005) Short review: the mechanisms of radiocaesium uptake by *Arabidopsis* roots. Nukleonika 50:S3–S8

Herdel K, Schmidt P, Feil R, Mohr A, Schurr U (2001) Dynamics of concentrations and nutrient fluxes in the xylem of *Ricinus communis* – diurnal course, impact of nutrient availability and nutrient uptake. Plant Cell Environ 24:41–52

Hermans C, Hammond JP, White PJ, Verbruggen N (2006) How do plants respond to nutrient shortage by biomass allocation? Trends Plant Sci 11:610–617

Hirsch RE, Lewis BD, Spalding EP, Sussman MR (1998) A role for the AKT1 potassium channel in plant nutrition. Science 280:918–921

Huang CX, van Steveninck RFM (1989) Longitudinal and transverse profiles of K^+ and Cl^- concentration in 'low-' and 'high-salt' barley roots. New Phytol 112:475–480

Ivashikina N, Becker D, Ache P, Meyerhoff O, Felle HH, Hedrich R (2001) K^+ channel profile and electrical properties of *Arabidopsis* root hairs. FEBS Lett 508:463–469

James RA, Munns R, von Caemmerer S, Trejo C, Miller C, Condon AG (2006) Photosynthetic capacity is related to the cellular and subcellular partitioning of Na^+, K^+ and Cl^- in salt-affected barley and durum wheat. Plant Cell Environ 29:2185–2197

Jaquinod M, Villiers F, Kieffer-Jaquinod S, Hugouvieux V, Bruley C, Garin J, Bourguignon J (2007) A proteomics dissection of *Arabidopsis thaliana* vacuoles isolated from cell culture. Mol Cell Proteomics 63:394–412

Jeschke WD (1983) Cation fluxes in excised and intact roots in relation to specific and varietal differences. Plant Soil 72:197–212

Jeschke WD (1984) Effects of transpiration on potassium and sodium fluxes in root cells and the regulation of ion distribution between roots and shoots of barley seedlings. J Plant Physiol 117:267–285

Jeschke WD, Pate JS (1991) Cation and chloride partitioning through xylem and phloem within the whole plant of *Ricinus communis* L. under conditions of salt stress. J Exp Bot 42:1105–1116

Jeschke WD, Wolf O, Hartung W (1992) Effect of NaCl salinity on flows and partitioning of C, N, and mineral ions in whole plants of white lupin, *Lupinus albus* L. J Exp Bot 43:777–788

Jeschke WD, Klagges S, Hilpert A, Bhatti AS, Sarwar G (1995) Partitioning and flows of ions and nutrients in salt-treated plants of *Leptochloa fusca* L. Kunth. I. Cations and chloride. New Phytol 130:23–35

Jeschke WD, Kirkby EA, Peuke AD, Pate JS, Hartung W (1997) Effects of P deficiency on assimilation and transport of nitrate and phosphate in intact plants of castor bean (*Ricinus communis* L.). J Exp Bot 48:75–91

Johansson I, Wulfetange K, Porée F, Michard E, Gajdanowicz P, Lacombe B, Sentenac H, Thibaud J-B, Mueller-Roeber B, Blatt MR, Dreyer I (2006) External K^+ modulates the activity of the *Arabidopsis* potassium channel SKOR via an unusual mechanism. Plant J 46:269–281

Johnson CR (1973) Symptomatology and analyses of nutrient deficiencies produced on flowering annual plants. Commun Soil Sci Plant Anal 4:185–196

Jungk A (2001) Root hairs and the acquisition of plant nutrients from soil. J Plant Nutr Soil Sci 164:121–129

Jungk A, Claassen N (1997) Ion diffusion in the soil-root system. Adv Agron 61:53–110

Kaplan B, Sherman T, Fromm H (2007) Cyclic nucleotide-gated channels in plants. FEBS Lett 581:2237–2246

Karley AJ, White PJ (2009) Moving cationic minerals to edible tissues: potassium, magnesium, calcium. Curr Opin Plant Sci 12:291–298

Keunecke M, Lindner B, Seydel U, Schulz A, Hansen U-P (2001) Bundle sheath cells of small veins in maize leaves are the location of uptake from the xylem. J Exp Bot 52:709–714

Kochian LV, Lucas WJ (1988) Potassium transport in roots. Adv Bot Res 15:93–178

Laohavisit A, Mortimer JC, Demidchik V, Coxon KM, Stancombe MA, Macpherson N, Brownlee C, Hofmann A, Webb AAR, Miedema H, Battey NH, Davies JM (2009) *Zea mays* annexins modulate cytosolic free Ca^{2+} and generate a Ca^{2+}-permeable conductance. Plant Cell 21:479–493

Latz A, Becker D, Hekman M, Müller T, Beyhl D, Marten I, Eing C, Fischer A, Dunkel M, Bertl A, Rapp UR, Hedrich R (2007) TPK1, a Ca^{2+}-regulated Arabidopsis vacuole two-pore K^+ channel is activated by 14–3–3 proteins. Plant J 52:449–459

Lebaudy A, Hosy E, Simonneau T, Sentenac H, Thibaud J-B, Dreyer I (2008) Heteromeric K^+ channels in plants. Plant J 54:1076–1082

Lebaudy A, Véry A-A, Sentenac H (2007) K^+ channel activity in plants: genes, regulations and functions. FEBS Lett 581:2357–2366

Leigh RA, Storey R (1993) Intercellular compartmentation of ions in barley leaves in relation to potassium nutrition and salinity. J Exp Bot 44:755–762

Leigh RA, Wyn Jones RG (1984) A hypothesis relating critical potassium concentrations for growth to the distribution and functions of this ion in the plant cell. New Phytol 97:1–13

Lu YX, Li CJ, Zhang FS (2005) Transpiration, potassium uptake and flow in tobacco as affected by nitrogen forms and nutrient levels. Ann Bot 95:991–998

Maathuis FJM, Sanders D (1993) Energization of potassium uptake in *Arabidopsis thaliana*. Planta 191:302–307

Macduff JH, Dhanoa MS (1996) Diurnal and ultradian rhythms in K^+ uptake by *Trifolium repens* under natural light patterns: evidence for segmentation at different root temperatures. Physiol Plant 98:298–308

Macduff JH, Bakken AK, Dhanoa MS (1997) An analysis of the physiological basis of commonality between diurnal patterns of NH_4^+, NO_3^- and K^+ uptake by *Phleum pratense* and *Festuca pratensis*. J Exp Bot 48:1691–1701

Malone M, Herron M, Morales M-A (2002) Continuous measurement of macronutrient ions in the transpiration stream of intact plants using the meadow spittlebug coupled with ion chromatography. Plant Physiol 130:1436–1442

Marschner H, Kirkby EA, Engels C (1997) Importance of cycling and recycling of mineral nutrients within plants for growth and development. Bot Acta 110:265–273

Martínez-Cordero MA, Martínez V, Rubio F (2005) High-affinity K^+ uptake in pepper plants. J Exp Bot 56:1553–1562

Memon AR, Saccomani M, Glass ADM (1985) Efficiency of potassium utilization by barley varieties: the role of subcellular compartmentation. J Exp Bot 36:1860–1876

Mengel K, Haeder H-E (1977) Effect of potassium supply on the rate of phloem sap exudation and the composition of phloem sap of *Ricinus communis*. Plant Physiol 59:282–284

Mengel K, Kirkby EA, Kosegarten H, Appel T (2001) Principles of plant nutrition. Kluwer, Dordrecht

Moran N (2007) Osmoregulation of leaf motor cells. FEBS Lett 581:2337–2347

Morris J, Tian H, Park S, Sreevidya CS, Ward JM, Hirschi KD (2008) AtCCX3 is an *Arabidopsis* endomembrane H^+-dependent K^+ transporter. Plant Physiol 148:1474–1486

Mortimer JC, Laohavisit A, Macpherson N, Webb A, Brownlee C, Battey NH, Davies JM (2008) Annexins: multifunctional components of growth and adaptation. J Exp Bot 59:533–544

Mouline K, Very A-A, Gaymard F, Boucherez J, Pilot G, Devic M, Bouchez D, Thibaud J-B, Sentenac H (2002) Pollen tube development and competitive ability are impaired by disruption of a Shaker K^+ channel in *Arabidopsis*. Gene Dev 16:339–350

Munns R (1985) Na^+, K^+ and Cl^- in xylem sap flowing to shoots of NaCl-treated barley. J Exp Bot 36:1032–1042

Nieves-Cordones M, Miller AJ, Alemán F, Vicente Martínez V, Rubio F (2008) A putative role for the plasma membrane potential in the control of the expression of the gene encoding the tomato high-affinity potassium transporter HAK5. Plant Mol Biol 68:521–532

Padmanaban S, Chanroj S, Kwak JM, Li X, Ward JM, Sze H (2007) Participation of endomembrane cation/H^+ exchanger AtCHX20 in osmoregulation of guard cells. Plant Physiol 144:82–93

Pardo JM, Cubero B, Leidi EO, Quintero FJ (2006) Alkali cation exchangers: roles in cellular homeostasis and stress tolerance. J Exp Bot 57:1181–1199

Peiter E, Maathuis FJM, Mills LN, Knight H, Pelloux J, Hetherington AM, Sanders D (2005) The vacuolar Ca^{2+}-activated channel TPC1 regulates germination and stomatal movement. Nature 434:404–408

Pettersson S, Jensén P (1979) Regulation of rubidium uptake in sunflower roots. Physiol Plant 45:83–87

Pettersson S, Jensén P (1983) Variation among species and varieties in uptake and utilization of potassium. Plant Soil 72:231–237

Peuke AD, Jeschke WD (1993) The uptake and flow of C, N and ions between roots and shoots in *Ricinus communis* L. 1. Grown with ammonium or nitrate as nitrogen source. J Exp Bot 44:1167–1176

Peuke AD, Rokitta M, Zimmermann U, Schrieber L, Haase A (2001) Simultaneous measurement of water flow velocity and solute transport in xylem and phloem of adult plants of *Ricinus communis* over a daily time course by nuclear magnetic resonance spectrometry. Plant Cell Environ 24:491–503

Peuke AD, Jeschke WD, Hartung W (2002) Flows of elements, ions and abscisic acid in *Ricinus communis* and site of nitrate reduction under potassium limitation. J Exp Bot 53:241–250

Philippar K, Fuchs I, Lüthen H, Hoth S, Bauer CS, Haga K, Thiel G, Ljung K, Sandberg G, Böttger M, Becker D, Hedrich R (1999) Auxin-induced K^+ channel expression represents an essential step in coleoptile growth and gravitropism. Proc Natl Acad Sci U S A 96:12186–12191

Pilot G, Gaymard F, Mouline K, Chérel I, Sentenac H (2003) Regulated expression of *Arabidopsis* Shaker K^+ channel genes involved in K^+ uptake and distribution in the plant. Plant Mol Biol 51:773–787

Qi Z, Hampton CR, Shin R, Barkla BJ, White PJ, Schachtman DP (2008) The high affinity K^+ transporter AtHAK5 plays a physiological role *in planta* at very low K^+ concentrations and provides a caesium uptake pathway in *Arabidopsis*. J Exp Bot 59:595–607

Ranf S, Wünnenberg P, Lee J, Becker D, Dunkel M, Hedrich R, Scheel D, Dietrich P (2008) Loss of the vacuolar cation channel, AtTPC1, does not impair Ca^{2+} signals induced by abiotic and biotic stresses. Plant J 53:287–299

Reintanz B, Szyroki A, Ivashikina N, Ache P, Godde M, Becker D, Palme K, Hedrich R (2002) AtKC1, a silent *Arabidopsis* potassium channel α-subunit modulates root hair K^+ influx. Proc Natl Acad Sci USA 99:4079–4084

Rengel Z, Damon PM (2008) Crops and genotypes differ in efficiency of potassium uptake and use. Physiol Plant 133:624–636

Roy SJ, Gilliham M, Berger B, Essah PA, Cheffings C, Miller AJ, Davenport RJ, Liu L-H, Skynner MJ, Davies JM, Richardson P, Leigh RA, Tester M (2008) Investigating glutamate receptor-like gene co-expression in *Arabidopsis thaliana*. Plant Cell Environ 31: 861–871

Rubio F, Nieves-Cordones M, Alemán F, Martínez V (2008) Relative contribution of AtHAK5 and AtAKT1 to K^+ uptake in the high-affinity range of concentrations. Physiol Plant 134: 598–608

Schachtman DP, Shin R (2007) Nutrient sensing and signaling: NPKS. Annu Rev Plant Biol 58:47–69

Schurr U, Schulze E-D (1995) The concentration of xylem sap constituents in root exudate, and in sap from intact, transpiring castor bean plants (*Ricinus communis* L.). Plant Cell Environ 18:409–442

Shin R, Schachtman DP (2004) Hydrogen peroxide mediates plant root cell response to nutrient deprivation. Proc Natl Acad Sci U S A 101:8827–8832

Siddiqi MY, Glass ADM (1983a) Studies of the growth and mineral nutrition of barley varieties. I. Effect of potassium supply on the uptake of potassium and growth. Can J Bot 61:671–678

Siddiqi MY, Glass ADM (1983b) Studies of the growth and mineral nutrition of barley varieties. II. Potassium uptake and its regulation. Can J Bot 61:1551–1558

Siddiqi MY, Glass ADM, Hsiao AI, Minjas AN (1987) Genetic differences among wild oat lines in potassium uptake and growth in relation to potassium supply. Plant Soil 99:93–105

Siebrecht S, Herdel K, Schurr U, Tischner R (2003) Nutrient translocation in the xylem of poplar - diurnal variations and spatial distribution along the shoot axis. Planta 217:783–793

Spear SN, Asher CJ, Edwards DG (1978a) Response of cassava, sunflower, and maize to potassium concentration in solution. I. Growth and plant potassium concentration. Field Crop Res 1:347–361

Spear SN, Asher CJ, Edwards DG (1978b) Response of cassava, sunflower, and maize to potassium concentration in solution. II. Potassium absorption and its relation to growth. Field Crop Res 1:363–373

Steingrobe B, Claassen N (2000) Potassium dynamics in the rhizosphere and K efficiency of crops. J Plant Nutr Soil Sci 163:101–106

Stephens NR, Qi Z, Spalding EP (2008) Glutamate receptor subtypes evidenced by differences in desensitization and dependence on the *GLR3.3* and *GLR3.4* genes. Plant Physiol 146:529–538

Sze H, Padmanaban S, Cellier F, Honys D, Cheng N-H, Bock KW, Conéjéro G, Li X, Twell D, Ward JM, Hirschi KD (2004) Expression patterns of a novel *AtCHX* gene family highlight potential roles in osmotic adjustment and K^+ homeostasis in pollen development. Plant Physiol 136:2532–2547

Talke IN, Blaudez D, Maathuis FJM, Sanders D (2003) CNGCs: prime targets of plant cyclic nucleotide signalling? Trends Plant Sci 8:286–293

Tapken D, Hollmann M (2008) *Arabidopsis thaliana* glutamate receptor ion channel function demonstrated by ion pore transplantation. J Mol Biol 383:36–48

Trehan SP (2005) Nutrient management by exploiting genetic diversity of potato - a review. Potato J 32:1–15

Urquhart W, Gunawardena AHLAN, Moeder W, Ali R, Berkowitz GA, Yoshioka K (2007) The chimeric cyclic nucleotide-gated ion channel ATCNGC11/12 constitutively induces programmed cell death in a Ca^{2+} dependent manner. Plant Mol Biol 65:747–761

Voelker C, Schmidt D, Mueller-Roeber B, Czempinski K (2006) Members of the Arabidopsis AtTPK/KCO family form homomeric vacuolar channels *in planta*. Plant J 48:296–306

Walker DJ, Leigh RA, Miller AJ (1996) Potassium homeostasis in vacuolate plant cells. Proc Natl Acad Sci U S A 93:10510–10514

Walker DJ, Black CR, Miller AJ (1998) Role of cytosolic potassium and pH in the growth of barley roots. Plant Physiol 118:957–964

Wang T-B, Gassmann W, Rubio F, Schroeder JI, Glass ADM (1998) Rapid up-regulation of *HKT1*, a high-affinity potassium transporter gene, in roots of barley and wheat following withdrawal of potassium. Plant Physiol 118:651–659

Watanabe T, Broadley MR, Jansen S, White PJ, Takada J, Satake K, Takamatsu T, Tuah SJ, Osaki M (2007) Evolutionary control of leaf element composition in plants. New Phytol 174:516–523

White PJ (1993) Relationship between the development and growth of rye (*Secale cereale* L.) and the potassium concentration in solution. Ann Bot 72:349–358

White PJ (1997a) Cation channels in the plasma membrane of rye roots. J Exp Bot 48:499–514

White PJ (1997b) The regulation of K^+ influx into roots of rye (*Secale cereale* L.) seedlings by negative feedback via the K^+ flux from shoot to root in the phloem. J Exp Bot 48:2063–2073

White PJ (2000) Calcium channels in higher plants. Biochim Biophys Acta 1465:171–189

White PJ, Broadley MR (2003) Calcium in plants. Ann Bot 92:487–511

White PJ, Clarkson DT, Earnshaw MJ (1987) Acclimation of potassium influx in rye (*Secale cereale*) to low root temperatures. Planta 171:377–385

White PJ, Earnshaw MJ, Clarkson DT (1991) Effects of growth and assay temperatures on unidirectional K^+ fluxes in roots of rye (*Secale cereale*). J Exp Bot 42:1031–1041

White PJ, Bowen HC, Demidchik V, Nichols C, Davies JM (2002) Genes for calcium-permeable channels in the plasma membrane of plant root cells. Biochim Biophys Acta 1564:299–309

White P, Bowen H, Broadley M, Hammond J, Hampton C, Payne K (2004) The mechanisms of cesium uptake by plants. In: Inabe J, Tsukada H, Takeda A (eds) Proceeduings of the international symposium on radioecology and environmental dosimetry, Rokkasho, Aomori, Japan, October 22–24, 2003. Institute for Environmental Sciences, Aomori, Japan, pp 255–262

White PJ, Broadley MR, Greenwood DJ, Hammond JP (2005) Proceedings of the International Fertiliser Society 568. Genetic modifications to improve phosphorus acquisition by roots. IFS, York, UK

Wild A, Skarlou V, Clement CR, Snaydon RW (1974) Comparison of potassium uptake by four plant species grown in sand and in flowing solution culture. J Appl Ecol 11:801–812

Wrona AF, Epstein E (1985) Potassium and sodium absorption kinetics in roots of two tomato species. Plant Physiol 79:1064–1067

Zhang Z, Tian X, Duan L, Wang B, He Z, Li Z (2007) Differential responses of conventional and Bt-transgenic cotton to potassium deficiency. J Plant Nutr 30:659–670

Zhao J, Cheng N-H, Motes CM, Blancaflor EB, Moore M, Gonzales N, Padmanaban S, Sze H, Ward JM, Hirschi KD (2008) AtCHX13 is a plasma membrane K^+ transporter. Plant Physiol 148:796–807

Zimmermann S, Chérel I (2005) Potassium. In: Broadley MR, White PJ (eds) Plant nutritional genomics. Blackwell, Oxford, pp 26–65

Selenium Metabolism in Plants

Elizabeth A.H. Pilon-Smits and Colin F. Quinn

Abstract Selenium (Se) is an essential nutrient for many organisms, but also toxic at higher levels. While certain algae require Se to make selenoproteins, no such requirement has been shown for higher plants. Still, plants readily take up and assimilate Se using sulfur (S) transporters and biochemical pathways, and can also volatilize methylated Se. Some plants can even hyperaccumulate Se to levels around 1% of plant dry weight, in the form of methyl-selenocysteine, probably as a defense mechanism. Plants may be used both to provide dietary Se in areas of Se deficiency, and to clean up Se pollution from seleniferous areas. These applications benefit from better insight into the genetic and biochemical mechanisms that control plant Se tolerance and accumulation. Here we give a review of plant Se metabolism, and present new insights into plant Se tolerance and hyperaccumulation mechanisms. Moreover, we summarize research on the ecological aspects of plant Se accumulation.

1 Introduction

The element selenium (Se) is chemically similar to sulfur (S), and as a result, plants and other organisms readily take up and metabolize Se via S transporters and pathways. Since replacement of S by Se in proteins and other S compounds disrupts the function of these molecules, Se is toxic at elevated levels to most organisms. For instance, a diet containing 1 mg/kg DW Se may lead to chronic Se poisoning in humans and animals, and one-time ingestion of plant material containing 1,000 mg/kg DW Se can lead to acute Se poisoning and death (Draize and Beath 1935; Rosenfeld and Beath 1964; Wilber 1980). Both chronic and acute Se poisoning are serious

E.A.H. Pilon-Smits (✉) and C.F. Quinn
Biology Department, Colorado State University, Anatomy/Zoology Building, Fort Collins, CO 80523, USA
e-mail: epsmits@lamar.colostate.edu; cquinn@lamar.colostate.edu

R. Hell and R.-R. Mendel (eds.), *Cell Biology of Metals and Nutrients*, Plant Cell Monographs 17,
DOI 10.1007/978-3-642-10613-2_10,

problems in seleniferous areas such as in the Western USA, where Se is naturally present in soils derived from Cretaceous shale rock (Ohlendorf et al. 1986; Harris 1991; Kabata-Pendias 1998; Terry et al. 2000). On the other hand, Se is an essential trace element for many organisms including mammals, many bacteria, and certain green algae (Stadtman 1990, 1996; Fu et al. 2002). These organisms contain the so-called selenoproteins that contain selenocysteine (SeCys) in their active site. Interestingly, this SeCys is encoded by an opal stopcodon, which when in the right mRNA context, encodes SeCys instead. Se-requiring organisms can also contain seleno-tRNAs (Mihara and Esaki 2002). Selenoproteins invariably have antioxidant functions, including the scavenging of reactive oxygen species. In relation to this, Se deficiency is associated with an elevated probability of developing cancers or viral infections, as well as male infertility (Ellis et al. 2004; Diwadkar-Navsariwala et al. 2006; White and Broadley 2009). Extreme Se deficiency can lead to a type of heart disease termed white muscle disease in livestock and Keshan disease in people, after a province in China where this disease is common (Whanger 1989). The difference between the amount of Se required as a nutrient and the amount that is toxic is small; as a consequence both Se deficiency and toxicity are common problems worldwide (Terry et al. 2000). For higher plants, Se is known to be a beneficial nutrient but it has not been shown to be essential (Hartikainen 2005; Lyons et al. 2009; Pilon-Smits et al. 2009). Plant homologues of genes encoding selenoproteins in other organisms, such as glutathione peroxidase (GPX), were shown to encode a cysteine (Cys) instead of SeCys in the active site (Novoselov et al. 2002). Based on these *in silico* analyses it has been hypothesized that essential Se metabolism is a primitive trait that has been lost in evolution in higher plants and in other groups that do not require Se. It cannot be excluded, however, that some plants can post translationally convert an amino acid (e.g. serine) into selenocysteine, and thus produce selenoproteins differently. In this context it is interesting to note that Se treatment has been reported to enhance glutathione peroxidase activity in plants, and to reduce lipid peroxidation (Cartes et al. 2005; Djanaguiraman et al. 2005; Hartikainen 2005). Moreover, plants may have seleno-tRNAs; this has not been investigated.

While higher plants do not appear to require Se, they readily take it up from their environment and incorporate it into organic compounds using S assimilation enzymes, as depicted in Fig. 1. and, presented in more detail below. In short, inorganic selenate is reduced and assimilated into organic Se. The first organic form of Se produced is SeCys. This amino acid can be nonspecifically incorporated into proteins instead of Cys, leading to toxicity. An alternative fate of SeCys is to be converted to selenomethionine (SeMet), which also can be misincorporated into proteins, with less harmful effects. SeMet can also be converted to volatile dimethylselenide (DMSe), offering a release valve for excess Se from the plant (Lewis et al. 1966). SeCys can also be converted in plants to elemental Se and alanine (Pilon et al. 2003). Elemental Se is relatively innocuous; many bacteria use a similar Se detoxification mechanism. Furthermore, SeCys can be methylated, to form methyl-SeCys. This form of Se can safely be accumulated since it is not incorporated into proteins (Neuhierl et al. 1999). Methyl-SeCys can also act as

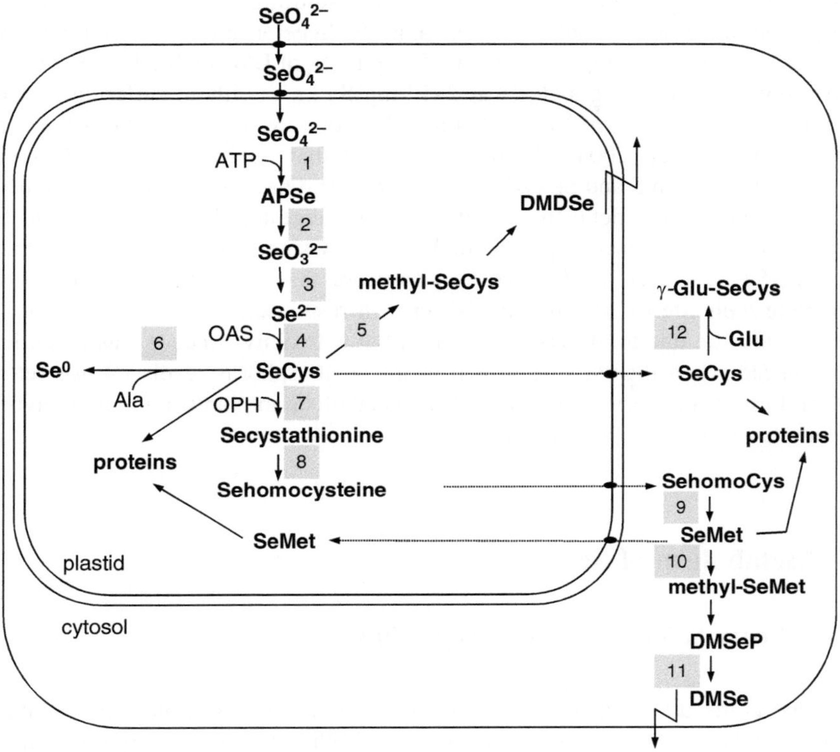

Fig. 1 Schematic overview of Se metabolism in plants. *APSe* adenosine phospho selenate, *SAT* serine acetyl transferase, *OAS O*-acetylserine, *OPH O*-phosphohomoserine, *SeCys* selenocysteine, *SeMet* selenomethionine, *DMSeP* dimethylselenoproprionate, *DMSe* dimethylselenide, *DMDSe* dimethyldiselenide. Numbers denote known enzymes. (1) ATP sulfurylase, (2) adenosine phosphosulfate reductase, (3) sulfite reductase (or glutathione), (4) OAS thiol lyase, (5) SeCys methyltransferase, (6) SeCys lyase, (7) cysthathionine-γ-synthase, (8) cysthathionine-β-lyase, (9) methionine synthase, (10) methionine methyltransferase, (11) DMSP lyase, (12) γ-glutamylcysteine synthetase

a precursor for the production of another form of volatile Se, dimethyldiselenide (DMDSe) (Terry et al. 2000; Sors et al. 2005).

In general, plants that have a tendency to accumulate high levels of S compounds, such as members of the Brassica genus (mustards and cabbages), also are good Se accumulators. Such S-loving species that accumulate Se to fairly high levels (0.1% of dry weight (DW), or 1,000 mg Se kg^{-1} DW) when supplied with adequate external Se levels are called accumulator species. Accumulator species likely do not have any Se-specific pathways but take up and metabolize Se and S indiscriminately, simply at elevated rates compared to nonaccumulators. Another category of plant species are the Se hyperaccumulators. These plants, from the families Brassicaceae, Fabaceae, and Asteraceae, are only found on seleniferous soils and typically accumulate Se to levels 100-fold higher than the surrounding vegetation in the field (Beath et al. 1939a, b). Se hyperaccumulators preferentially take up Se over S, and hyperaccumulate Se up to1% of DW, or 10,000 mg Se kg^{-1} DW from soil containing as little as 2–10 mg Se kg^{-1} without suffering toxicity (Neuhierl and Böck 1996; Neuhierl et al. 1999; Persans and Salt 2000; Ellis et al. 2004; LeDuc et al. 2004). There is evidence that Se hyperaccumulators can distinguish between S and Se (White et al. 2007) and have Se-specific metabolism, as discussed in more detail below. It has been suggested that Se is essential for hyperaccumulators, since hyperaccumulators grow significantly better in the presence of Se. However, to date there is no proof that these plants require Se to complete their life cycle. The positive growth response of hyperaccumulators to Se may also be due to alleviation of phosphorus toxicity, since it was much less pronounced when plants were grown at lower phosphorus levels (Broyer et al. 1972). Below, an overview is given of Se metabolism in plants, both in nonhyperaccumulators and hyperaccumulators.

2 Metabolism of Se

2.1 From Selenate to Selenocysteine

Selenate is the predominant form of bioavailable Se in oxic soils and selenite is more abundant in anoxic wetland conditions. Both forms are readily taken up by plants. Selenate is taken up and distributed by means of sulfate-proton cotransporters (Smith et al. 1995). All of the sulfate transporters in plants likely can transport selenate as well (Leustek 1996; Yoshimoto et al. 2002, 2003; Hawkesford 2003; Maruyama-Nakashita et al. 2004). Selenate assimilation takes place predominantly in the leaf chloroplasts (Pilon-Smits et al. 1999). The reduction of selenate to selenite appears to be a rate-limiting step in the Se assimilation pathway, since most plants supplied with selenate accumulate predominantly selenate, while plants supplied with selenite accumulate organic Se (de Souza et al. 1998). The conversion of selenate to selenite involves the consecutive action of two enzymes (Fig. 1). ATP sulfurylase (APS) couples selenate to ATP, forming adenosine phosphoelenate (APSe) (Wilson and Bandurski 1958). This is subsequently reduced to selenite by APS reductase (APR). There are isozymes for APS and APR in both chloroplast and cytosol, but most of the selenate reduction likely takes place in the chloroplast. The further reduction of selenite to selenide may happen exclusively in the chloroplast if it is mediated by sulfite reductase, in analogy with sulfite reduction. However, it has also been suggested that nonenzymatic reduction by reduced glutathione (GSH) may play a significant role in selenite reduction (Anderson 1993; Terry et al. 2000). Selenide can subsequently be coupled to *O*-acetylserine (OAS) to form SeCys, by means of OAS thiol lyase (also called cysteine synthase). This enzyme activity is found in cytosol, chloroplasts, and mitochondria. OAS is synthesized by the enzyme serine acetyl transferase, and functions as a signal molecule that upregulates the activity of sulfate transporters and sulfate assimilation enzymes.

2.2 From Selenocysteine to Other Selenocompounds

SeCys to SeMet and DMSe – SeCys can be converted to SeMet via the action of three enzymes (Fig. 1). The first, cystathionine-γ-synthase (CγS), couples SeCys to *O*-phosphohomoserine to form Se-cystathionine. The second enzyme, cystathionine-β-lyase, converts Se-cystathionine into Se-homocysteine. These first two enzymes are thought to be chloroplastic. The next step, however, occurs in the cytosol. Se-homocysteine is converted to SeMet via the action of Met synthase. SeMet has multiple possible fates, one of which is to be methylated to methyl-SeMet via the enzyme methionine methyltransferase. Methyl-SeMet can be further metabolized to volatile DMSe, which is cleaved off of the intermediate, dimethylselenopropionate (DMSeP), by DMSeP lyase.

SeCys to Se(0) – SeCys can be converted to elemental Se (Se(0)), via the action of a selenocysteine lyase (SL). NifS-like enzymes with SL activity have been found in both chloroplasts and mitochondria (Pilon et al. 2003). In organisms that require Se, SL enzymes provide elemental Se for selenoproteins and Se-tRNAs (Mihara and Esaki 2002). Overexpression of the chloroplastic plant SL (called cpNifS) was shown to reduce incorporation of Se into proteins as well as to enhance Se accumulation (Van Hoewyk et al. 2005). Whether this SL activity has any function *in vivo* is questionable. The main function of the NifS-like enzymes in plants is probably to act as Cys desulfurases in S metabolism, providing elemental S for iron–sulfur cluster formation (Van Hoewyk et al. 2007).

SeCys to Methyl-SeCys and DMDSe – SeCys can be methylated to form methyl-SeCys, via the action of SeCys methyltransferase (SMT). SMT enzyme activity is particularly pronounced in hyperaccumulators, and as a result, these species accumulate Se predominantly in the form of methyl-SeCys when supplied with selenate, while most other species accumulate selenate (de Souza et al. 1998; Freeman et al. 2006b). Since Methyl-SeCys does not enter proteins, it can be safely accumulated, explaining in part the tolerance of hyperaccumulators to Se. Recently, a Brassica species (*B. oleracea*, broccoli) was also shown to have an SMT enzyme, which was only expressed in the presence of Se (Lyi et al. 2005). Methyl-SeCys can be further converted to volatile DMDSe, the predominant volatile form of Se produced by Se hyperaccumulators (Terry et al. 2000; Kubachka et al. 2007). Hyperaccumulators also have been found to couple glutamate to methyl-SeCys, to form γ-glutamyl-methyl-SeCys, a major storage form of Se in hyperaccumulator seeds (Freeman et al. 2007; Kubachka et al. 2007). The enzyme mediation of this reaction is likely to be γ-glutamylcysteine synthetase (ECS). In S metabolism, this same enzyme functions in glutathione production (Glu–Cys–Gly). Reduced glutathione (GSH) has many redox functions in cells, and also is a negative regulator that downregulates S uptake and assimilation.

3 Genetic Engineering of Plant Se Metabolism

3.1 Results Obtained from Various Transgenic Approaches

As described above, all plants can take up inorganic selenate and selenite and assimilate them to SeCys and other organic selenocompounds, including volatile forms. Hyperaccumulators of Se may have additional metabolic pathways for Se, particularly methylation of SeCys and the conversion of methyl-SeCys to volatile DMDSe. To further enhance plant Se accumulation, tolerance, and volatilization, various transgenic approaches have been used.

One approach involved upregulation of key genes involved in S/Se assimilation and volatilization. First, overexpression of the first gene involved in selenate-to-selenite conversion, ATP sulfurylase (APS) in *Brassica juncea* (Indian mustard) resulted in enhanced selenate reduction, judged from the observation that the transgenic APS plants accumulated an organic form of Se when supplied with selenate, while wildtype (untransformed) controls accumulated selenate (Pilon-Smits et al. 1999). The APS transgenics accumulated two to threefold more Se than wild type, and 1.5-fold more S. The APS plants tolerated the accumulated Se better than wild type, perhaps because of the different form of Se accumulated. Se volatilization rate was not affected in the APS transgenics.

Second, overexpression in *B. juncea* of the first enzyme in the conversion of SeCys to SeMet, cystathionine-γ-synthase (CgS), resulted in two to threefold higher volatilization rates compared to untransformed plants (Van Huysen et al. 2003). Probably as a result of their enhanced volatilization, the CgS transgenics accumulated 40% less Se in their tissues than wild type. The CgS transgenics were also more Se tolerant than wildtype plants, probably due to their lower tissue Se levels.

Another genetic engineering approach to manipulate plant Se metabolism targeted SeCys, and particularly the prevention of the toxic process of its nonspecific incorporation into proteins. A mouse SL was expressed in *A. thaliana* and *B. juncea* (Pilon et al. 2003; Garifullina et al. 2003). This enzyme specifically breaks down SeCys into alanine and elemental Se. The SL transgenics showed reduced Se incorporation into proteins. When expressed in the cytosol of *A. thaliana*, the mouse SL enhanced plant Se tolerance, but when expressed in the chloroplast Se tolerance was reduced (Pilon et al. 2003). Perhaps the produced elemental Se interfered with iron–sulfur cluster formation in this compartment, which uses elemental S. All the transgenic SL plants showed enhanced Se accumulation, up to twofold compared to wildtype plants. Later, similar results were obtained when an *A. thaliana* homologue of the mouse SL (called CpNifS) was discovered and overexpressed: the CpNifS transgenics showed less Se incorporation in proteins, twofold enhanced Se accumulation, as well as enhanced Se tolerance (Van Hoewyk et al. 2005).

In another approach to prevent SeCys incorporation into proteins, SeCys methyltransferase (SMT) from the Se hyperaccumulator *A. bisulcatus* was overexpressed in *A. thaliana* and *B. juncea* (Ellis et al. 2004; LeDuc et al. 2004). The SMT transgenics showed enhanced Se accumulation, in the form of methyl-SeCys, as well as enhanced Se tolerance. The expression of SMT also resulted in increased rates of Se volatilization, with more volatile Se produced in the form of DMDSe.

While the expression of SMT enhanced Se tolerance, accumulation, and volatilization, the effects were more pronounced when the plants were supplied with selenite as opposed to selenate. Thus, the conversion of selenate to selenite appeared to be a rate-limiting step for the production of SeCys. To overcome this rate-limitation, APS and SMT transgenics were crossed to create double-transgenic plants that overexpress both APS and SMT (APSxSMT plants). The APS x SMT double transgenics accumulated up to nine times higher Se levels than wild type (LeDuc et al. 2006). Most of the Se in the double transgenics was in the form of methyl-SeCys: the APSxSMT plants accumulated up to eightfold more methyl-SeCys than wild type and nearly twice as much as the SMT transgenics. Se tolerance was similar in the single and double transgenics.

3.2 Obtained Insight into Rate-controlling Steps and Se Detoxification Mechanisms

From the genetic engineering studies we can conclude that the sulfate assimilation and volatilization pathway is capable of selenate assimilation and volatilization as well. The enzyme APS appears to be rate-limiting for the assimilation of selenate to organic Se, while CgS is rate-limiting for DMSe volatilization. Enhanced APS expression also appears to trigger selenate uptake and Se and S accumulation, likely due to upregulation of sulfate transporter expression.

The results from the SL and CpNifS transgenics show that the specific breakdown of SeCys can reduce nonspecific incorporation of Se into proteins. As long as the elemental Se does not interfere with cellular processes, this enhances Se tolerance. As mentioned above, whether CpNifS in plants functions in Se tolerance in nature is unknown; it's most important function is likely in synthesis of iron–sulfur clusters (Van Hoewyk et al. 2007). Since overexpression of SL or CpNifS led to enhanced, Se accumulation it appears that introduction of this new sink for Se upregulated Se and S uptake. The results from the SMT transgenics show that SMT is a key enzyme for Se hyperaccumulation, conferring enhanced Se tolerance and accumulation when expressed in nonhyperaccumulators. However, for optimal Se assimilation and detoxification, APS needs to be overexpressed together with SMT. APS x SMT double transgenics combine the ability to reduce selenate to selenite and SeCys with the ability to methylate SeCys and thus to detoxify the increased pool of internal Se.

3.3 Testing the Potential of the Transgenics for Phytoremediation, and as Fortified Foods

As described above, several different transgenics have been obtained that showed enhanced Se tolerance, accumulation, and assimilation from inorganic to organic

Se, and volatilization. Se accumulation was up to ninefold higher and volatilization up to threefold faster, under laboratory conditions. These properties may be useful for cleanup of excess levels of Se in the environment (phytoremediation), and also as fortified foods to prevent Se deficiency. Accumulators of MetSeCys would be particularly useful for the latter purpose, since this form of Se is particularly anticarcinogenic (Unni et al. 2005). In a first step to assess the transgenics' potential for phytoremediation or as Se-fortified food, they were tested for their capacity to accumulate Se from naturally seleniferous soil and from Se-contaminated sediment.

When grown on naturally seleniferous soil in a greenhouse pot experiment, the APS transgenics accumulated Se to threefold higher levels than wildtype *B. juncea*, and the CgS transgenics contained 40% lower Se levels than wild type (Van Huysen et al. 2004). These results are in agreement with the laboratory results. Plant growth was the same for all plant types in this experiment. Subsequently, a field experiment was carried out on Se (selenate)-contaminated sediment in the San Joaquin Valley (CA, USA) by Gary Banuelos and coworkers (Bañuelos et al. 2005). The APS transgenics accumulated Se to fourfold higher levels than wildtype *B. juncea*, which is similar to the laboratory and greenhouse results. In a second field experiment on the same Se-polluted sediment, the cpSL and SMT transgenics showed twofold higher Se accumulation than wildtype *B. juncea*, also in agreement with earlier laboratory experiments (Bañuelos et al. 2007). In both field experiments, biomass was comparable for the different plant types. Thus, the results obtained from the different transgenics using naturally seleniferous or Se-contaminated soils in a greenhouse or field are similar to those obtained under controlled laboratory conditions. The various transgenics showed enhanced Se accumulation, volatilization and/or tolerance, all promising traits for use as Se-fortified foods or for phytoremediation.

4 New Insights into Plant Se Responses and Tolerance Mechanisms

4.1 *Results Using the Model Nonaccumulator Species Arabidopsis thaliana*

Comparative studies have been performed using relatively Se-tolerant *A. thaliana* accessions versus nontolerant accessions, with the aim to reveal new insight into the genes that control Se uptake, (hyper) accumulation, and volatilization. Several quantitative trait loci (QTL) were identified that cosegregated with the higher selenate tolerance in *A. thaliana* accession Columbia compared with accession Landsberg erecta, using a population of recombinant inbred lines (Zhang et al. 2006a). Several genes involved in S assimilation are located in one of the identified chromosomal regions, which may be responsible for the Se tolerance conferred by this QTL. Other genetic and biochemical studies using 3 and 19 *A. thaliana*

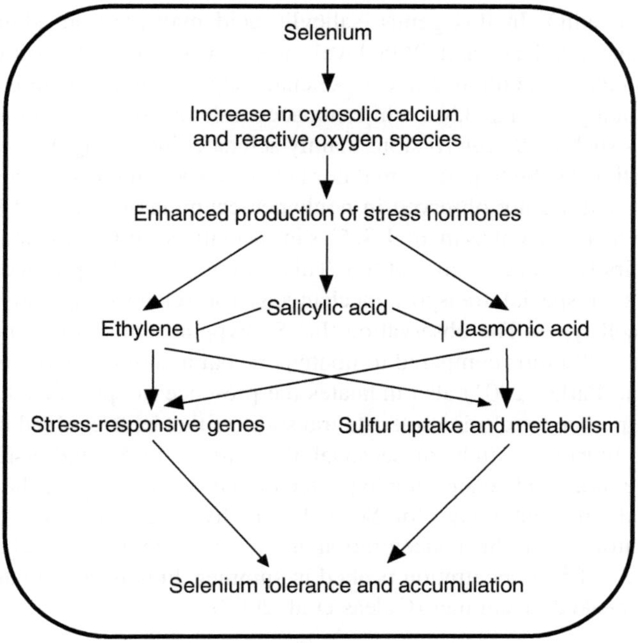

Fig. 2 Working model of Se responses contributing to Se tolerance in Arabidopsis (from data presented by Tamaoki et al. 2008)

accessions, respectively, have given further insight into Se tolerance and accumulation mechanisms in this species (Zhang et al. 2006b,c). Tolerance to selenate and selenite appear to be controlled at least in part by different loci, and tolerance and accumulation are not correlated. Genomic, genetic, and biochemical studies using Arabidopsis accessions differing in selenite or selenate tolerance revealed an important role for the plant hormones jasmonic acid (JA) and ethylene (Tamaoki et al. 2008; Van Hoewyk et al. 2008). Reactive oxygen species (ROS) may also have a signaling role, and the resistance mechanism appears to involve enhanced sulfate uptake and reduction, which may serve to prevent Se from replacing S in proteins and other S compounds. A working model for Se responses and tolerance is shown in Fig. 2.

4.2 *Results Using Se Hyperaccumulators and Related Nonhyperaccumulators*

In comparative studies using hyperaccumulator and nonhyperaccumulator species from the genus *Stanleya*, indications of similar Se tolerance mechanisms were found to those described above for Arabidopsis (Freeman, Tamaoki and Pilon-Smits, unpublished results). In this genus, salicylic acid may play an additional role, besides JA and ethylene, and ROS levels were lower rather than higher in the Se-tolerant taxa. As mentioned above, selenate-supplied hyperaccumulators store Se predominantly as MeSeCys, while nonaccumulators such as Arabidopsis and accumulators such as *B. juncea* store mainly selenate, indicating differences in Se metabolic pathways. Se hyperaccumulator taxa also show interesting Se sequestration patterns that are not observed in nonhyperaccumulators. Around 90% of the accumulated Se is present as methyl-SeCys in specialized cells in the leaf epidermis or in leaf hairs (Freeman et al. 2006a). This may indicate the presence in hyperaccumulators of special transport mechanisms for selenocompounds into these specialized cell types. The observation that Se hyperaccumulators generally have much higher Se/S ratios compared to nonhyperaccumulators growing on the same soil (Feist and Parker 2001) also indicates the presence of specialized Se-specific transporters, perhaps exclusive selenate transporters that have evolved from sulfate transporters. Indeed, a study of seasonal fluctuations in Se and S levels in Se hyperaccumulators and related nonhyperaccumulators growing on the same field site indicated different fluxes for Se and S in hyperaccumulators, but not for nonaccumulators. Leaf Se concentration in hyperaccumulators peaked in early spring, while leaf S concentration peaked in summer. In nonaccumulators both Se and S levels peaked in summer (Galeas et al. 2007).

5 Ecological Aspects of Plant Se Accumulation

5.1 Contribution of Microbes to Se Uptake and Volatilization

The fluxes and metabolic conversions of Se through plants are summarized in Fig. 3. So far, studies on plant Se metabolism have generally been done using nonsterile plants. Since all plants live in symbiosis with a host of bacteria and fungi, and since these microbes can metabolize and volatilize Se as well, (Thompson-Eagle et al. 1989; de Souza and Terry 1997; Pankiewicz et al. 2006), an important question is: what role do plant-associated microbes play in plant Se accumulation and volatilization? Different groups of plant-associated microbes may affect Se uptake and volatilization in plants: rhizosphere microbes, living in the area that is under the influence of the plant root, endophytic microbes that live within plant tissues, as well as microbes that live on leaf surfaces. The microbes may be envisioned to help plants take up Se via different mechanisms, or to help them metabolize it and volatilize it.

There is convincing evidence that bacteria contribute to plant Se uptake and volatilization. In broccoli (*B. oleracea*) 95% of Se root volatilization was inhibited when roots were treated with the antibiotics chlortetracycline and penicillin (Zayed and Terry 1994). Similarly, Indian mustard (*B. juncea*) plants treated with the antibiotic ampicillin volatilized 30% less Se and accumulated 70% less Se than

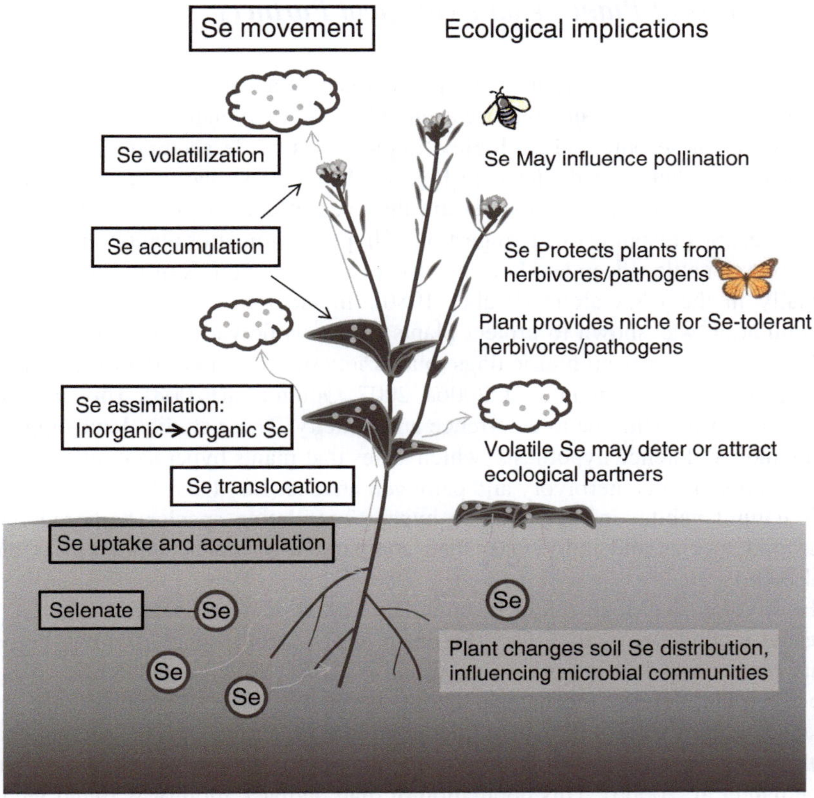

Fig. 3 Overview of the movement and metabolic conversion of Se by plants (*left side*) and their ecological implications (*right side*)

untreated plants. In addition, Indian mustard plants grown from surfaced-sterilized seeds that were subsequently inoculated with rhizospheric bacteria accumulated fivefold more Se and volatilized fourfold more Se than control plants from surface sterilized seeds that were not inoculated with bacteria. The mechanism for the stimulatory effect by the bacteria appeared to be both stimulation of root growth, and stimulation of S/Se uptake and assimilation. Plants inoculated with rhizospheric bacteria had an increased root surface area and the culture media contained ninefold higher serine levels than control plants. OAS is known to stimulate S uptake and assimilation (de Souza et al. 1999).

Less is known about a possible role for plant-associated fungi in plant Se uptake and volatilization. In one study, the nonaccumulator ryegrass accumulated less Se when treated with the mycorrhizal fungus *Glomus mosseae* compared to controls lacking in mycorrhizal fungus (Munier-Lamy et al. 2007). Also virtually nothing is known about the role endophytic microbes may play in Se uptake and volatilization. These will be interesting areas for further study.

5.2 *Effects of Plant Se on Ecological Partners*

The high Se levels in hyperaccumulators likely play an important role in the ecology of these plants, and even Se accumulated in crop plants may have ecological effects. These potential ecological implications of Se accumulation and volatilization by plants are depicted in Fig. 3. Most well known effects of plant Se hyperaccumulation on other species are the toxic effects of plant Se on livestock herbivores. Se poisoning due to ingestion of hyperaccumulator plants is responsible for losses of cattle, sheep and horses to the extent of hundreds of millions of US $ annually in the USA alone (Wilber 1980). In laboratory and field studies, Se accumulation was shown to protect plants from a wide variety of herbivores and pathogens, ranging from prairie dogs to a variety of arthropods and fungi (Hanson et al. 2003, 2004; Freeman et al. 2006a, 2007; Quinn et al. 2008). This protective effect of Se was both due to deterrence and toxicity. These studies lend support to the elemental defence hypothesis, which states that plants hyperaccumulate metals as protection against herbivory and pathogen attacks (Boyd and Martens 1993). In their natural habitat in the field, Se hyperaccumulating species harbored fewer arthropod species and individuals than comparable Se nonaccumulators (Galeas et al 2008).

Herbivores that ingest hyperaccumulator plant material readily convert the ingested methyl-SeCys to SeCys (Freeman et al. 2006b), which is toxic because of its inadvertent incorporation into proteins. Thus, Se hyperaccumulation is an effective plant defense mechanism against herbivory. Like all plant defense, over time, some herbivores will evolve tolerance. Indeed, a population of diamondback moth living in a seleniferous area was shown to have evolved Se tolerance (Freeman et al. 2006b). This population of diamondback moth was found feeding primarily on the Se hyperaccumulator *S. pinnata*. The mechanism for Se tolerance appears to be metabolic: the Se-tolerant moth accumulated Se in the ingested form, methyl-SeCys, which is not incorporated into proteins, while a control population of diamondback moth from a nonseleniferous habitat was Se-sensitive and converted the ingested methyl-SeCys to SeCys (Freeman et al. 2006b).

Another mechanism herbivores may utilize to minimize Se toxicity is to avoid Se-rich plant tissue. Se is not distributed evenly throughout Se hyperaccumulating plants, and Se levels fluctuate over the growing season. Leaf Se concentrations peak in early spring, and are much higher in young leaves and reproductive tissues than in older leaves (Galeas et al. 2007). Also, within the flowers of *S. pinnata*, the stamens and pistils have a higher Se concentration than the petals and sepals (Quinn et al unpublished results). Furthermore, hyperaccumulators preferentially allocate Se to the periphery of the leaf, unlike nonhyperaccumulators. In *S. pinnata*, Se is stored primarily in specialized cells in the epidermis, and in *A. bisulcatus* Se is stored primarily in leaf hairs (Freeman et al. 2006b). Therefore, it appears that hyperaccumulators preferentially allocate Se to their most valuable tissues, for protection from herbivores and pathogens. Sequestration in the epidermis may also contribute to Se tolerance. Depending on herbivore feeding mode,

Se hyperaccumulation may be more or less effective against different herbivores. In view of the particularly high Se levels in the flowers,Se may also play a role in pollination ecology. This will be an interesting area of further research

The elevated Se levels in and around hyperaccumulator plants likely also affects local microbial communities. Soil around Se hyperaccumulating plants has a ~tenfold higher Se concentration than the surrounding bulk soil, and there is evidence that rhizosphere and saprophytic fungi from seleniferous areas have evolved enhanced Se tolerance (Wangeline and Pilon-Smits, unpublished results; Quinn and Pilon-Smits, unpublished results).

6 Conclusions and Future Prospects

Building on the genomic and biochemical studies described above, follow-up research may reveal key genes that trigger the cascade of responses that together provide Se tolerance and accumulation in model plants and hyperaccumulators. Also, genes may be found that encode specific transporters of selenocompounds into and within hyperaccumulators. Such key genes will be the ultimate candidates for overexpression studies, with the potential of transferring the complete Se hyperaccumulator profile into high-biomass species.

Recent research has elucidated many important ecological interactions involving Se in plants. This research has helped identify important areas for future research. Particularly, more research is warranted on the role microbes play in plant Se uptake and volatilization, and the movement of Se through the food chain via Se hyperaccumulators or Se-fortified crop plants. The role of Se in below-ground ecological interactions with microbes and other organisms is also a fairly unexplored area. In addition to effects of Se on root–microbe interactions, Se may protect plants from root feeding herbivores, and selenocompounds released from hyperaccumulator roots may be toxic to surrounding vegetation. Similarly, the effects of Se on pollination ecology will be an interesting field of further study.

Better knowledge of the processes involved in plant metabolism of Se, the limiting factors involved, the contributions of ecological partners and the effects of Se on ecological partners are all useful for minimizing potential harmful effects of Se while benefiting from the positive effects of plant Se on animal and human health.

The capacity of plants to accumulate and volatilize Se will be very useful for the phytoremediation of Se-contaminated soils and waters (Bañuelos and Meek 1990; Hansen et al. 1998). When plant Se accumulation is managed well, this offers an efficient and cost-effective way to remove Se from the environment. Since plants are an effective source of dietary Se, Se-enriched plant material from phytoremediation or other sources can be considered fortified food. After being grown on Se-contaminated soil or being irrigated with Se-contaminated water, the Se-laden plant material may be used as a feed supplement for livestock, or as a biofuel. If successful, the potential of this strategy may be further enhanced by the use of selected transgenic lines. Of course, any use of Se-accumulating wildtype or transgenic plants will need to be accompanied by careful risk assessment, to avoid escape of transgenes and any adverse ecological effects of the accumulated Se.

Acknowledgements The writing of this manuscript was supported by National Science Foundation grant # IOS-0817748 to EAHPS. We thank Wiebke Tapken for creative design of figures. We thank all researchers who contributed to the results described here and apologize to those whose work we could not cover due to space limitation.

References

Anderson JW (1993) Selenium interactions in sulfur metabolism. In: De Kok LJ (ed) Sulfur nutrition and assimilation in higher plants – regulatory, agricultural and environmental aspects. SPB Academic, The Netherland, pp 49–60

Bañuelos GS, Meek DW (1990) Accumulation of selenium in plants grown on selenium-treated soil. J Environ Qual 19:772–777

Bañuelos G, Terry N, LeDuc DL, Pilon-Smits EAH, Mackey B (2005) Field trial of transgenic Indian mustard plants shows enhanced phytoremediation of selenium contaminated sediment. Environ Sci Technol 39:1771–1777

Bañuelos G, LeDuc DL, Pilon-Smits EAH, Tagmount A, Terry N (2007) Transgenic Indian mustard overexpressing selenocysteine lyase, selenocysteine methyltransferase, or methionine methyltransferase exhibit enhanced potential for selenium phytoremediation under field conditions. Environ Sci Technol 41:599–605

Beath OA, Gilbert CS, Eppson HF (1939a) The use of indicator plants in locating seleniferous areas in Western United States. I. General. Am J Bot 26:257–269

Beath OA, Gilbert CS, Eppson HF (1939b) The use of indicator plants in locating seleniferous areas in Western United States. II. Correlation studies by states. Amer J Bot 26:296–315

Boyd RS, Martens SN (1993) The raison d'etre for metal hyperaccumulation by plants. In: Baker AJM, Proctor J, Reeves RD (eds) The vegetation of ultramafic (serpentine) soils. Intercept, Andover, UK, 279–289

Broyer TC, Huston RP, Johnson CM (1972) Selenium and nutrition of Astragalus. 1. Effects of selenite or selenate supply on growth and selenium content. Plant Soil 36:635–649

Cartes P, Gianfreda L, Mora ML (2005) Uptake of selenium and its antioxidant activity in ryegrass when applied as selenate and selenite forms. Plant Soil 276:359–367

de Souza MP, Terry N (1997) Selenium volatilization by rhizosphere bacteria. Abstr Gen Meet Am Soc Microbiol 97:499

de Souza MP, Pilon-Smits EAH, Lytle CM, Hwang S, Tai JC, Honma TSU, Yeh L, Terry N (1998) Rate-limiting steps in selenium volatilization by *Brassica juncea*. Plant Physiol 117:1487–1494

de Souza MP, Chu D, Zhao M, Zayed AM, Ruzin SE, Schichnes D, Terry N (1999) Rhizosphere bacteria enhance selenium accumulation and volatilization by Indian mustard. Plant Physiol 119:565–574

Djanaguiraman M, Durga Devi D, Shanker AK, Sheeba JA, Bangarusamy U (2005) Selenium – an antioxidative protectant in soybean during selenscence. Plant Soil 272:77–86

Diwadkar-Navsariwala V, Prins GS, Swanson SM, Birch LA, Ray VH, Hedayat S, Lantvit DL, Diamond AM (2006) Selenoprotein deficiency accelerates prostate carcinogenesis in a transgenic model. Proc Natl Acad Sci USA 103:8179–8184

Draize JH, Beath OA (1935) Observation on the pathology of "blind staggers" and "alkali disease". Am Vet Med Assoc J 86:53–763

Ellis DR, Sors TG, Brunk DG, Albrecht C, Orser C, Lahner B, Wood KV, Harris HH, Pickering IJ, Salt DE (2004) Production of Se-methylselenocysteine in transgenic plants expressing selenocysteine methyltransferase. BMC Plant Biol 4:1–11

Feist LJ, Parker DR (2001) Ecotypic variation in selenium accumulation among populations of *Stanleya pinnata*. New Phytol 149:61–69

Freeman JL, Quinn CF, Marcus MA, Fakra S, Pilon-Smits EAH (2006a) Selenium tolerant diamondback moth disarms hyperaccumulator plant defense. Current Biol 16:2181–2192

Freeman JL, Zhang LH, Marcus MA, Fakra S, McGrath SP, Pilon-Smits EAH (2006b) Spatial imaging, speciation and quantification of selenium in the hyperaccumulator plants *Astragalus bisulcatus* and *Stanleya pinnata*. Plant Physiol 142:124–134

Freeman JL, Lindblom SD, Quinn CF, Fakra S, Marcus MA, Pilon-Smits EAH (2007) Selenium accumulation protects plants from herbivory by orthoptera due to toxicity and deterrence. New Phytol 175:490–500

Fu L-H, Wang X-F, Eyal Y, She Y-M, Donald LJ, Standing KG, Ben-Hayyim G (2002) A selenoprotein in the plant kingdom: Mass spectrometry confirms that an opal codon (UGA) encodes selenocysteine in *Chlamydomonas reinhardtii* glutathione peroxidase. J. Biol Chem 277:25983–25991

Galeas ML, Zhang LH, Freeman JL, Wegner M, Pilon-Smits EAH (2007) Seasonal fluctuations of selenium and sulfur accumulation in selenium hyperaccumulators and related non-accumulators. New Phytol 173:517–525

Galeas ML, Klamper EM, Bennett LE, Freeman JL, Kondratieff BC, Pilon-Smits EAH (2008) Selenium hyperaccumulation affects plant arthropod load in the field. New Phytol 177:715–724

Garifullina GF, Owen JD, Lindblom S-D, Tufan H, Pilon M, Pilon-Smits EAH (2003) Expression of a mouse selenocysteine lyase in *Brassica juncea* chloroplasts affects selenium tolerance and accumulation. Physiol Plant 118:538–544

Hansen D, Duda PJ, Zayed A, Terry N (1998) Selenium removal by constructed wetlands: role of biological volatilization. Environ Sci Technol 32:591–597

Hanson BR, Garifullina GF, Lindblom SD, Wangeline A, Ackley A, Kramer K, Norton AP, Lawrence CB, Pilon-Smits EAH (2003) Selenium accumulation protects *Brassica juncea* from invertebrate herbivory and fungal infection. New Phytol 159:461–469

Hanson BR, Lindblom SD, Loeffler ML, Pilon-Smits EAH (2004) Selenium protects plants from phloem-feeding aphids due to both deterrence and toxicity. New Phytol 162:655–662

Harris T (1991) Death in the Marsh. Island Press, Washington, DC

Hartikainen H (2005) Biogeochemistry of selenium and its impact on food chain quality and human health. J Trace Elem Med Biol 18:309–318

Hawkesford MJ (2003) Transporter gene families in plants: the sulphate transporter gene family – redundancy or specialization? Physiol Plant 117:155–163

Kabata-Pendias A (1998) Geochemistry of selenium. J Environ Pathol Toxicol Oncol 17:173–177

Kubachka KM, Meija J, LeDuc DL, Terry N, Caruso JA (2007) Environ Sci Technol 41:1863–1869

LeDuc DL, Tarun AS, Montes-Bayon M, Meija J, Malit MF, Wu CP, AbdelSamie M, Chiang C-Y, Tagmount A, deSouza MP, Neuhierl B, Bock A, Caruso JA, Terry N (2004) Overexpression of selenocysteine methyltransferase in *Arabidopsis* and indian mustard increases selenium tolerance and accumulation. Plant Physiol 135:377–383

LeDuc DL, AbdelSamie M, Montes-Bayón M, Wu CP, Reisinger SJ, Terry N (2006) Overexpressing both ATP sulfurylase and selenocysteine methyltransferase enhances selenium phytoremediation traits in Indian mustard. Environ Pollut 144:70–76

Leustek T (1996) Molecular genetics of sulfate assimilation in plants. Physiol Plant 97:411–419

Lewis BG, Johnson CM, Delwiche CC (1966) Release of volatile selenium compounds by plants: collection procedures and preliminary observations. J Agric Food Chem 14:638–640

Lyi SM, Heller LI, Rutzke M, Welch RM, Kochian LV, Li L (2005) Molecular and biochemical characterization of the selenocysteine Se-methyltransferase gene and Se-methylselenocysteine synthesis in broccoli. Plant Physiol 138:409–420

Lyons GH, Genc Y, Soole K, Stangoulis JCR, Liu F, Graham RD (2009) Selenium increases seed production in *Brassica*. Plant Soil 318:73–80

Maruyama-Nakashita A, Nakamura Y, Yamaya T, Takahashi H (2004) A novel regulatory pathway of sulfate uptake in *Arabidopsis* roots: implication of CRE1/WOL/AHK4-mediated cytokinin-dependent regulation. Plant J 38:779–789

Mihara H, Esaki N (2002) Bacterial cysteine desulfurases: their function and mechanisms. Appl Microbiol Biotechnol 60:12–23

Munier-Lamy C, Deneux-Mustin S, Mustin C, Merlet D, Berthelin J, Leyval C (2007) Selenium bioavailability and uptake as affected by four different plants in a loamy clay soil with particular attention to mycorrhizae inoculated ryegrass. J Environ Radioact 97:148–158

Neuhierl B, Böck A (1996) On the mechanism of selenium tolerance in selenium accumulating plants. Purification and characterization of a specific selenocysteine methyltransferase from cultured cells of *Astragalus bisulcatus*. Eur J Biochem 239:235–238

Neuhierl B, Thanbichler M, Lottspeich F, Böck A (1999) A family of S-methylmethionine dependent thiol/selenol methyltransferases. Role in selenium tolerance and evolutionary relation. J Biol Chem 274:5407–5414

Novoselov SV, Rao M, Onoshko NV, Zhi H, Kryukov GV, Xiang Y, Weeks DP, Hatfield DL, Gladyshev VN (2002) Selenoproteins and selenocysteine insertion system in the model plant system, *Chlamydomonas reinhardtii*. EMBO J 21:3681–3693

Ohlendorf HM, Hoffman DJ, Salki MK, Aldrich TW (1986) Embryonic mortality and abnormalities of aquatic birds: apparent impacts of selenium from irrigation drain water. Sci Total Environ 52:49–63

Pankiewicz U, Jamroz J, Schodziñski A (2006) Optimization of selenium accumulation in *Rhodotorula rubra* cells by treatment of culturing medium with pulse electric field. Int Agrophysics 20:147–152

Persans MW, Salt DE (2000) Possible molecular mechanisms involved in nickel, zinc and selenium hyperaccumulation in plants. Biotechnol Genet Eng Rev 17:389–413

Pilon M, Owen JD, Garifullina GF, Kurihara T, Mihara H, Esaki N, Pilon-Smits EAH (2003) Enhanced selenium tolerance and accumulation in transgenic *Arabidopsis thaliana* expressing a mouse selenocysteine lyase. Plant Physiol 131:1250–1257

Pilon-Smits EAH, Hwang S, Lytle CM, Zhu Y, Tai JC, Bravo RC, Chen Y, Leustek T, Terry N (1999) Overexpression of ATP sulfurylase in Indian mustard leads to increased selenate uptake, reduction, and tolerance. Plant Physiol 119:123–132

Pilon-Smits EAH, Quinn CF, Tapken W, Malagoli M, Schiavon M (2009) Physiological functions of beneficial elements. Curr Opin Plant Biol, in press

Quinn CF, Freeman JF, Galeas ML, Klamper EM, Pilon-Smits EAH (2008) Selenium protects plants from prairie dog herbivory – implications for the functional significance and evolution of Se hyperaccumulation. Oecologia 155:267–275

Rosenfeld I, Beath OA (1964) Selenium, geobotany, biochemistry, toxicity, and nutrition. Academic, New York

Smith FW, Ealing PM, Hawkesford MJ, Clarkson DT (1995) Plant members of a family of sulfate transporters reveal functional subtypes. Proc Natl Acad Sci USA 92:9373–9377

Sors TG, Ellis DR, Salt DE (2005) Selenium uptake, translocation, assimilation and metabolic fate in plants. Photosynth Res 86:373–389

Stadtman TC (1990) Selenium biochemistry. Annu Rev Biochem 59:111–127

Stadtman TC (1996) Selenocysteine. Annu Rev Biochem 65:83–100

Tamaoki M, Freeman JL, Pilon-Smits EAH (2008) Cooperative ethylene and jasmonic acid signaling regulates selenite resistance in *Arabidopsis thaliana*. Plant Physiol 146:1219–1230

Terry N, Zayed AM, de Souza MP, Tarun AS (2000) Selenium in higher plants. Ann Rev Plant Physiol Plant Mol Biol 51:401–432

Thompson-Eagle ET, Frankenberger WT Jr, Karlson U (1989) Volatilization of selenium by *Alternaria alternata*. Appl Environ Microbiol 55:1406–1413

Van Hoewyk D, Garifullina GF, Ackley AR, Abdel-Ghany SE, Marcus MA, Fakra S, Ishiyama K, Inoue E, Pilon M, Takahashi H, Pilon-Smits EAH (2005) Overexpression of AtCpNifS enhances selenium tolerance and accumulation in *Arabidopsis*. Plant Physiol 139:1518–1528

Van Hoewyk D, Abdel-Ghany SE, Cohu C, Herbert S, Kugrens P, Pilon M, Pilon-Smits EAH (2007) The *Arabidopsis* cysteine desulfurase CpNifS is essential for maturation of iron-sulfur cluster proteins, photosynthesis, and chloroplast development. Proc Natl Acad Sci USA 104:5686–5691

Van Hoewyk D, Takahashi H, Hess A, Tamaoki M, Pilon-Smits EAH (2008) Transcriptomeand biochemical analyses give insights into selenium-stress responses and selenium tolerance mechanisms in *Arabidopsis*. Physiol Plant 132:236–253

Van Huysen T, Abdel-Ghany S, Hale KL, LeDuc D, Terry N, Pilon-Smits EAH (2003) Overexpression of cystathionine-γ-synthase in indian mustard enhances selenium volatilization. Planta 218:71–78

Van Huysen T, Terry N, Pilon-Smits EAH (2004) Exploring the Selenium phytoremediationpotential of transgenic *Brassica juncea* overexpressing ATP sulfurylase or cystathionine γ-synthase. Int J Phytoremed 6:111–118

Unni E, Koul D, Alfred Yung W-K, Sinha R (2005) Se-methylselenocysteine inhibits phosphatidylinositol 3-kinase activity of mouse mammary epithelial tumor cells in vitro. Breast Cancer Res 7:R699–R707

Whanger PD (1989) China, a country with both selenium deficiency and toxicity: some thoughts and impressions. J Nutr 119:1236–1239

White PJ, Bowen HC, Marshall B, Broadley MR (2007) Extraordinarily high leaf selenium to sulfur ratios define 'Se-accumulator' plants. Ann Bot 100:111–118

White PJ, Broadley MR (2009) Biofortification of crops with seven mineral elements often lacking in human diets – iron, zinc, copper, calcium, magnesium, selenium and iodine. New Phytol 182:49–84

Wilber CG (1980) Toxicology of selenium: a review. Clin Toxicol 17:171–230

Wilson LG, Bandurski RS (1958) Enzymatic reactions involving sulfate, sulfite, selenate and molybdate. J Biol Chem 233:975–981

Yoshimoto N, Takahashi H, Smith FW, Yamaya T, Saito K (2002) Two distinct high affinity sulfate transporters with different inducibilities mediate uptake of sulfate in *Arabidopsis* roots. Plant J 29:465–473

Yoshimoto N, Inoue E, Saito K, Yamaya T, Takahashi H (2003) Phloem-localizing sulfate transporter, Sultr1;3, mediates re-distribution of sulfur from source to sink organs in *Arabidopsis*. Plant Physiol 131:1511–1517

Zayed AM, Terry N (1994) Selenium volatilization in roots and shoots: effects of shoot removal and sulfate level. J Plant Physiol 143:8–14

Zhang L, Byrne PF, Pilon-Smits EAH (2006a) Mapping quantitative trait loci associated with selenate tolerance in *Arabidopsis thaliana*. New Phytol 170:33–42

Zhang L-H, Ackley AR, Pilon-Smits EAH (2006b) Variation in selenium tolerance and accumulation among nineteen *Arabidopsis* ecotypes. J Plant Physiol 164:327–336

Zhang L-H, Abdel-Ghany SE, Freeman JL, Ackley AR, Schiavon M, Pilon-Smits EAH (2006c) Investigation of Selenium tolerance mechanisms in *Arabidopsis thaliana*. Physiol Plant 128:212–223

Cellular Biology of Sulfur and Its Functions in Plants

Rüdiger Hell, M. Sayyar Khan, and Markus Wirtz

Abstract Sulfur is one of the most versatile elements in life. It functions in fundamental processes such as electron transport, structure, and regulation. In plants, additional roles have developed with respect to photosynthetic oxygen production, abiotic and biotic stress resistance and secondary metabolism. Sulfate uptake, reductive assimilation, and integration into cysteine and methionine are the central processes that direct oxidized and reduced forms of organically-bound sulfur into its various functions. These steps are distributed between several cellular compartments and tightly regulated by supply, demand, and environmental factors in a network with assimilation of carbon and nitrogen. Signaling cues such as sulfate availability and thiol-based redox homeostasis via glutathione and their integrating by sensing systems will be presented in this chapter and analyzed.

1 Sulfur is an Essential Mineral Element

1.1 Physiological Functions of Sulfur

Sulfur is of elemental importance for life due to its versatility and reactivity in different oxidation and reduction states. In phototrophic organisms in general, the redox properties of sulfur in proteins and in sulfur-containing metabolites are important as mediators between the reductive assimilation processes of

R. Hell (✉) and M. Wirtz
Heidelberg Institute for Plant Sciences, University of Heidelberg, Im Neuenheimer Feld 360, 69120 Heidelberg, Germany
e-mail: rhell@hip.uni-heidelberg.de

M.S. Khan
Department of Plant Breeding and Genetics, NWFP Agricultural University, Peshawar, Pakistan

R. Hell and R.-R. Mendel (eds.), *Cell Biology of Metals and Nutrients*,
Plant Cell Monographs 17,
DOI 10.1007/978-3-642-10613-2_11, © Springer-Verlag Berlin Heidelberg 2010

photosynthesis and reactive oxygen species that arise as byproducts of electron transport chains. Reduced sulfur compounds have been assumed as early energy donors and, together with iron, may have contributed to early electron transport processes in proto-life (Martin et al. 2003). The reductive conditions before the invention of oxygenic photosynthesis probably allowed the evolution of processes based on the redox properties of sulfur. These sulfur-based reduction systems remained when the atmospheric environment became oxidative as a consequence of the effective development of photosynthesis in bacteria and later in algae and plants. It was suggested that this transition promoted the evolution of oxygen-dependent pathways as basis of subsequently complex organismal development (Falkowski 2006).

In addition to iron-sulfur clusters, the catalytic abilities of reduced sulfur are documented in the cofactors and ligands coenzyme A, biotin, thiamin, lipoic acid, and others. They are of fundamental importance to all cells since the very origin of life and have essential functions as vitamins in mammals. In proteins, the structural role of disulfide bridges is well established and may be renewed in this chapter by the discovery of two functionally important disulfide bridges in plant γ-glutamylcysteine ligase (GSH1), the first enzyme of GSH biosynthesis (Jez et al. 2004; Hothorn et al. 2006; Gromes et al. 2008). Regulatory functions of disulfide bridges in plants are prominent in enzymes of the Calvin Cycle but are found in increasing numbers in control switches such as NPR1 in pathogen defence (Tada et al. 2008) and transcription factors (Ndamukong et al. 2007). Monothiol- and dithiol mediated transfer processes are essential in these highly specific and directed processes that include GSH, NO, thioredoxins and glutaredoxins (Meyer and Hell 2005; Meyer et al. 2008).

These primary cellular functions are, in plants, complemented by the so-called secondary sulfur compounds. Many of them have roles in plant defence processes against phytopathogenic microbes and pests, such as the phytoalexin camalexin, sulfur-rich peptides of the thionin and defensin groups, and glucosinolates in the Brassicaceae family (Rausch and Wachter 2005). Lack of these compounds reduces the defence capabilities, while their formation upon attack goes along with an activation of primary sulfur metabolism. Since the synthesis of these defence compounds seems to partially depend on optimal sulfate supply, the term "sulfur-enhanced defence" has been coined (Kruse et al. 2007). The many aspects of sulfur function in phototrophic organisms have recently been comprehensively reviewed (Dahl et al. 2008).

1.2 Symptoms of Sulfur Deficiency

The numerous functions of sulfur in plants give rise to a characteristic long-term deficiency phenotype: chlorosis of interveinal sections of young leaves. The lack of reduced sulfur for iron-sulfur clusters in photosynthesis is presumably one reason that gives rise to chlorophyll oxidation; degradation of photosynthetic proteins to regain reduced sulfur is another (Ferreira and Teixeira 1992; Gilbert et al. 1997). In contrast to nitrate metabolism, where mature (source) leaves under nitrogen starvation turn chlorotic because they export nitrate to young sink leaves for growth, the young (sink) leaves react first and strongest to prolonged sulfate deficiency. The reason for this difference is seen in a slow release of sulfate stored in the vacuoles of mature leaves (Bell et al. 1994). In comparison, the mobility of nitrate between vacuole and cytosol upon demand is much higher (Miller et al. 2009). The activation rate of sulfate seems to be connected to nitrogen availability and growth rate (Blake-Kalff et al. 1998; Dubousset et al. 2009). This physiological feature gives rise to problems with early diagnosis of sulfur deficiency in agriculture. Determination of free sulfate in leaves is not indicative of the sulfate supply of a crop. Leaf sulfate contents may still be high, but are of only limited use, while the plant already suffers from sulfate deficiency in the soil solution (Blake-Kalff et al. 1998). As observed for many nutrient deficiency responses, the shoot:root ratio decreases during prolonged sulfate starvation. Interestingly, the morphological phenotype also includes changes in root architecture, where reactive oxygen species are produced in response to sulfate deprivation (Schachtman and Shin 2007). The precise developmental response depends on the plant species, but in Arabidopsis lateral root initiation is enhanced but not elongation (Kutz et al. 2002). At least Arabidopsis roots are also able to grow toward sulfate-rich zones as has been observed for Arabidopsis and several crop plants in response to nitrate, phosphate or potassium rich patches of soil (Robinson 1994).

Since sulfate belongs to the six essential plant nutrients, its availability has strong impact not only on plant growth and development, but also on crop yield and quality. Optimized sulfate fertilization that is in equilibrium with nitrogen application is of great concern in agriculture (Hell and Hillebrand 2001; Howarth et al. 2008; Dubousset et al. 2009). Since the decline of atmospheric intake of SO_2 due to clean air acts in the 1980s, the recognition of sulfate deficiency and precise fertilization regimes for sulfate has not only become a requirement in high yield agriculture, in particular for oilseed rape, but also cereals and sugar beet. Today, addition of 20–40 kg S/ha are regular measures to optimize yield in sulfur deficient soils. The timing of application is important, for example, before the onset of seed filling, because, in marked difference to nitrogen metabolism, the current availability of stored sulfate and less its reactivation defines the availability to the seeds (Walker and Boothe 2003).

1.3 Acclimatory Responses to Sulfur Starvation

In addition to morphological changes in response to long-term sulfate starvation, plants also respond to short-term deprivation at several levels. Marine algae live in an environment with about 30 mM sulfate and hardly experience a deficiency situation. However, sweet water algae and land plants retrieve sulfate from solutions in the micromolar range (Giordano et al. 2005). They are equipped with high-affinity sulfate transporters that operate with half-maximal activity in the low micromolar range and are rapidly de-repressed upon sulfate removal from the solution (see Sect. 2). The short-term response consists of characteristic adaptive reactions in plants and algae: de-repression of genes of sulfate uptake and increased uptake rates at the plasmalemma (Hawkesford and De Kok 2006); induction of genes of sulfate assimilation and several marker genes (*Nit3*, Kutz et al. 2002; *UP9*, Wawrzynska et al. 2005; *Sdi1*, Howarth et al. 2009); decreased levels of cysteine and glutathione (GSH) and enhanced concentrations of *O*-acetylserine (OAS), the intermediate of cysteine synthesis (see Sect. 4). This characteristic response has been widely investigated (Lewandowska and Sirko 2008) and is the subject of comprehensive bioinformatics and system biology approaches (reviewed in Hirai and Saito 2008; Hoefgen and Nikiforova 2008; Amtmann and Armengaud 2009; Amtmann and Blatt 2009); After about 24–72 h of depletion, depending on the species and growth regime, the long-term response develops (Nikiforova et al. 2005): secondary sulfur compounds such as glucosinolates are degraded as well as proteins of the photosynthetic apparatus, genes of hormone synthesis and signalling begin to change the morphology. If the generative stage is affected by sulfur deficiency the seed developmental program is changed. Again depending on the current nitrogen status (Howarth et al. 2009), seed storage protein composition is shifted by transcriptional and posttranscriptional changes in gene expression of sulfur-rich proteins (zein, PA1) and sulfur-poor proteins (vicilin, β-conglycinin; Kim et al. 1999; Tabe et al. 2002). This shift is enhanced by posttranslational processes including proteolysis in ripening seed to optimize seed viability for germination (Higashi et al. 2006).

2 The Acquisition and Allocation of Sulfur Compounds

2.1 Sulfate Acquisition by Plant Roots

Plants take up sulfur from the soil mainly in the form of sulfate (SO_4^{2-}), although they are also able to use reduced sulfur compounds from the atmosphere such as sulfur dioxide or hydrogen sulfide (Leustek et al. 2000). After being actively transported into the roots, sulfate is distributed throughout the plant. The suggested mechanism for sulfate transport is a coupled H^+ co-transport with a probable $3H^+:1SO_4^{2-}$ stoichiometry as shown in *Lemna gibba* (Lass and Ullrich-Eberius 1983) and is driven by a proton gradient maintained by P-type ATPase. Plant sulfate transporters have been analyzed for a number of species, including crop plants and described in several recent reviews (Buchner et al. 2004a; Hawkesford and De Kok 2006; Hawkesford 2008; Miller et al. 2009). In general sulfate transporters are encoded by gene families with more than ten members, which are subdivided into five groups according to amino acid similarity and function. The best investigated organism in this respect is *Arabidopsis thaliana* with 14 putative sulfate transporter (*Sultr*) genes.

The size of the encoded proteins of the SulP type transporters ranges from 500 to 700 amino acids. The secondary structure consensus predicts 10–12 transmembrane spanning helices that constitute the catalytic part of the protein. A short linker joins a C-terminal region with similarity to bacterial anti-sigma factor antagonists such as the *Bacillus subtilis* SpoIIAA. This domain is termed STAS (sulfate transporter and antisigma factor antagonist) and thought to enable protein-protein interactions. All plant sulfate transporter types possess this domain except group 5 transporters (Hawkesford 2008). Expression of chimaeric transmembrane and STAS domain constructs had a deleterious effect on transport kinetics and deletions of the STAS domain prevented trafficking to the plasma membrane (Shibagaki and Grossman 2004). Mutations of amino acids in the *AtSultr1;2* STAS domain with analogy to the phosphorylated serine of SpoIIAA resulted in a complete loss of activity of sulfate transport (Rouached et al. 2005). Thus, the STAS domain is likely to contribute to sulfate transport control but the precise mechanism of action is unclear.

Sulfate transporter group 1 is best characterized and encodes plasmalemma transporters with high affinity for sulfate ($K_m = 1.5–10\ \mu M$; Hawkesford 2003). They are expressed predominately in roots and responsible for uptake of sulfate from soil solution into the root cells (Shibagaki et al. 2002; Yoshimoto et al. 2002). The steady levels of their mRNA increases rapidly, upon sulfur starvation, leading to an increased capacity of the roots for sulfate uptake. Detailed comparison of *AtSultr1;1* and *AtSultr1;2* showed similar expression patterns in the cortex, epidermis, and root hairs based on GFP fusion approaches (Yoshimoto et al. 2002). However, *Atsultr1;1* is also expressed specifically in leaf hydathodes and *AtSultr1;2* in guard cells, suggesting individual roles in these specialized cells. It has been suggested that *Atsultr1;2* is responsible for constitutive uptake, whereas *AtSultr1;1* is inducible under stress (Yoshimoto et al. 2002).

Group 2 comprises low-affinity sulfate transporters (K_m for AtSultr2;1 = 0.41 mM and AtSultr2; 2 $\geq$ 1.2; Takahashi et al. 2000) that are responsible for translocation of sulfate within the plant. After the uptake of sulfate into the symplast, it is transported in the cytoplasm from peripheral root cells radially through the cortex and endodermis into the root stele. The delivery of sulfate from the symplast to the xylem vessel apoplast is thought to occur as efflux of sulfate from neighboring xylem parenchyma cells (Smith and Diatlof 2005). Apoplastic sulfate may be picked up by Sultr2;1, since its expression was found in xylem parenchyma cells and pericycles of roots, and xylem parenchyma and phloem cells of leaves, leading to the conclusion that it is responsible for uptake from the apoplasm within the vascular bundle and therefore involved in root to shoot transport (Takahashi et al. 2000). Once reached to the fine leaf veins, sulfate is again taken up into the leaf symplasm by sulfate transporters present in the bundle sheath cells that surround the vascular bundles. *Sultr2;1* and *Sultr2;2* are expressed throughout the plant (Buchner et al. 2004a), but tend to localize in vascular tissues (Takahashi et al. 2000). It is therefore likely that both contribute to translocation of sulfate between cells of the vascular tissue. AtSultr2;1 may also be involved in the transport of sulfur to the seed according to expression in the base of the silique and in the funiculus (Awazuhara et al. 2005).

Group 3 represents a diverse group that may be further subdivided into three or possibly four small clusters (Hawkesford 2008), each containing both Arabidopsis and rice examples and indicating relatively ancient gene duplications. Evidence for sulfate uptake using yeast complementation assays is missing in group 3. However, AtSultr3;5 has been co-localized with AtSultr2;1 in xylem parenchyma and pericycle cells in roots (Kataoka et al. 2004a). Sulfate uptake was hardly detectable with Sultr3;5 alone in the yeast sulfate uptake mutant system, but cells coexpressing both Sultr2;1 and Sultr3;5 showed three-times enhanced uptake activity (v_{max}) compared to Sultr2;1 expression alone, suggesting a role in root-to-shoot transport *in planta* (Kataoka et al. 2004a).

Group 4 sulfate transporters, although first reported to be plastid localized (Takahashi et al. 1999), appear to be tonoplast located according to GFP fusion studies (Kataoka et al. 2004b). In Arabidopsis and Brassica the two group 4 isoforms show highest expression in roots (Buchner et al. 2004b; Kataoka et al. 2004b) and were inducible by sulfur deficiency. Analysis of Arabidopsis double knockout plants indicated their role in sulfate efflux from the vacuole tissue. Nevertheless, import of sulfate into plastids for assimilatory reduction is absolutely necessary. It is believed to be constituted by a minor activity of the triose-phosphate translocator (Gross et al. 1990) or possibly of an ATP binding cassette transporter system in the envelope (Hawkesford 2008). In the moss *Marchantia polmorpha* and green alga *Chlamydomonas reinhardtii* genes encoding sulfate permeases in the chloroplast membrane have been identified. However no such genes have been found in vascular plants (Melis and Chen 2005).

Group 5 sulfate transporters are quite different from the rest with respect to primary sequence and the absence of the C-terminal STAS domain There are typically two isoforms for any given species which are also quite distinct from one another. Group 5 transporters may be functionally different since Sultr5;2 appears to be involved in molybdate metabolism. Natural variation across Arabidopsis ecotypes associated low expression of *Sultr5.2* gene with low moldybdate content while the protein was shown to be localized in the mitochondrial envelope (Baxter et al. 2008). Plasmalemma uptake of molybdate using a yeast sulfate uptake mutant assay was also shown for the same gene (Tomatsu et al. 2007) and the orthologue from *Chlamydomonas*. Affinity for molybdate was in the low nanomolar range (Tejada-Jimenez et al. 2007), resulting in renaming of Sultr5;2 into Mot1. However, the latter reports used GFP fusions to locate Mot1 in the endomembrane system or indirectly in the plasmalemma, leaving the exact function and cellular localization of this protein unsettled. Plasmalemma uptake of molybdate may in fact be a side function of group 1 transporters as shown for SHST1 from *Stylosanthes hamata* in the yeast sulfate uptake mutant assay (Fitzpatrick et al. 2008). No positive indications of sulfate or molybdate transport were observed for Sultr5;1 (see also the chapter on molybdate by Bittner and Mendel in this book).

2.2 Whole Plant Allocation of Sulfur Compounds

Sulfate is the major transport form of sulfur in vascular plants. Its allocation is orchestrated by the tissue-specific expression of the *Sultr* family as described above (Miller et al. 2009). Once sulfate has been taken up by the plant and reached the leaf symplast as described above, it is moved via plasmodesmata to the leaf mesophyll cells to the sites for reductive assimilation. Sulfate constitutes 70%–90% of total sulfur in mature leaves of oilseed rape (*Brassica napus* L.), while less than 1% is found in GSH and glucosinolates. In contrast, free sulfate (42%) is less dominant in young leaves of oilseed rape, where approximately 50% of the total sulfur was incorporated into insoluble (protein) sulfur fraction, 2% into GSH, and 6% into glucosinolates (Blake-Kalff et al. 1998). From shoot to root, sulfate is mobile in the phloem (Rennenberg et al. 1979; Lappartient and Touraine 1996), to redistribute either excess sulfate or stored sulfate pools in the vacuoles under long term sulfur stress, albeit too slow to efficiently support new growth (Clarkson et al. 1983; Bell et al. 1995). In Arabidopsis, Sultr1;3 has been shown to be localized in the phloem (Yoshimoto et al. 2003) where it probably plays a role to gather and keep sulfate within the phloem tissues for long distance recycling to other organs of the plant.

The smaller part of sulfur is allocated or moved between cells in reduced forms. In maize leaves assimilatory sulfate reduction is restricted to bundle-sheath cells, whereas the formation of GSH takes place predominantly in the mesophyll cells (Kopriva and Koprivova 2005). Here, cysteine acts as the transport metabolite of reduced sulfur between the two cell types (Burgener et al. 1998). However, maize may be a special case, since this is the only plant where a direct role of cysteine as metabolic repressor of sulfate uptake has been reported (Bolchi et al. 1999). GSH has long been assumed to be the main reduced long-distance sulfur compound (Rennenberg 1976; Brunold and Rennenberg 1997). GSH and/or its precursor γ-glutamylcysteine (γ-EC) have been shown to be transported over long-distance along the phloem to the roots (Fig. 1, Lappartient and Touraine 1996; Li et al. 2006). GSH must also be transported from maternal tissue into the embryo. The embryo is symplastically isolated and elimination of glutathione synthesis in the embryo causes lethality (Cairns et al. 2006).

More recently S-Methylmethionine (SMM) has also been shown to contribute up to 50% of the sulfur moving to developing grains in wheat, albeit the contribution was less in other species (Bourgis et al. 1999). SMM can be formed by methionine methyltransfease (MMT) and converted to methionine by homocysteine *S*-metyltransferase (Fig. 1). Both would form a futile SMM cycle if expressed at the same time and place, leaving its function at least in part unclear. SMM arrives in the phloem at a level of ~2 mol% of amino acids and could readily provide the methionine needed to synthesize grain proteins with an average methionine content of ~2 mol%. The claim that SMM is a source of sulfur in

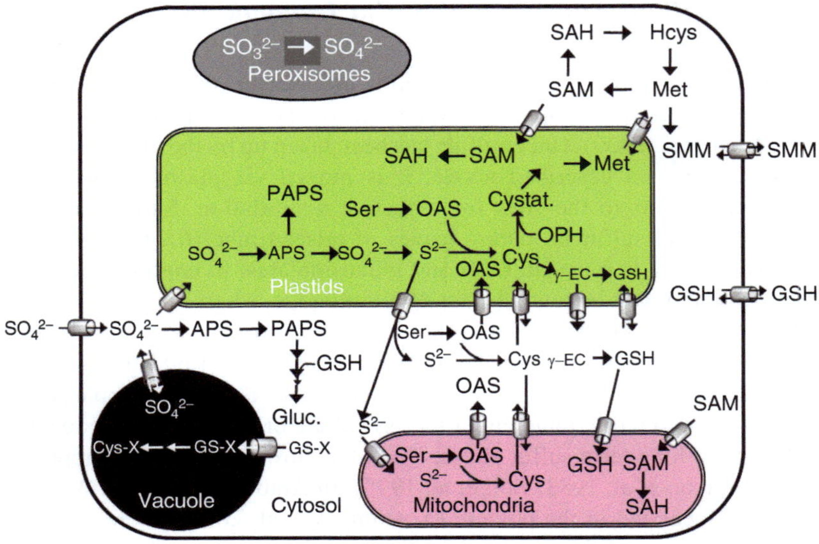

Fig. 1 *Transport of sulfur metabolites within plant cells.* Sulfur containing metabolites are actively transported between sub cellular compartments of a plant cell to separate reduction or oxidation of sulfur. E.g. sulfate is exclusively reduced in plastids will be used in its reduced form (sulfide, S^{2-}) for synthesis of cysteine in plastids, mitochondria and cytosol, but its oxidation is restricted to peroxisomes. A network of transporters (barrels) links production and use of key metabolites for primary and secondary sulfur metabolism like SAM. Finally, long distance transport of reduced or oxidized sulfur compounds crossing the plasmalemma takes place to meet demand of source and sink tissues

the seeds is based on the indirect evidence that it is essentially absent from mature grains in wheat and must be metabolized (Pimenta et al. 1998), and conversion to methionine is its only likely fate (Bourgis et al. 1999). However, the role of SMM in methylation or transport of sulfur has been investigated in *mmt* null mutants of Arabidopsis and maize that lacked the capacity to produce SMM (Kocsis et al. 2003). The normal growth and seeds sulfur contents in Arabidopsis *mmt* plants rule out the indispensable role for SMM in sulfur transport but increased methylation ratio. Thus, sulfate and GSH might play a compensatory role in long distance sulfur transport, or perhaps SMM is a minor form of pholem sulfur present in these plants or the SMM cycle function in methylation. Interestingly, not only primary sulfur compounds but also secondary sulfur compounds can be transported between cells. An example is glucosinolates. Feeding experiments with radiolabeled *p*-hydroxybenzyl-glucosinolate (*p*-OHBG) showed long distance phloem transport in Arabidopsis. When applied to the tip of detached leaves the labeled glucosinolate was rapidly transported from the application site into the whole plant and intact *p*-OHBG recovered from phloem sap and different tissues (Chen et al. 2001).

2.3 Cellular Distribution of Sulfur-containing Compounds

Secondary sulfur compounds have a very complex cell biology with respect to distribution of biosynthetic pathways between subcellular compartments, accumulation inside and outside the cell and, accordingly, transmembrane transport. Often taxonomic specialization of these compounds is observed, as for the alliins in the Alliaceae, while others such as defensins are truly ubiquitous. The reader is referred to specialized reviews with respect to sulfate-containing peptides (phytosulfokines; Fukuda et al. 2007), sulfur-rich proteins (thionins an defensins; Kruse et al. 2005) and sulfur-containing defence compounds such as phytoalexins (Glawischnig 2007), glucosinolates (Halkier and Gershenzon 2006), alliins (Burow et al. 2008) and elemental sulfur (Cooper and Williams 2004).

In this context it should be noted that many of these pathways finish with the transfer of sulfate from the activated form phosphoadenosine phosphosulfate to hydroxylgroups of the final biologically activating compound, e.g. glucosinolates or phytosulfokines. Such sulfation reactions are carried out by sulfotransferases (SULTs), a family of enzymes that is absent in basal land plants and *C. reinhardtii* (Kopriva et al. 2007a).Their substrates and physiological functions are predicted to be very diverse (Hernández-Sebastià et al. 2008). Small organic molecules such as flavonoids, steroids, glucosinolates, and hydroxyjasmonates are sulfonated by cytosolic SULTs, whereas membrane-associated SULTs sulfonate larger biomolecules such as peptides, proteins, and complex carbohydrates (Hernández-Sebastià et al. 2008).

The exception from this complexity is the sulfolipids, both with respect to pathway and distribution. Sulfolipids are ubiquitous in oxygenic photosynthetic organisms and in eukaryotes are always associated with plastids. They are sulfoquinosovyl-diacylgycerides and form a substantial part of polar lipids in plastid membranes. Their synthesis includes two steps and introduces the unusual substrate sulfite into the sugar moiety (Benning et al. 2008). Apart from many less confirmed assumptions their only known function is the replacement of phospholipids during phosphate deficiency stress.

3 Reductive Sulfate Assimilation

3.1 Subcellular Organization of Reactions

Assimilatory reduction of sulfate is predominantly a plastid localized process. However, branching points of the pathway often mark parallel enzymatic activities in other cellular compartments. A comparison of sequenced plant genomes shows that the presence of paralogous genes related to sulfur metabolism is quite common (Kopriva 2006; Kopriva et al. 2007a). After uptake of sulfate at the plasmalemma the pathway is initiated by ATP dependent activation of sulfate to adenosine

5′-phosphosulfate (APS) that is catalyzed by ATP sulfurylase (ATPS). In Arabidopsis and other plants nuclear–encoded isoforms of ATPS reside in plastids and the cytosol (Klonus et al. 1994; Rotte and Leustek 2000). All four ATPS forms in Arabidopsis carry transit peptides but which form provides in addition cytosolic activity via dual targeting is not known. In contrast to the endosymbiont theory ATPS proteins in organisms with primary plastids have been suggested to be of eukaryotic (host) origin (Patron et al. 2008). Further activation of APS with ATP is catalyzed by APS kinase (APK) and yields 3′-phosphoadenosyl-5′-phosphosulfate (PAPS). APK is also present in plastids and the cytosol to provide PAPS in both compartments for sulfation reactions (Mugford et al. 2009; see Sect. 2.3).

In plastids, APK and the assimilatory reduction pathway compete for APS/PAPS. This pathway operates exclusively in plastids in photoautotrophic as well as in heterotrophic tissues and has long been disputed (for review, see Schmidt and Jäger 1992; Leustek 1996; Hell 1997; Leustek et al. 2000; Kopriva and Koprivova 2004; Kopriva 2006). The first option refers to the so-called "bound pathway" of sulfate reduction. It starts from APS and requires the activities of APS sulfotransferase, a thiol-bound intermediate (possibly GSH) and thiosulfonate reductase (TSR) to release sulfide for cysteine synthesis. The bound pathway would avoid the production of free sulfite. Sulfite ions (HSO_3^- and SO_3^{2-}) are strong nucleophiles that can deleteriously react with a wide variety of cellular components. The bound pathway was first described for green algae (Hodson and Schiff 1971) and evidence for APS sulfotransferase activity and regulation was reported from numerous vascular plants (Schmidt 1975; Brunold and Suter 1990). Activity of TSR was described from the green alga *Chlorella* as well (Schmidt 1973).

The second option suggests a "free pathway" that begins with PAPS and would proceed as found in enterobacteria via thioredoxin-dependent PAPS reductase to free sulfite and via sulfite reductase (SiR) to sulfide. Both enzyme activities were found in *E. coli*, *Salmonella typhimurium,* and *Saccharomyces cerevisiae* and also in vascular plants (Schwenn and Kemena 1984; Schwenn 1989). Progress toward bound and free pathway hypotheses suffered from the reactive chemistry of sulfur in its different oxidation states and a number of biochemical ambiguities. The corresponding enzymes were never purified to homogeneity from plants. However, search for genes encoding PAPS reductase activity by functional complementation of an *E. coli cysH* mutant strain that lacked this enzyme led to the breakthrough discovery of APS reductase (APR) in Arabidopsis (Gutierrez-Marcos et al. 1996; Setya et al. 1996), favoring SiR-dependent reduction over TSR-mediated electron transfer to sulfide. The APR protein consists of a plastid transit sequence, a catalytic domain carrying a 4S–4Fe cluster with sequence homology to CysH and a glutaredoxin-like C-terminus that was shown to mediate electron transfer from GSH for reduction (Bick et al. 1998). APR from Arabidopsis and other plants strongly prefers APS instead of PAPS as substrate, its expression responds to sulfate and nitrate availability, and a number of stress factors result in activation of its activity (Leustek et al. 2000, for review). In addition, flux analysis using ^{35}S-labeled sulfate hinted that APR, after sulfate uptake, exerts strongest control over flux through the sulfate reduction pathway in Arabidopsis (Vauclare et al. 2002) and is responsible for genetically determined variation in sulfate content in Arabidopsis ecotypes (Loudet et al. 2007). To add to the controversy, the identification of APR led to the discovery that many bacterial taxa, but not the model organism *E. coli*, prefer APS as substrate for the free reduction pathway and possess a bacterial thioredoxin-dependent APR without glutaredoxin domain (Bick et al. 2000). Furthermore, the moss *Physcomitrella patens* was shown to carry two genes encoding plant-type APR and bacteria-like APR-B. The latter lacks the plant-specific iron-sulfur cluster as ligand and the glutaredoxin domain. It preferentially reduces not only APS but also PAPS although at a slow rate (Kopriva et al. 2007b). Targeted knockout lines of the *AprA* locus in Physcomitrella were still viable, demonstrating that under nonstressed conditions APR-B was sufficient for normal growth (Koprivova et al. 2002). Phylogenetic comparisons suggest that vascular plant APR with Fe-S cluster is of endosymbiontic origin although the cyanobacterial heritage is not entirely clear, while APR-B is derived from γ-proteobacterial PAPR lacking the Fe-S cluster (Patron et al. 2008). The amino acid sequence from a native purified APS sulfotransferase of *Lemna minor* confirmed, that this protein and APR are, the same enzyme (Suter et al. 2000). This apparently solves the controversy in favor of a free reduction pathway via APR and sulfite. Nevertheless a cDNA from *Arabidopsis thaliana* encoding APS kinase was reported to confer also APS sulfotransferase activity in a nonphysiological side reaction (Schiffmann and Schwenn 1994), documenting the many difficulties of sulfur biochemistry.

The second enzyme of the free reduction pathway, SiR, is exclusively localized in plastids and consists of two 65 kDa subunits. It contains a single siroheme and (4Fe–4S) cluster as prosthetic groups, has a high affinity for sulfite ($K_m^{sulfite}$ ~10 μM; Krueger and Siegel 1982; Nakayama et al. 2000) and releases sulfide that is then integrated into cysteine in a step that is comparable to ammonia fixation into glutamine (see Sect. 4). Ferredoxin acts as physiological donor of six electrons required for sulfite reduction, while bacterial SiR uses NADPH (Yonekura-Sakakibara et al. 2000). The structure, sequence and ligands of SiR in bacteria, archea and eukaryotes are similar to nitrite reductase, which catalyzes an equivalent reduction step in nitrate assimilation, i.e. a six electron reduction of nitrite to ammonia (Crane et al. 1995; Swamy et al. 2005). SiR is able to reduce nitrite as well and substrate preference can be converted by a single amino acid mutation (Nakayama et al. 2000). Physiological relevance of SiR had thus been questioned and sulfite as a substrate was regarded as an artifact, since a supposedly SiR deficient mutant strain of *Chlorella* was reported to carry TSR instead (Schmidt et al. 1974; Schmidt 1976). Nitrite reductase may react similarly, adding to the uncertainty of the free pathway as indispensable reduction pathway (Schmidt and Jäger 1992). Phylogenetic analysis showed that both SiR and NIR arose from an ancient gene-duplication in eubacteria, before the primary endosymbiosis that gave rise to plastids (Patron et al. 2008). It should be noted that SiR was observed in association with nucleoids in plastids of pea, maize and soybean. SiR was suggested to compact nucleoids and to repress DNA synthesis (Cannon et al. 1999) and transcription (Sekine et al. 2002). It cannot be ruled out that SiR has a second "moonlighting" function, although it appears completely soluble in biochemical experiments.

3.2 *Signal Mechanisms and Homeostasis of Uptake and Reductive Assimilation*

The characteristic short- and long-term responses to sulfate deficiency described in Sect. 2.2 are based on the entire inventory of regulatory mechanisms in eukaryotic cells (Amtmann and Blatt 2009; Miller et al. 2009). Induction of *Sultr1;1* expression was suggested to involve phosphorylation/dephosphorylation as regulatory part of root sulfate uptake (Maruyama-Nakashita et al. 2004). A remarkable mechanism is the enhanced sulfate transport capacity based on membrane protein interaction. The almost identical expression patterns of *Sultr2;1* and *Sultr3;5* in the vascular tissue (see Sect. 2.1) prompted co-expression analysis of both proteins in yeast and showed increased sulfate uptake rates compared to SULTR2;1 or SULTR3;5 alone. The function of this interaction was indirectly confirmed by reduced root-shoot sulfate transport in a *sultr3;5* null mutant (Kataoka et al. 2004a).

A recently discovered regulatory component in plants is miRNAs that respond to nutrient deprivation (Pant et al. 2009). In Arabidopsis miRNA395 targets *ATPS* genes *1*, *3* and *4*. and *Sultr2;1*. When miRNA395 increases during sulfate deprivation, the abundance of ATPS transcripts decreases (Kawashima et al. 2009). However, expression of genes encoding ATPS are also known to respond with moderately enhanced mRNA contents to sulfate starvation and feeding with GSH (Logan et al. 1996; Lappartient et al. 1999). Whether ATPS has a limiting role for flux control in the reduction pathway is unresolved. Over-expression experiments in tobacco cells and in *B. juncea* plants were contradictive (Hatzfeld et al. 1998; Pilon-Smits et al. 1999). The role of miRNA395 that is encoded from several loci in the Arabidopsis genome is also not clear at present and experiments with respect to short-term deprivation and long-term starvation need to be conducted to resolve these contrasting results. Interestingly, *mi395* expression is controlled by transcription factor SLIM1 (Maruyama-Nakashita et al. 2006), but not the mi395 target *Sultr2;1* due to cell-type specific patterns (Kawashima et al. 2009).

Once inside the cell, APR is in command of flux for sulfate reduction (Vauclare et al. 2002), although mutant analysis of the semi-constitutive SiR gene suggests a potentially limiting role at high flux conditions (Khan, Wirtz, Hell, unpublished). Transcriptional regulation of APR genes in response to sulfate and nitrate deficiency, environmental stresses and during day-night cycle has long been known (Brunold and Suter 1990; Kopriva 2006). Support for the major control function comes from constitutive overexpression of bacterial APR from *Pseudomonas aeruginosa* in Arabidopsis and maize that resulted in massive deregulation of primary sulfur metabolism (Tsakraklides et al. 2002; Martin et al. 2005). In addition, posttranslational control by redox processes was shown in vitro and in vivo, treatment with ozone enhanced APR activity without changes in protein abundance and in the presence of transcription or translation inhibitors (Bick et al. 2001).

Availability and demand are discussed as driving forces for regulation of sulfate uptake and reduction. Most of the conditions and metabolites tested affect the steps of the pathway similarly, although to varying extent. Under normal conditions the genes of primary sulfur metabolism are under the same control. The rate of sulfate uptake by the root more or less equals the sulfur requirement for growth of the plant (Hawkesford and De Kok 2006). The most rapid response occurs in the classical sulfate deprivation experiment where plants are grown under regular sulfate supply and are transferred to low or zero sulfate concentrations in the growth medium. Genes encoding sulfate transporters respond first, usually within less than 1 h, suggesting sensing of sulfate outside or in the cytsol of root surface cells, although evidence for this is rather indirect at present (Rouached et al. 2008; Amtmann and Blatt 2009). A catabolite repression system as proposed for control of nitrate uptake and assimilation may operate via GSH or possibly cysteine (Hell and Hillebrand 2001; Hell et al. 2002). Feeding of cysteine repressed sulfate uptake in maize seedlings (Bolchi et al. 1999), whereas only GSH but not cysteine down-regulated uptake in Arabidopsis roots (Lappartient et al. 1999), suggesting species specific differences in regulation. OAS may have a special role, because it links nitrogen and carbon metabolism with sulfur. Feeding of OAS rapidly triggers not only *Sultr* and *APR* genes in different plants, but induces numerous genes, effectively imitating the transcriptome deficiency response (Hirai et al. 2003; Hirai and Saito 2008; Hoefgen and Nikiforova 2008). These studies also revealed network links to jasmonate signalling and auxin metabolism. Over-expression of SAT to enhance internal OAS concentrations in potato enhanced *Sultr* transcription in the roots, but in long-term studies no correlation between OAS levels and sulfate uptake activity was observed (Hopkins et al. 2005). Whether OAS accumulation during sulfate deprivation is purely a secondary process that regulates the fine-tuning of cellular cysteine synthesis (see Sect. 4) or mediates actually long-term responses remains to be elucidated.

Signal transduction research in the sulfate deficiency response identified a seven base pair *cis*-element (SURE; sulfur-responsive element) in the promoters of the group 1 transporters and other sulfur-related genes. Since the SURE element is present also in nonresponsive genes more *cis*-elements are required for a sulfur-specific response (Maruyama-Nakashita et al. 2005). A genetic screen exploited a chimeric promoter of an OAS responsive *cis*-element of the bean β-conglycinin promoter with a 35S minimal promoter fused to GFP in transgenic Arabidopsis. This approach led to the isolation of several genes none of which revealed a direct regulatory function (Ohkama-Ohtsu et al. 2004; Kasajima et al. 2007). A similar screen used the promoter of Arabidopsis *Sultr1;1* fused to GFP to isolate mutants that are nonresponsive to sulfate deficiency. This identified the transcription factor SLIM1 (sulfur-limitation1) that belongs to the family of ethylene-insensitive such as trans-factors (Maruyama-Nakashita et al. 2006). SLIM1 appears to be a major regulator and was shown to be involved in the control of the sulfur deficiency response of numerous genes including degradation of glucosinolates, but not of *APR* genes. It is further puzzling that SLIM1 seems to be exclusively expressed in the vascular tissue and thus far away from Sultr1;1 occurrence in the rhizodermal cells (Maruyama-Nakashita et al. 2006). Other transcription factors were described for regulation of primary and secondary sulfur metabolism. R2R3 type Myb transcription factors were observed to be up-regulated during sulfate deficiency in microarray experiments but have not definitively been verified (Nikiforova et al. 2003). Glucosinolate biosynthesis and degradation are known to be regulated in response to biotic and abiotic stress (Malitsky et al. 2008). Several Myb factors seem to specifically share control of promoters of genes of aliphatic and indol glucosinolates (Gigolashvili et al. 2007; Hirai et al. 2007). It remains to be analyzed if some of these factors are able to address promoters of genes of primary sulfur metabolism to achieve coordination of assimilatory reduction and downstream demand.

4 Regulation of Sulfur Amino Acids Biosynthesis

4.1 Regulation of Cysteine Biosynthesis

Synthesis of cysteine is the entry point of reduced sulfur into metabolism. It can be subdivided into three steps: (1) the assimilatory sulfate reduction for provision of sulfide (Sect. 3), (2) the synthesis of the carbon and nitrogen containing backbone of cysteine, and (3) the incorporation of reduced sulfur into the organic backbone. The released cysteine is the pivotal compound in sulfur metabolism and the starting point for production of all compounds containing reduced sulfur like methionine and GSH (Hell 1997; Saito 2004).

As in bacteria, the synthesis of the carbon-nitrogen backbone of cysteine is catalyzed by serine acetyltransferase (SAT, EC 2.3.1.30), which transfers an acetyl-moiety from acetyl-coenzyme A to serine, leading to *O*-acetylserine (OAS) formation (Kredich and Tomkins 1966; Hell et al. 2002; Droux 2004). Then *O*-acetylserine(thiol)lyase (OAS-TL, EC 2.5.1.47) converts OAS into cysteine in the presence of sulfide (Hell 1997; Droux 2003). SAT and OAS-TL form a multienzyme complex called cysteine synthase complex (CSC; (Kredich et al. 1969; Bogdanova and Hell 1997), in which protein–protein interactions regulate activities of both enzymes (Droux et al. 1998; Wirtz et al. 2001).

SAT and OAS-TL are encoded by small nuclear gene families. Their members are ubiquitously transcribed and the gene products localized in the plastid, the mitochondria and the cytosol (Hell et al. 2002; Kawashima et al. 2005). In contrast to sulfate transporters, ATPS and APR (Sect. 1), SAT and OAS-TL isoforms are hardly regulated at their transcript levels in response to sulfur availability or exogenous application of OAS, a key regulator for cysteine synthesis (Hirai et al. 2003; Maruyama-Nakashita et al. 2003; Nikiforova et al. 2004; Kawashima et al.

2005). Nevertheless, transcription of OAS-TLs and SATs respond to abiotic stress conditions, which are known to perturb redox (Freeman et al. 2004; Dominguez-Solis et al. 2008; Lehmann et al. 2009). Most likely cysteine synthesis is up-regulated under these conditions to cope with the higher demand of cysteine for synthesis of GSH that is used for detoxification of reactive oxygen species (Sect. 5).

It is currently unknown why SAT and OAS-TL activities are unequally abundant in their subcellular compartments (Lunn et al. 1990; Rolland et al. 1992; Ruffet et al. 1995; Kuske et al. 1996). Purification of OAS-TL proteins in combination with reverse genetics approaches for all OAS-TL and SAT genes in Arabidopsis revealed that in leaves 90% of OAS-TL activity and protein is present in cytosol and plastids, while only a minor fraction is found in the mitochondria (Heeg et al. 2008; Lopez-Martin et al. 2008; Watanabe et al. 2008a). In contrast, 80% of total SAT activity is associated with mitochondria, while the residual activity is found equally distributed in cytosol and plastids (Watanabe et al. 2008b). None of the single knock out lines for SAT and OAS-TL show a lethal phenotype indicating that OAS and cysteine can be transported sufficiently between the subcellular compartments of the plant cell (see Fig. 2). However, the mitochondrial SAT has a pace making function in OAS net synthesis, which limits total cysteine synthesis and growth of Arabidopsis (Haas et al. 2008). The importance of mitochondrial OAS synthesis is in agreement with provision of substrates for SAT reaction, acetyl-coenzyme A and

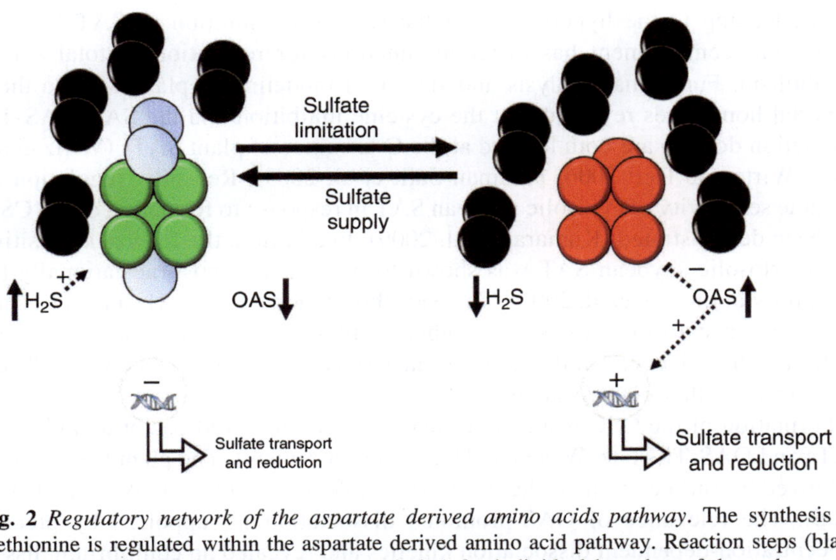

Fig. 2 *Regulatory network of the aspartate derived amino acids pathway*. The synthesis of methionine is regulated within the aspartate derived amino acid pathway. Reaction steps (black arrows) and key regulatory enzymes (grey circles) for individual branches of the pathway are highlighted. The positive (blue) and negative (red) regulatory feedback loops within the pathway are marked by dotted lines. Aspartate kinase (AK), cystathionine β-lyase (CBL), cystathionine γ-synthase (CGS), methionine γ-lyase (MGL), methionine synthase (MS), SAM-dependant methyl-transferase (MT), SAM synthase (SAMS), threonine synthase (TS)

serine, by citric acid cycle and photorespiration and the high SAT activity found in this compartment (Wirtz and Droux 2005). It also provides an elegant explanation for the unexpected growth phenotype of the mitochondrial OAS-TL knockout (Heeg et al. 2008). Only knockout of the minor abundant mitochondrial OAS-TL activity (10% of total activity) but not of plastidic (~45% of total activity) and cytosolic OAS-TL (~45% of total activity) results in reduced growth, most likely because mitochondrial OAS-TLs regulates total OAS net synthesis via formation of the CSC (see below) in mitochondria. Sulfide, which is exclusively produced in the plastid (Sect. 3), meets the OAS in the cytosol, which led to the hypothesis that under normal sulfur supply cysteine is mainly produced in cytosol (see Fig. 2). In agreement with this hypothesis the knockout of cytosolic OAS-TL most strongly affects cysteine net synthesis (Heeg et al. 2008; Watanabe et al. 2008a).

The strong limitation of cysteine synthesis by provision of OAS is further supported by the 300-fold lower activity of SAT in comparison to OAS-TL in leaves (Ruffet et al. 1994) and the successful attempts to increase cysteine contents by over-expression of SAT (Blaszczyk et al. 1999; Harms et al. 2000; Sirko et al. 2004). In contrast, OAS-TL over-expression approaches in cytosol and plastids increase the total cysteine content only marginally because OAS is limiting (Saito et al. 1994a; Saito et al. 1994b; Sirko et al. 2004).

Consequently cysteine shows no feedback-inhibition of OAS-TL which produces cysteine, but SAT that produces the limiting precursor, OAS (Noji et al. 1998). SAT activity is inhibited by cysteine in an isoform-specific manner in plants, leading to the hypothesis that the feedback inhibition of SAT in each subcellular compartment has different functions for regulation of total sulfur metabolism. Functional analysis and structural modeling of plant SAT to their bacterial homologes revealed that the cysteine inhibition and the SAT-OAS-TL interaction domain are both located at the C-terminus of plant SATs (Wirtz et al. 2001; Wirtz and Hell 2006; Feldman-Salit et al. 2009). Recently, regulation of cysteine sensitivity of cytosolic soybean SAT in response to formation of the CSC has been demonstrated (Kumaran et al. 2009). In addition, the feedback sensitivity of cytosolic soybean SAT was shown to be regulated posttranslationally by phosphorylation (Liu et al. 2006). The phosphorylation site of soybean SAT is not conserved in cytosolic SATs from other plant species, which adds a note of caution to the universal validity of posttranslationally regulated cysteine feedback sensitivity in other plant systems.

Formation of the CSC is a general feature of all analyzed bacterial and plant SATs and OAS-TLs (see Wirtz and Hell 2006 for review). The plant CSC can be stabilized by sulfide in vitro whereas OAS is able to dissociate it (Wirtz and Hell 2006). The relevance of OAS-promoted dissociation is supported by precise concentration-dependent dissociation kinetics and revealed an equilibrium dissociation constant of 57 μM OAS (Berkowitz et al. 2002). These findings suggest that the equilibrium of association/dissociation can be effectively shifted like a switch, since fluctuations of cellular OAS concentration in this range have been observed in response to sulfate and nitrate availability (Kim et al. 1999). Inside the CSC SAT is activated, while OAS-TL is efficiently inactivated by the C-terminus of SAT that binds in the active site of OAS-TL (Droux et al. 1998; Wirtz et al. 2001; Bonner et al. 2005; Francois et al. 2006). As a consequence, OAS leaves the CSC and will be converted by free OAS-TL to cysteine, if sulfide is present. Most likely substrate channeling is actively prohibited in the CSC to allow free OAS to act as a regulator of the transcriptional response toward sulfur availability (Kim et al. 1999; Hirai et al. 2003). During sulfate deficiency the OAS level increases as a result of missing sulfide. This leads to the dissociation of the complex by (1) the destabilizing effect of OAS and (2) the missing stabilization by sulfide. The reason for dual control of complex formation by OAS and sulfide could be the need for complex stabilization at very high cysteine synthesis rates. As a result of the strong dependence of cysteine synthesis for OAS, the most efficient way to increase net cysteine synthesis is a higher OAS level. Recently, the importance of CSC formation for regulation of cysteine synthesis *in planta* was demonstrated by over-expression of active and an inactive SAT, which was still able to enter the CSC in the cytosol of transgenic tobacco plants (Wirtz and Hell 2007).

4.2 *Catabolism, Storage and Transport of Cysteine*

Breakdown of cysteine can be catalyzed by two classes of enzymes. Class 1 enzymes use the reduced sulfur in cysteine to incorporate it in other metabolites, while the enzymes of class 2 break down cysteine for catabolism or release the reduced sulfur of cysteine. Cysteine desulfurases of the NifS-type (EC 2.8.1.7), which provide elemental sulfur for molybdenum cofactor synthesis in the cytosol and iron-sulfur cluster formation in mitochondria and plastids belong to class 1. Typical class 2 enzymes are: cysteine desulfhydrases (L-CDes, EC 4.4.1.15), cystine desulfhydrase (cystine lyase, EC 4.4.1.13), which breakdown cysteine or cysteine to ammonia and pyruvate (Jones et al. 2003; Bloem et al. 2004), and β-cyanoalanine synthases (EC 4.4.1.9) that use cysteine for detoxification of cyanide (Meyers and Ahmad 1991). For an update of catabolic fate of cysteine and the biosynthesis of iron sulfur clusters the reader is referred to Papenbrock et al. (2007) and Balk and Lobreaux (2005).

Long distance allocation of cysteine is mainly accomplished by phloem-specific transport of GSH (Herschbach et al. 2000, Sect. 2). The efficiency of the transport allows restoring of GSH mediated tolerance against heavy metals in roots of plants with a genetically engineered shoot-specific synthesis of GSH (Li et al. 2006). GSH can also serve as a transient storage of cysteine, which can be efficiently remobilized by degradation of GSH (see Sect. 5). Nonetheless, the multiple functions of GSH make it an unfavorable storage compound. Long term storage of cysteine is therefore achieved in proteins like the 2S-albumins in seeds (see Fujiwara et al. 2002 for review).

4.3 Biosynthesis of Methionine

Research interest in the regulation of methionine biosynthesis is driven at least in part by its economic value. Especially in legumes seeds and potato tubers, methionine is low-abundant and limits nutritional quality for feedstuff (Müntz et al. 1998; Stiller et al. 2007). For that reason, biosynthesis of methionine was first analyzed in crops like maize, potato, soybean and wheat (Müntz et al. 1998; Wang et al. 2001; Zeh et al. 2001), although recent innovations were achieved in *Arabidopsis thaliana*. In contrast to cysteine, methionine levels respond only marginally to sulfur deprivation (Hirai et al. 2004), indicating that methionine synthesis is not primarily limited by the supply of reduced sulfur. Indeed, methionine formation seems to be regulated within the regulatory network of the aspartate-derived amino acids pathway (ADAAP). The complex network for biosynthesis of lysine, methionine, isoleucine and threonine from the common precursor aspartate is depicted in Fig. 3.

One cannot overestimate the importance of allosteric regulation of key enzymes by end products of each branch of the pathway for regulation of the entire network

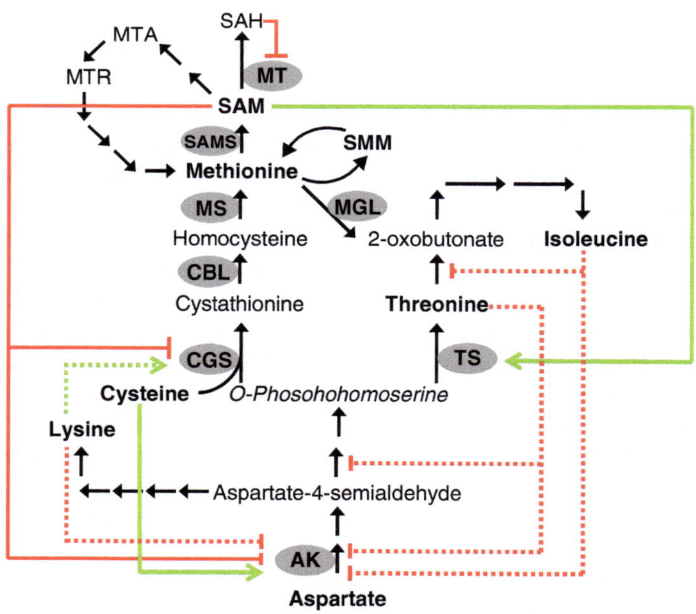

Fig. 3 *Regulatory model for the cysteine synthase complex*. Cellular sulfur availability adjusts the equilibrium of CSC, which regulates the activity of SAT and OAS-TL. Solid arrows indicate fluctuation of metabolites in response to sulfur availability, while dashed arrows represent positive (+) or negative (−) regulatory effects of these metabolites on CSC formation or gene expression. *Green circles*: active SAT, *red circles*: inactive SAT, *blue circles*: inactive OAS-TL, *black circles*: active OAS-TL

and the dependence of fluxes between the different branches. For example, the highest accumulation of methionine so far observed in plants (250-fold of wild type) is not achieved by over-expression of the key enzyme for the methionine branch of ADAAP (Cystathionine γ-synthase, CGS, EC 2.5.1.48), but by specific down-regulation of threonine synthase (TS, EC 4.2.3.1), which forces flux of aspartate into the methionine branch of the pathway (Bartlem et al. 2000; Zeh et al. 2001). TS is allosterically activated by S-adensoylmethionine (SAM) to direct flux to threonine and isoleucine formation, if methionine steady state levels are adequate (Curien et al. 1998; Laber et al. 1999; Mas-Droux et al. 2006a). Consequently, a forward genetic screen for methionine over-accumulating mutants (*mto*) using the resistance to ethionine identified *mto1-1* and *mto2-1*, which are affected in CGS and TS, respectively (Inaba et al. 1994; Bartlem et al. 2000). In summary, one can conclude that the competition of TS and CGS for their common substrate O-phosphohomoserine (OPH) is a strong determinant for methionine synthesis (Amir et al. 2002).

SAM controls not only TS activity but also methionine synthesis itself by (1) allosteric inhibition of aspartate kinase (AK, EC 2.7.2.4) and (2) down-regulation of CGS activity (Chiba et al. 1999; Chiba et al. 2003; Mas-Droux et al. 2006b; Curien et al. 2007). AK activates aspartate under consumption of ATP for further use in the ADAAP. One member of the AK protein family in Arabidopsis is strongly stimulated by presence of cysteine, which allows sufficient synthesis of OPH for methionine synthesis, if reduced sulfur is available (Curien et al. 2005). CGS catalyses the first committed step of methionine synthesis by formation of cystathionine from OPH and cysteine (Fig. 3). Cystathionine β-lyase (CBL, EC 4.4.1.8) subsequently cleaves cystathionine to release homocysteine, which is the substrate for methylation by methionine synthase (MS, EC 2.1.1.14). The activity of CGS is rate limiting and can be regulated by proteolytic removal of the N-terminal regulatory domain, which results in accumulation of methionine and SAM (Hacham et al. 2002; Hacham et al. 2006; Loizeau et al. 2007). In contrast to CGS and CBL, which are exclusively localized in plastids, MS is present in plastids and the cytosol (Fig. 2). Solely in the latter, S-adenosylmethionine synthase (SAMS, EC 2.5.1.6) combines methionine with ATP to produce SAM (Ravanel et al. 2004). Although SAM can be transported into plastids and mitochondria (Ravanel et al. 2004; Palmieri et al. 2006), it is the cytosolic SAM pool that regulates plastidic CGS activity by posttranscriptional destabilization of the CGS mRNA, at least in Arabidopsis (Chiba et al. 1999; Chiba et al. 2003). The latter observation points toward a complex regulatory circuit controlling methionine synthesis that includes communication between pool sizes of the key effectors, SAM and methionine, in the plastid and the cytosol (Fig. 1). Reduction of SAMS activity in the *mto3-1* and *mto3-2* mutants led to a 200-fold increase of methionine levels in Arabidopsis, which is probably caused by reduced incorporation of methionine in SAM and loss of feedback control via SAM (Goto et al. 2002; Shen et al. 2002). SAM levels can also be reduced by exogenous application of lysine, which down-regulates SAMS transcription and activity, allowing an efficient communication between the methionine and lysine branch of ADAAP (Hacham et al. 2007).

4.4 Catabolism, Storage and Transport of Methionine

SAMS directs about 80% of the metabolic flux of methionine to SAM, which is used by SAM-dependent methyltransferases (MTs) to methylate nucleic acids, proteins and cell wall components like lipids, lignins and pectins (see Lu 2000 for review; Shen et al. 2002; Yang et al. 2006). SAM is also substrate for nicotianamine synthase (EC 2.5.1.43), SAM-decarboxylase (EC 4.1.1.50) and 1-aminocyclopropane-1-carboxylate synthase (EC 4.4.1.14) to produce nicotianamine, polyamines and ethylene, respectively (Hesse et al. 2004). These results demonstrate that SAM is (1) the second most frequently used cofactor in nature, after ATP (Cantoni 1975; Lu 2000) and (2) mandatory for proper development of plants by influencing metal homeostasis and hormone function (Burstenbinder et al. 2007; Klatte et al. 2009). Interestingly, SAMS1 activity can be efficiently inhibited by S-nitrosylation by the natural NO donor S-nitrosoglutathione, which may mediate cross-talk between NO and ethylene signaling pathways in plants (Lindermayr et al. 2005)

Two cytosolic cycles regenerate methionine and SAM. First, the transfer of the methyl group from SAM generates S-adenosylhomocysteine (SAH), a potent inhibitor of MTs. To mitigate the toxic effects of SAH on MT activity and to recycle methionine, SAH is removed by SAH hydrolase (EC 3.3.1.1) in a reaction generating adenosine and homocysteine (Hesse et al. 2004). The latter can be used as building block for synthesis of methionine by MS as described above. Second, recycling of methionine at high rates of ethylene production is achieved by the Yang- or Met-cycle (Adams and Yang 1977, Fig. 2), which converts methylthioadenosine, the byproduct of SAM dependent ethylene formation in four steps to methionine. Recently, analysis of methylthioribose kinase (EC 2.7.1.100) revealed the significance of methionine recycling under sulfur limiting conditions in Arabidopsis (Burstenbinder et al. 2007).

Besides consumption of methionine in form of SAM, methionine can be catabolized in the cytosol by the activity of methionine γ-lyase (MGL, EC 4.4.1.11), which produces ammonia, 2-oxobutanoate and methanthiol (Rebeille et al. 2006). Methanethiol can be incorporated into cysteine, while 2-oxobutanoate can serve as a precursor for isoleucine synthesis (Rebeille et al. 2006; Goyer et al. 2007). Under normal growth condition a knockout of methionine γ-lyase activity in Arabidopsis results in no visible phenotype, but a tenfold accumulation of methionine under sulfate-limiting conditions (Goyer et al. 2007).

Long distance transport of methionine is achieved via the phloem after conversion of methionine to S-methylmethionine (SMM, see Sect. 2.2), which is also assumed to function as temporary storage of methionine in leaves (Bourgis et al. 1999). Methionine methyltransferase (EC 2.1.1.12) uses SAM as a methyl donor to form SMM from methionine and is unique in plants although it is nonessential (Bourgis et al. 1999; Ranocha et al. 2000; Kocsis et al. 2003).

5 Roles of GSH in Redox Homeostasis and Detoxification

5.1 GSH Biosynthesis and Functions

The tripeptide GSH (γ-glutamylcysteinylglycine) is ubiquitous in cells and organisms, among the few exceptions are trypanosomes that contain GSH-derived compounds (i.e., trypanothione). In plants, members of the Poaceae and Fabaceae carry GSH variants with C-terminal amino acids other than glycine (references in Meyer and Hell 2005). In all cases GSH biosynthesis is a two step, ATP-dependent process that is catalyzed by γ-glutamylcysteine ligase (GSH1) and GSH synthetase (GSH2). After the evolutionary invention of photosynthesis about 2.8 million years ago, cells encountered the problem of an increasingly oxidizing environment. The rapid oxidation of cysteine required the development of a reduced internal redox state. Indeed comparison of protein composition derived from sequenced genomes shows that today cysteine is one of the most rarely used amino acids (Pe'er et al. 2004). For maintenance of these thiol groups and cellular redox state, GSH may have evolved. The presence of GSH opened the additional possibility to develop redox-based sensing mechanisms that mediate between cell and environment. An emerging part of this role is the formation of mixed disulfides between GSH and thiols of cysteine residues in proteins. This S-glutathionylation or thiolation has originally been seen as reversible protection against the irreversible oxidation of protein thiols to sulfinic or sulfonic acids or reaction with NO to S-nitrosylated proteins, but is now considered as posttranslational modification for redox-driven signal transduction (Rinalducci et al. 2008). GSH itself freely reacts with NO to form S-nitrosoglutathione that is able to modify protein thiol groups by both protein S-nitrosylation and S-glutathionylation. The reader is referred to specialized reviews on this aspect (Foyer and Noctor 2005; Meyer and Hell 2005; Meyer 2008; Rouhier et al. 2008).

In addition, GSH acquired numerous other functions that are often specific for organism types. In plants, GSH has been suggested as a long-distance transport form of reduced sulfur, as an intermediate storage for reduced sulfur and scavenger of xenobiotics and reactive oxygen species (Fig. 1, May et al. 1998). The superfamily of glutathione-*S*-transferases (GSTs) with multiple activities toward S–C conjugation, peroxidation and secondary metabolism transport is indicative of these additional roles (Dixon et al. 2009). Among those, detoxification of xenobiotica by is one of the most widespread tasks of GSTs in eukaryotic organisms. Detoxification phase I is catalyzed by P450 enzymes, followed by phase II GSTs that form glutathione-S-conjugates to xenobiotics for either removal from potential susceptible target sites of intoxication or metabolic degradation or both. Reactive oxygen species are detoxified either by direct thiol oxidation or within the ascorbate-GSH cycle (Foyer et al. 2009b). GSH also serves as substrate for the synthesis of phytochelatins, γ-glutamylcysteine polymers with a terminal glycine residue, with high affinity to heavy metals such as cadmium via metal-sulfide chelation (Grill et al. 1985). They had long been suspected as principal heavy metal detoxification system in plants but more recently are believed to contribute homeostasis of metal micronutrients such as zinc (Tennstedt et al. 2009).

GSH is essential for survival of plants (Cairns et al. 2006; Pasternak et al. 2008). It is present in major plant cell compartments except the vacuole. However, its biosynthesis takes place only in plastids and the cytosol. Using aqueous biochemical fractionation of pea and spinach leaves GSH1 72% and 61%, respectively, of activity was found in chloroplasts and the rest in the cytosol but not mitochondria (Hell and Bergmann 1990). GSH2 activity of 24% was demonstrated in chloroplasts of photoheterotrophic tobacco cells (Hell and Bergmann 1988) and, after nonaqueous fractionation, between 47% and 64% in chloroplasts of pea leaves (Klapheck et al. 1987). In contrast, *Arabidopsis thaliana* was reported to have single genes encoding each of the enzymes with GSH1 only present in plastids and GSH2, via differential splicing, in both compartments (Fig. 1). From immunolocalization an estimated less than 10% of GSH2 protein was present in plastids and more than 90% in the cytosol (Wachter et al., 2005). In Brassicaceae and several other taxa all *GSH1* genes so far have been found to encode proteins with predicted plastid transit peptide (Gromes et al. 2008). In view of these apparent discrepancies major taxon-specific differences cannot be excluded.

In Arabidopsis this distribution implies transport of γ-EC and/or GSH across the plastid envelope. Pasternak et al. (2008) used a T-DNA null mutant of GSH2 to show that plants with only γ-EC are seedling lethal. Since *gsh1* mutant Arabidopsis plants are lethal at the embryo stage, this suggests that γ-EC compensates to a very limited extent for GSH in early developmental stages. Complementation of *gsh2* mutant plants with cytosol-specific GSH2 produced phenotypic wild type-like plants. This result implies that, due to the exclusive localization of GSH1 in plastids, γ-EC can leave plastids so as to supply the cytosol with the precursor of GSH synthesis and that GSH itself can be imported into plastids. These and other data also show that feedback inhibition of GSH1 is an important regulatory mechanism for GSH synthesis (Cairns et al. 2006; Pasternak et al. 2008).

Maintenance of redox homeostasis of GSH requires continuous re-reduction of oxidized GSH (GSSG) during the detoxification of reactive oxygen species by the ascorbic acid-GSH cycle (Foyer et al. 2009a). In Arabidopsis this is achieved by two genes encoding NADPH-dependent GSH reductases (GR). GR1 is localized in the cytosol and GR2 has a bipartite transit peptide for plastid and mitochondrial localization (Chew et al. 2003). Under nonstress conditions, GR keeps GSH mainly in the reduced form with only nanomolar concentrations present as GSSG. These observations were decisively promoted by the application of an engineered reduction-oxidation sensitive green fluorescent protein (roGFP) to plants (Meyer et al. 2007). roGFP also revealed that loss of cytosolic GR1 in an Arabidopsis *gr1* deficient mutant is partly compensated by a cytosolic backup system. NADPH-dependent thioredoxin reductase constitutes an efficient electron transfer from NADPH via cytosolic thioredoxin to GSSG to rescue *gr1* mutants at least under nonstressed conditions (Marty et al. 2009).

Since mitochondria harbor no GSH biosynthetic enzymes they must be able to import GSH (Fig. 1). Analysis of a null mutant of the ATP binding cassette (ABC) type transporter Atm1 in yeast and Atm3 (=Sta1) in Arabidopsis showed that this membrane protein contributes to maturation of iron-sulfur clusters in the cytosol (Kushnir et al. 2001). It was suggested that GSH acts as a carrier for precursors of such clusters (Balk and Lobreaux 2005). The endoplasmatic reticulum and peroxisomes have been assumed to contain GSH. In these compartments, the organelles and other possible cellular locations GSH in its reduced and/or oxidized form must be transported by a so far unknown mechanism. Membrane protein families capable of transport of GSH or GSH conjugates with rather broad substrate range have been identified by complementation of a yeast mutant deficient in plasmalemma GSH transport (Cagnac et al. 2004; Zhang et al. 2004). Whether these proteins are true GSH transporters and whether some family members mediate also intracellular glutathione transport, remains to be investigated. Clear evidence for low specificity transport of GSH and other conjugates comes from members of the superfamily of multidrug resistance-associated proteins. Several ABC transporters of this group localize to the tonoplast and are active in conjugate transport (Rea 2007).

5.2 GSH Degradation and Detoxification of Xenobiotics

Degradation of GSH in animals is part of the γ-glutamyl cycle where extracytoplasmic GSH is cleaved into its constituent amino acids that are absorbed by plasmalemma transporters followed by re-synthesis of GSH inside the cell (Meister 1995). GSH is degraded by the sequential reaction of γ-glutamyl transpeptidase, (GGT), γ-glutamyl cyclotransferase, and 5-oxoprolinase to yield glutamate and cysteinylglycine that is cleaved by peptidase. These enzyme activities are also found in plants, but studies focused on GGT protein family in Arabidopsis revealed no conclusive evidence for a plant γ-glutamyl cycle (Grzam et al. 2007; Martin et al. 2007; Ohkama-Ohtsu et al. 2007b; Ohkama-Ohtsu et al. 2007a). Three of the four *GGT* genes in Arabidopsis encode functional GGT proteins while *GGT3* apparently is a pseudogene. GGT1 and GGT2 are plasmalemma-associated and most likely exposed to the apoplast while GGT4 is tonoplast-associated and exposed to the vacuole. GGT1 is involved in the oxidative stress response by degradation of extracellular GSSG while GGT2 may contribute to GSH transport into siliques (Ohkama-Ohtsu et al. 2007b). GGT4 cleaves GSH conjugates that have been produced by GSTs in the cytosol and rapidly imported into the vacuole by ABC type transporters (Rea 2007). In vivo labeling of GSH with a fluorescent dye imitates the xenobiotics conjugation response of phase II detoxification and allows life cell imaging of conjugate transport and breakdown (Meyer and Fricker 2002; Meyer and Rausch 2008). This approach showed that Arabidopsis mutants of GGT4 were unable to initiate breakdown and accumulated GSH conjugates in the vacuole (Grzam et al. 2006; Grzam et al. 2007; Ohkama-Ohtsu et al. 2007a). These observations are supported by localization and feeding experiments with barley roots. (Ferretti et al. 2009) concluded from their results that apolastic GGT activity is a component of the system for retrieving exogenous GSH, possibly in connection with the oxidative state of the apoplast and as part of a plant γ-glutamylcycle between apoplast and cytoplasm.

Similar reactions that recognize γ-glutamylpeptides can be carried out by other enzymes in the plant cytosol. Glucosinolate engineering recently identified glutathione as a donor for reduced sulfur in biosynthesis of benzyl glucosinolates and a γ-glutamyl intermediate as product of a γ-glutamyl peptidase with a glutamine amidotransferase domain (Fig. 1, Geu-Flores et al. 2009). Phytochelatin synthase, a metal cofactor-dependent cytosolic enzyme that synthesizes γ-glutamylcysteinyl-polymers from GSH (phytochelatins) is capable of hydrolysis of the C-terminal glycine residue of GSH conjugates as a first step of conjugate degradation (Blum et al. 2007). The in vivo function of this reaction is not fully understood since the cytosol cannot further metabolize the resulting γ-glutamtylcysteine conjugates and in comparison the GSH conjugates are very efficiently imported in the vacuole by ABC transporters in the first place and fully degraded (Grzam et al. 2006; Grzam et al. 2007).

New support for operation of a γ-glutamylcycle in plants comes from the analysis of Arabidopsis T-DNA double mutants which are deficient in both genes encoding cytosolic 5-oxoprolinase, but in a GGT independent way (Ohkama-Ohtsu et al. 2008). Such plants accumulate 5-oxoprolin from GSH and have less glutamate compared to the wild-type, suggesting a rate-limiting role of 5-oxoprolinase for cytosolic GSH degradation. From metabolite data obtained for a *ggt1/ggt4/oxp1* triple mutant without detectable GGT activity in leaves it was suggested that GGTs have no major role in GSH degradation, but that γ-glutamycyclotransferase in the cytosol is the major source of 5-oxoproline formation from GSH (Ohkama-Ohtsu et al. 2008). The molecular identification of γ-glutamycyclotransferase is thus the missing link for a complete γ-glutamylcycle in plants. With respect to turnover, plasmalemma GGTs seem to contribute mostly to GSH degradation in the apoplast and vacuolar GGT4 serves in conjugate metabolization but not degradation of free GSH.

References

Adams DO, Yang SF (1977) Methionine metabolism in apple tissue: implication of S-adenosylmethionine as an intermediate in the conversion of methionine to ethylene. Plant Physiol 60:892–896

Amir R, Hacham Y, Galili G (2002) Cystathionine γ-synthase and threonine synthase operate in concert to regulate carbon flow towards methionine in plants. Trends Plant Sci 7:153–156

Amtmann A, Armengaud P (2009) Effects of N, P, K and S on metabolism: new knowledge gained from multi-level analysis. Curr Opin Plant Biol 12:275–283

Amtmann A, Blatt MR (2009) Regulation of macronutrient transport. New Phytol 181:35–52

Awazuhara M, Fujiwara T, Hayashi H, Watanabe-Takahashi A, Takahashi H, Saito K (2005) The function of SULTR2;1 sulfate transporter during seed development in *Arabidopsis thaliana*. Physiol Plant 125:95–105

Balk J, Lobreaux S (2005) Biogenesis of iron-sulfur proteins in plants. Trends Plant Sci 10:324–331

Bartlem D, Lambein I, Okamoto T, Itaya A, Uda Y, Kijima F, Tamaki Y, Nambara E, Naito S (2000) Mutation in the threonine synthase gene results in an over-accumulation of soluble methionine in Arabidopsis. Plant Physiol 123:101–110

Baxter I, Muthukumar B, Park HC, Buchner P, Lahner B, Danku J, Zhao K, Lee J, Hawkesford MJ, Guerinot ML, Salt DE (2008) Variation in molybdenum content across broadly distributed populations of *Arabidopsis thaliana* is controlled by a mitochondrial molybdenum transporter (MOT1). PLoS Genet 4:e1000004

Bell CI, Clarkson DT, Cram WJ (1995) Sulphate supply and its regulation of transport in roots of a tropical legume *Macroptilium atropurpureum* cv. Siratro J Exp Bot 282:65–71

Bell CS, Cram WJ, Clarkson DT (1994) Compartmental analysis of $^{35}SO_4^{2-}$ exchange kinetics in roots and leaves of a tropical legume *Macroptilium atropurpureum* cv. Siratro. J Exp Bot 45:879–886

Benning C, Garavito RM, Shimojima M (2008) Sulfolipid biosynthesis and function in plants. In: Hell R, Dahl C, Leustek T (eds) Sulfur Metabolism in phototrophicorganisms. Springer, Dordrecht, The Netherlands, pp 185–200

Berkowitz O, Wirtz M, Wolf A, Kuhlmann J, Hell R (2002) Use of biomolecular interaction analysis to elucidate the regulatory mechanism of the cysteine synthase complex from *Arabidopsis thaliana*. J Biol Chem 277:30629–30634

Bick JA, Aslund F, Chen Y, Leustek T (1998) Glutaredoxin function for the carboxyl-terminal domain of the plant-type 5′-adenylylsulfate reductase. Proc Natl Acad Sci USA 95:8404–8409

Bick JA, Dennis JJ, Zylstra GJ, Nowack J, Leustek T (2000) Identification of a new class of 5′-adenylylsulfate (APS) reductases from sulfate-assimilating bacteria. J Bacteriol 182:135–142

Bick JA, Setterdahl AT, Knaff DB, Chen Y, Pitcher LH, Zilinskas BA, Leustek T (2001) Regulation of the plant-type 5′-adenylyl sulfate reductase by oxidative stress. Biochemistry 40:9040–9048

Blake-Kalff MM, Harrison KR, Hawkesford MJ, Zhao FJ, McGrath SP (1998) Distribution of sulfur within oilseed rape leaves in response to sulfur deficiency during vegetative growth. Plant Physiol 118:1337–1344

Blaszczyk A, Brodzik R, Sirko A (1999) Increased resistance to oxidative stress in transgenic tobacco plants overexpressing bacterial serine acetyltransferase. Plant J 20:237–243

Bloem E, Riemenschneider A, Volker J, Papenbrock J, Schmidt A, Salac I, Haneklaus S, Schnug E (2004) Sulphur supply and infection with *Pyrenopeziza brassicae* influence L-cysteine desulphydrase activity in *Brassica napus* L. J Exp Bot 55:2305–2312

Blum R, Beck A, Korte A, Stengel A, Letzel T, Lendzian K, Grill E (2007) Function of phytochelatin synthase in catabolism of glutathione-conjugates. Plant J 49:740–749

Bogdanova N, Hell R (1997) Cysteine synthesis in plants: protein-protein interactions of serine acetyltransferase from *Arabidopsis thaliana*. Plant J 11:251–262

Bolchi A, Petrucco S, Tenca PL, Foroni C, Ottonello S (1999) Coordinate modulation of maize sulfate permease and ATP sulfurylase mRNAs in response to variations in sulfur nutritional status: stereospecific down-regulation by L-cysteine. Plant Mol Biol 39:527–537

Bonner ER, Cahoon RE, Knapke SM, Jez JM (2005) Molecular basis of cysteine biosynthesis in plants: structural and functional analysis of *O*-acetylserine sulfhydrylase from *Arabidopsis thaliana*. J Biol Chem 280:38803–38813

Bourgis F, Roje S, Nuccio ML, Fisher DB, Tarczynski MC, Li C, Herschbach C, Rennenberg H, Pimenta MJ, Shen TL, Gage DA, Hanson AD (1999) S-methylmethionine plays a major role in phloem sulfur transport and is synthesized by a novel type of methyltransferase. Plant Cell 11:1485–1498

Brunold C, Suter M (1990) Adenosine-5-phosphosulfate sulfotransferase. In Methods Plant Biochem., P. Lea, ed (London: Academic Press), pp. 339–343

Brunold C, Rennenberg H (1997) Regulation of sulfur metabolism in plants: first molecular approaches. Progr Bot 58:164–186

Buchner P, Takahashi H, Hawkesford MJ (2004a) Plant sulphate transporters: co-ordination of uptake, intracellular and long-distance transport. J Exp Bot 55:1765–1773

Buchner P, Stuiver CEE, Westerman S, Wirtz M, Hell R, Hawkesford MJ, De Kok LJ (2004b) Regulation of sulfate uptake and expression of sulfate transporter genes in *Brassica oleracea* as affected by atmospheric H_2S and pedospheric sulfate nutrition. Plant Physiol 136:3396–3408

Burgener M, Suter M, Jones S, Brunold C (1998) Cyst(e)ine is the transport metabolite of assimilated sulfur from bundle-sheath to mesophyll cells in maize leaves. Plant Physiol 116:1315–1322

Burow M, Wittstock U, Gershenzon J (2008) Sulfur-containing secondary metabolites and their role in plant defense. In: Hell R, Dahl C, Leustek T (eds) Sulfur metabolism in phototrophic organisms. Springer, Dordrecht, The Netherlands, pp 205–226

Burstenbinder K, Rzewuski G, Wirtz M, Hell R, Sauter M (2007) The role of methionine recycling for ethylene synthesis in *Arabidopsis*. Plant J 49:238–249

Cagnac O, Bourbouloux A, Chakrabarty D, Zhang M-Y, Delrot S (2004) AtOPT6 transports glutathione derivatives and is induced by primisulfuron. Plant Physiol 135:1378–1387

Cairns NG, Pasternak M, Wachter A, Cobbett CS, Meyer AJ (2006) Maturation of *Arabidopsis* seeds is dependent on glutathione biosynthesis within the embryo. Plant Physiol 141:446–455

Cannon G, Ward L, Case C, Heinhorst S (1999) The 68 kDa DNA compacting nucleoid protein from soybean chloroplasts inhibits DNA synthesis in vitro. Plant Mol Biol 39:835–845

Cantoni GL (1975) Biological methylation: selected aspects. Annu Rev Biochem 44:435–451

Chen S, Petersen BL, Olsen CE, Schulz A, Halkier BA (2001) Long-distance phloem transport of glucosinolates in *Arabidopsis*. Plant Physiol 127:194–201

Chew O, Whelan J, Millar AH (2003) Molecular definition of the ascorbate-glutathione cycle in *Arabidopsis* mitochondria reveals dual targeting of antioxidant defenses in plants. J Biol Chem 278:46869–46877

Chiba Y, Sakurai R, Yoshino M, Ominato K, Ishikawa M, Onouchi H, Naito S (2003) S-adenosyl-L-methionine is an effector in the posttranscriptional autoregulation of the cystathionine gamma-synthase gene in *Arabidopsis*. Proc Natl Acad Sci USA 100:10225–10230

Chiba Y, Ishikawa M, Kijima F, Tyson RH, Kim J, Yamamoto A, Nambara E, Leustek T, Wallsgrove RM, Naito S (1999) Evidence for autoregulation of cystathionine γ-synthase mRNA stability in Arabidopsis. Science 286:1371–1374

Clarkson DT, Smith FW, Vanden Berg PJ (1983) Regulation of sulphate transport in a tropical legume, *Macroptilium atropurpureum*, cv. Siratro. J Exp Bot 34:1463–1483

Cooper RM, Williams JS (2004) Elemental sulphur as an induced antifungal substance in plant defence. J Exp Bot 55:1947–1953

Crane BR, Siegel LM, Getzoff ED (1995) Sulfite reductase structure at 1.6 Å evolution and catalysis for reduction of inorganic anions. Science 270:59–67

Curien G, Job D, Douce R, Dumas R (1998) Allosteric activation of Arabidopsis threonine synthase by S-adenosylmethionine. Biochemistry 37:13212–13221

Curien G, Ravanel S, Robert M, Dumas R (2005) Identification of six novel allosteric effectors of *Arabidopsis thaliana* aspartate kinase-homoserine dehydrogenase isoforms. Physiological context sets the specificity. J Biol Chem 280:41178–41183

Curien G, Laurencin M, Robert-Genthon M, Dumas R (2007) Allosteric monofunctional aspartate kinases from *Arabidopsis*. FEBS J 274:164–176

Dahl C, Hell R, Leustek T, Knaff D (2008) Introduction to sulfur metabolism in phototrophic organisms. In: Hell R, Dahl C, Leustek T (eds) Sulfur metabolism in phototrophic organisms. Springer, Dordrecht, The Netherlands, pp 1–16

Dixon DP, Hawkins T, Hussey PJ, Edwards R (2009) Enzyme activities and subcellular localization of members of the *Arabidopsis* glutathione transferase superfamily. J Exp Bot 60:1207–1218

Dominguez-Solis JR, He Z, Lima A, Ting J, Buchanan BB, Luan S (2008) A cyclophilin links redox and light signals to cysteine biosynthesis and stress responses in chloroplasts. Proc Natl Acad Sci USA 105:16386–16391

Droux M (2003) Plant serine acetyltransferase: new insights for regulation of sulphur metabolism in plant cells. Plant Physiol Biochem 41:619–627

Droux M (2004) Sulfur assimilation and the role of sulfur in plant metabolism: a survey. Photosynth Res 79:331–348

Droux M, Ruffet ML, Douce R, Job D (1998) Interactions between serine acetyltransferase and *O*-acetylserine (thiol) lyase in higher plants-structural and kinetic properties of the free and bound enzymes. Eur J Biochem 255:235–245

Dubousset L, Abdallah M, Desfeux AS, Etienne P, Meuriot F, Hawkesford MJ, Gombert J, Segura R, Bataille MP, Reze S, Bonnefoy J, Ameline AF, Ourry A, Le Dily F, Avice JC (2009) Remobilization of leaf S compounds and senescence in response to restricted sulphate supply during the vegetative stage of oilseed rape are affected by mineral N availability. J Exp Bot 60:3239–3253

Falkowski PG (2006) Evolution: tracing oxygen's imprint on earth's metabolic evolution. Science 311:1724–1725

Feldman-Salit A, Wirtz M, Hell R, Wade RC (2009) A mechanistic model of the cysteine synthase complex. J Mol Biol 386:37–59

Ferreira RM, Teixeira AR (1992) Sulfur starvation in Lemna leads to degradation of ribulose-bisphosphate carboxylase without plant death. J Biol Chem 267:7253–7257

Ferretti M, Destro T, Tosatto SCE, La Rocca L, Rascio N, Masi A (2009) γ-glutamyl transferase in the cell wall participates in extracellular glutathione salvage from the root apoplast. New Phytol 181:115–126

Fitzpatrick KL, Tyerman SD, Kaiser BN (2008) Molybdate transport through the plant sulfate transporter SHST1. FEBS Lett 582:1508–1513

Foyer CH, Noctor G (2005) Redox homeostasis and antioxidant signaling: a metabolic interface between stress perception and physiological responses. Plant Cell 17:1866–1875

Foyer CH, Bloom AJ, Queval G, Noctor G (2009a) Photorespiratory metabolism: genes, mutants, energetics, and redox signaling. Annu Rev Plant Biol 60:455–484

Foyer CH, Noctor G, Buchanan B, Dietz KJ, Pfannschmidt T (2009b) Redox regulation in photosynthetic organisms: signaling, acclimation, and practical implications. Antioxid Redox Signal 11:861–905

Francois JA, Kumaran S, Jez JM (2006) Structural basis for interaction of *O*-acetylserine sulfhydrylase and serine acetyltransferase in the *Arabidopsis* cysteine synthase complex. Plant Cell 18:3647–3655

Freeman JL, Persans MW, Nieman K, Albrecht C, Peer W, Pickering IJ, Salt DE (2004) Increased glutathione biosynthesis plays a role in nickel tolerance in *Thlaspi* nickel hyperaccumulators. Plant Cell 16:2176–2191

Fujiwara T, Nambara E, Yamagishi K, Goto DB, Naito S (2002) Storage proteins. The Arabidopsis Book, 1–12

Fukuda H, Hirakawa Y, Sawa S (2007) Peptide signaling in vascular development. Curr Opin Plant Biol 10:477–482

Geu-Flores F, Nielsen MT, Nafisi M, Moldrup ME, Olsen CE, Motawia MS, Halkier BA (2009) Glucosinolate engineering identifies a γ-glutamyl peptidase. Nat Chem Biol 5:575–577

Gigolashvili T, Berger B, Mock HP, Muller C, Weisshaar B, Flugge UI (2007) The transcription factor HIG1/MYB51 regulates indolic glucosinolate biosynthesis in *Arabidopsis thaliana*. Plant J 50:886–901

Gilbert SM, Clarkson DT, Cambridge M, Lambers H, Hawkesford MJ (1997) SO42-deprivation has an early effect on the content of ribulose-1, 5-bisphosphate carboxylase/oxygenase and photosynthesis in young leaves of wheat. Plant Physiol 115:1231–1239

Giordano M, Norici A, Hell R (2005) Sulfur and phytoplankton: acquisition, metabolism and impact on the environment. New Phytol 166:371–382

Glawischnig E (2007) Camalexin. Phytochem 68:401–406

Goto DB, Ogi M, Kijima F, Kumagai T, van Werven F, Onouchi H, Naito S (2002) A single-nucleotide mutation in a gene encoding S-adenosylmethionine synthetase is associated with methionine over-accumulation phenotype in *Arabidopsis thaliana*. Genes Genet Syst 77:89–95

Goyer A, Collakova E, Shachar-Hill Y, Hanson AD (2007) Functional characterization of a methionine γ-lyase in *Arabidopsis* and its implication in an alternative to the reverse trans-sulfuration pathway. Plant Cell Physiol 48:232–242

Grill E, Winnacker EL, Zenk MH (1985) Phytochelatins: the princial heavy-metal complexing peptides of higher plants. Science 230:674–676

Gromes R, Hothorn M, Lenherr ED, Rybin V, Scheffzek K, Rausch, T (2008) The redox switch of γ-glutamylcysteine ligase via a reversible monomer-dimer transition is a mechanism unique to plants. Plant J

Gross A, Brückner G, Heldt HW, Flügge U-I (1990) Comparison of the kinetic properties, inhibition and labelling of the phosphate translocators from maize and spinach mesophyll chloroplasts. Planta 180:262–271

Grzam A, Martin M, Hell R, Meyer A (2007) γ-Glutamyl transpeptidase GGT4 initiates vacuolar degradation of glutathione S-conjugates in Arabidopsis. FEBS Lett 581:3131–3138

Grzam A, Tennstedt P, Clemens,S, Hell R, Meyer AJ (2006) Vacuolar sequestration of glutathione S-conjugates outcompetes a possible degradation of the glutathione moiety by phytochelatin synthase. FEBS Lett 580:6384–6390

Gutierrez-Marcos JF, Roberts MA, Campbell EI, Wray JL (1996) Three members of a novel small gene-family from *Arabidopsis thaliana* able to complement functionally an *Escherichia coli* mutant defective in PAPS reductase activity encode proteins with a thioredoxin-like domain and "APS reductase" activity. Proc Natl Acad Sci USA 93:13377–13382

Haas FH, Heeg C, Queiroz R, Bauer A, Wirtz M, Hell R (2008) Mitochondrial serine acetyl-transferase functions as a pacemaker of cysteine synthesis in plant cells. Plant Physiol 148:1055–1067

Hacham Y, Avraham T, Amir R (2002) The N-terminal region of Arabidopsis cystathionine γ-synthase plays an important regulatory role in methionine metabolism. Plant Physiol 128:454–462

Hacham Y, Schuster G, Amir R (2006) An in vivo internal deletion in the N-terminus region of Arabidopsis cystathionine γ-synthase results in CGS expression that is insensitive to methionine. Plant J 45:955–967

Hacham Y, Song L, Schuster G, Amir R (2007) Lysine enhances methionine content by modulating the expression of S-adenosylmethionine synthase. Plant J 51:850–861

Halkier B, Gershenzon J (2006) Biology and biochemistry of glucosinolates. Annu Rev Plant Biol 57:303–333

Harms K, von Ballmoos P, Brunold C, Hofgen R, Hesse H (2000) Expression of a bacterial serine acetyltransferase in transgenic potato plants leads to increased levels of cysteine and glutathione. Plant J 22:335–343

Hatzfeld Y, Cathala N, Grignon C, Davidian J-C (1998) Effect of ATP sulfurylase overexpression in bright yellow 2 tobacco cells. Regulation of ATP sulfurylase and SO42-transport activities. Plant Physiol 116:1307–1313

Hawkesford MJ (2003) Transporter gene families in plants: the sulphate transporter gene family redundancy or specialization? Physiol Plant 117:155–163

Hawkesford MJ (2008) Uptake, distribution and subcellular transport of sulfate. In: Hell R, Dahl C, Leustek T (eds) Sulfur Metabolism in Phototrophic Organisms. Springer, Dordrecht, The Netherlands, pp 17–32

Hawkesford MJ, De Kok LJ (2006) Managing sulphur metabolism in plants. Plant Cell Environ 29:382–395

Heeg C, Kruse C, Jost R, Gutensohn M, Ruppert T, Wirtz M, Hell R (2008) Analysis of the *Arabidopsis O*-acetylserine(thiol)lyase gene family demonstrates compartment-specific differences in the regulation of cysteine synthesis. Plant Cell 20:168–185

Hell R (1997) Molecular physiology of plant sulfur metabolism. Planta 202:138–148

Hell R, Bergmann L (1988) Glutathione synthetase in tobacco suspension cultures: catalytic properties and localization. Physiol Plant 72:70–76

Hell R, Bergmann L (1990) γ-Glutamylcysteine synthetase in higher plants: catalytic properties and subcellular localization. Planta 180:603–612

Hell R, Hillebrand H (2001) Plant concepts for mineral acquisition and allocation. Curr Opin Biotechnol 12:161–168

Hell R, Jost R, Berkowitz O, Wirtz M (2002) Molecular and biochemical analysis of the enzymes of cysteine biosynthesis in the plant *Arabidopsis thaliana*. Amino Acids 22:245–257

Hernández-Sebastià C, Varin L, Marsolais F (2008) Sulfotransferases from plants, algae and phototrophic bacteria. In: Hell R, Dahl C, Leustek T (eds) Sulfur metabolism in phototrophic organisms. Springer, Dordrecht, The Netherlands, pp 111–130

Herschbach C, van Der Zalm E, Schneider A, Jouanin L, De Kok LJ, Rennenberg H (2000) Regulation of sulfur nutrition in wild-type and transgenic poplar over-expressing γ-glutamylcysteine synthetase in the cytosol as affected by atmospheric H_2S. Plant Physiol 124:461–473

Hesse H, Kreft O, Maimann S, Zeh M, Hoefgen R (2004) Current understanding of the regulation of methionine biosynthesis in plants. J Exp Bot 55:1799–1808

Higashi Y, Hirai MY, Fujiwara T, Naito S, Noji M, Saito K (2006) Proteomic and transcriptomic analysis of Arabidopsis seeds: molecular evidence for successive processing of seed proteins and its implication in the stress response to sulfur nutrition. Plant J 48:557–571

Hirai M, Fujiwara T, Awazuhara M, Kimura T, Noji M, Saito K (2003) Global expression profiling of sulfur-starved Arabidopsis by DNA macroarray reveals the role of *O*-acetyl-L-serine as a general regulator of gene expression in response to sulfur nutrition. Plant J 33:651–663

Hirai MY, Saito K (2008) Analysis of systemic sulfur metabolism in plants using integrated '-omics' strategies. Mol Biosyst 4:967–973

Hirai MY, Yano M, Goodenowe DB, Kanaya S, Kimura T, Awazuhara M, Arita M, Fujiwara T, Saito K (2004) Integration of transcriptomics and metabolomics for understanding of global responses to nutritional stresses in *Arabidopsis thaliana*. PNAS 101:10205–10210

Hirai MY, Sugiyama K, Sawada Y, Tohge T, Obayashi T, Suzuki A, Araki R, Sakurai N, Suzuki H, Aoki K, Goda H, Nishizawa OI, Shibata D, Saito K (2007) Omics-based identification of Arabidopsis Myb transcription factors regulating aliphatic glucosinolate biosynthesis. PNAS 104:6478–6483

Hodson RC, Schiff JA (1971) Studies of sulfate utilization by algae: 9. Fractionation of a cell-free system from *Chlorella* into two activities necessary for the reduction of adenosine 3′-phosphate 5′-phosphosulfate to acid-volatile radioactivity. Plant Physiol 47:300–305

Hoefgen R, Nikiforova VJ (2008) Metabolomics integrated with transcriptomics: assessing systems response to sulfur-deficiency stress. Physiol Plant 132:190–198

Hopkins L, Parmar S, Blaszczyk A, Hesse H, Hoefgen R, Hawkesford MJ (2005) *O*-acetylserine and the regulation of expression of genes encoding components for sulfate uptake and assimilation in potato. Plant Physiol 138:433–440

Hothorn M, Wachter A, Gromes R, Stuwe T, Rausch T, Scheffzek K (2006) Structural basis for the redox control of plant glutamate cysteine ligase. J Biol Chem 281:27557–27565

Howarth J, Parmar S, Barraclough P, Hawkesford M (2009) A sulphur deficiency-induced gene, *sdi1*, involved in the utilization of stored sulphate pools under sulphur-limiting conditions has potential as a diagnostic indicator of sulphur nutritional status. Plant Biotechnol J 7:200–209

Howarth JR, Parmar S, Jones J, Shepherd CE, Corol DI, Galster AM, Hawkins ND, Miller SJ, Baker JM, Verrier PJ, Ward JL, Beale MH, Barraclough PB, Hawkesford MJ (2008)

Co-ordinated expression of amino acid metabolism in response to N and S deficiency during wheat grain filling. J Exp Bot 59:3675–3689

Inaba K, Fujiwara T, Hayashi H, Chino M, Komeda Y, Naito S (1994) Isolation of an *Arabidopsis thaliana* mutant, *mto1*, that overaccumulates soluble methionine. Plant Physiol 104:881–887

Jez JM, Cahoon RE, Chen S (2004) *Arabidopsis thaliana* glutamate-cysteine ligase: functional properties, kinetic mechanism, and regulation of activity. J Biol Chem 279:33463–33470

Jones PR, Manabe T, Awazuhara M, Saito K (2003) A new member of plant CS-lyases. A cystine lyase from *Arabidopsis thaliana*. J Biol Chem 278:10291–10296

Kasajima I, Ohkama-Ohtsu N, Ide Y, Hayashi H, Yoneyama T, Suzuki Y, Naito S, Fujiwara T (2007) The BIG gene is involved in regulation of sulfur deficiency-responsive genes in *Arabidopsis thaliana*. Physiol Plant 129:351–363

Kataoka T, Hayashi N, Yamaya T, Takahashi H (2004a) Root-to-shoot transport of sulfate in *Arabidopsis*. Evidence for the role of SULTR3;5 as a component of low-affinity sulfate transport system in the root vasculature. Plant Physiol 136:4198–4204

Kataoka T, Watanabe-Takahashi A, Hayashi N, Ohnishi M, Mimura T, Buchner P, Hawkesford MJ, Yamaya T, Takahashi H (2004b) Vacuolar sulfate transporters are essential determinants controlling internal distribution of sulfate in *Arabidopsis*. Plant Cell 16:2693–2704

Kawashima CG, Berkowitz O, Hell R, Noji M, Saito K (2005) Characterization and expression analysis of a serine acetyltransferase gene family involved in a key step of the sulfur assimilation pathway in *Arabidopsis*. Plant Physiol 137:220–230

Kawashima CG, Yoshimoto N, Maruyama-Nakashita A, Tsuchiya YN, Saito K, Takahashi H, Dalmay T (2009) Sulphur starvation induces the expression of microRNA-395 and one of its target genes but in different cell types. Plant J 57:313–321

Kim H, Hirai MY, Hayashi H, Chino M, Naito S, Fujiwara T (1999) Role of *O*-acetyl-L-serine in the coordinated regulation of the expression of a soybean seed storage-protein gene by sulfur and nitrogen nutrition. Planta 209:282–289

Klapheck S, Latus C, Bergmann L (1987) Localization of glutathione synthetase and distribution of glutathione in leaf cells of *Pisum sativum* L. J Plant Physiol 131:123–131

Klatte M, Schuler M, Wirtz M, Fink-Straube C, Hell R, Bauer P (2009) The analysis of *Arabidopsis* nicotianamine synthase mutants reveals functions for nicotianamine in seed iron loading and iron deficiency responses. Plant Physiol 150:257–271

Klonus D, Hofgen R, Willmitzer L, Riesmeier JW (1994) Isolation and characterization of two cDNA clones encoding ATP-sulfurylases from potato by complementation of a yeast mutant. Plant J 6:105–112

Kocsis MG, Ranocha P, Gage DA, Simon ES, Rhodes D, Peel GJ, Mellema S, Saito K, Awazuhara M, Li C, Meeley RB, Tarczynski MC, Wagner C, Hanson AD (2003) Insertional inactivation of the methionine s-methyltransferase gene eliminates the s-methylmethionine cycle and increases the methylation ratio. Plant Physiol 131:1808–1815

Kopriva S (2006) Regulation of sulfate assimilation in *Arabidopsis* and beyond. Ann Bot 97:479–495

Kopriva S, Koprivova A (2004) Plant adenosine 5′-phosphosulphate reductase: the past, the present, and the future. J Exp Bot 55:1775–1783

Kopriva S, Koprivova A (2005) Sulfate assimilation and glutathione synthesis in C_4 plants. Photosynth Res 86:363–372

Kopriva S, Wiedemann G, Reski R (2007a) Sulfate assimilation in basal land plants – what does genomic sequencing tell us? Plant Biol 9:556–564

Kopriva S, Fritzemeier K, Wiedemann G, Reski R (2007b) The putative moss 3′-phosphoadenosine-5′-phosphosulfate reductase is a novel form of adenosine-5′-phosphosulfate reductase without an iron-sulfur cluster. J Biol Chem 282:22930–22938

Koprivova A, Meyer AJ, Schween G, Herschbach C, Reski R, Kopriva S (2002) Functional knockout of the adenosine 5′-phosphosulfate reductase gene in *Physcomitrella patens* revives an old route of sulfate assimilation. J Biol Chem 277:32195–32201

Kredich NM, Tomkins GM (1966) The enzymic synthesis of L-cysteine in *Escherichia coli* and *Salmonella typhimurium*. J Biol Chem 241:4955–4965

Kredich NM, Becker MA, Tomkins GM (1969) Purification and characterization of cysteine synthetase, a bifunctional protein complex, from *Salmonella typhimurium*. J Biol Chem 244:2428–2439

Krueger RJ, Siegel LM (1982) Evidence for siroheme-Fe_4S_4 interaction in spinach ferredoxin-sulfite reductase. Biochemistry 21:2905–2909

Kruse C, Jost R, Hillebrand H, Hell R (2005) Sulfur-rich proteins and their agrobiotechnological potential for resistance to plant pathogens. FAL Agricult Res 283:73–80

Kruse C, Jost R, Lipschis M, Kopp B, Hartmann M, Hell R (2007) Sulfur-enhanced defence: effects of sulfur metabolism, nitrogen supply, and pathogen lifestyle. Plant Biol 9:608–619

Kumaran S, Yi H, Krishnan HB, Jez JM (2009) Assembly of the cysteine synthase complex and the regulatory role of protein-protein interactions. J Biol Chem 284:10268–10275

Kushnir S, Babiychuk E, Storozhenko S, Davey MW, Papenbrock J, De Rycke R, Engler G, Stephan UW, Lange H, Kispal G, Lill R, Van Montagu M (2001) A mutation of the mitochondrial ABC transporter Sta1 leads to dwarfism and chlorosis in the *Arabidopsis* mutant *starik*. Plant Cell 13:89–100

Kuske CR, Hill KK, Guzman E, Jackson PJ (1996) Subcellular location of *O*-acetylserine sulfhydrylase isoenzymes in cell cultures and plant tissues of *Datura innoxia* Mill. Plant Physiol 112:659–667

Kutz A, Muller A, Hennig P, Kaiser WM, Piotrowski M, Weiler EW (2002) A role for nitrilase 3 in the regulation of root morphology in sulphur-starving *Arabidopsis thaliana*. Plant J 30:95–106

Laber B, Maurer W, Hanke C, Grafe S, Ehlert S, Messerschmidt A, Clausen T (1999) Characterization of recombinant *Arabidopsis thaliana* threonine synthase. Eur J Biochem 263:212–221

Lappartient AG, Touraine B (1996) Demand-driven control of root ATP sulfurylase activity and SO_4^{2-} uptake in intact Canola. Plant Physiol 111:147–157

Lappartient AG, Vidmar JJ, Leustek T, Glass AD, Touraine B (1999) Inter-organ signaling in plants: regulation of ATP sulfurylase and sulfate transporter genes expression in roots mediated by phloem-translocated compound. Plant J 18:89–95

Lass B, Ullrich-Eberius CI (1983) Evidence for proton/sulfate cotransport and its kinetics in *Lemna gibba* G1. Planta 161:53–60

Lehmann M, Schwarzlander M, Obata T, Sirikantaramas S, Burow M, Olsen CE, Tohge T, Fricker MD, Moller BL, Fernie AR, Sweetlove LJ, Laxa M (2009) The metabolic response of Arabidopsis roots to oxidative stress is distinct from that of heterotrophic cells in culture and highlights a complex relationship between the levels of transcripts, metabolites, and flux. Mol Plant 2:390–406

Leustek T (1996) Molecular genetics of sulfate assimilation in plants. Physiol Plant 97:411–419

Leustek T, Martin MN, Bick J-A, Davies JP (2000) Pathways and regulation of sulfur metabolism revealed through molecular and genetic studies. Ann Rev Plant Physiol Plant Mol Biol 51:141–165

Lewandowska M, Sirko A (2008) Recent advances in understanding plant response to sulfur-deficiency stress. Acta Biochim Pol 55:457–471

Li Y, Dankher OP, Carreira L, Smith AP, Meagher RB (2006) The shoot-specific expression of γ-glutamylcysteine synthetase directs the long-distance transport of thiol-peptides to roots conferring tolerance to mercury and arsenic. Plant Physiol 141:288–298

Lindermayr C, Saalbach G, Durner J (2005) Proteomic identification of S-nitrosylated proteins in *Arabidopsis*. Plant Physiol 137:921–930

Liu F, Yoo B-C, Lee J-Y, Pan W, Harmon AC (2006) Calcium-regulated phosphorylation of soybean serine acetyltransferase in response to oxidative stress. J Biol Chem 281:27405–27415

Logan HM, Cathala N, Grignon C, Davidian JC (1996) Cloning of a cDNA encoded by a member of the *Arabidopsis thaliana* ATP sulfurylase multigene family. Expression studies in yeast and in relation to plant sulfur nutrition. J Biol Chem 271:12227–12233

Loizeau K, Gambonnet B, Zhang GF, Curien G, Jabrin S, Van Der Straeten D, Lambert WE, Rebeille F, Ravanel S (2007) Regulation of one-carbon metabolism in Arabidopsis: the N-terminal regulatory domain of cystathionine γ-synthase is cleaved in response to folate starvation. Plant Physiol 145:491–503

Lopez-Martin MC, Becana M, Romero LC, Gotor C (2008) Knocking out cytosolic cysteine synthesis compromises the antioxidant capacity of the cytosol to maintain discrete concentrations of hydrogenperoxide in *Arabidopsis*. Plant Physiol 147:562–572

Loudet O, Saliba-Colombani V, Camilleri C, Calenge F, Gaudon V, Koprivova A, North K, Kopriva S, Daniel-Vedele F (2007) Natural variation for sulfate content in *Arabidopsis thaliana* is highly controlled by APR2. Nat Genet 39:896–900

Lu SC (2000) S-Adenosylmethionine. Int J Biochem Cell Biol 32:391–395

Lunn JE, Droux M, Martin J, Douce R (1990) Localization of ATP-sulfurylase and *O*-acetylserine (thiol)lyase in spinach leaves. Plant Physiol 94:1345–1352

Malitsky S, Blum E, Less H, Venger I, Elbaz M, Morin S, Eshed Y, Aharoni A (2008) The transcript and metabolite networks affected by the two clades of *Arabidopsis* glucosinolate biosynthesis regulators. Plant Physiol 148:2021–2049

Martin MN, Tarczynski MC, Shen B, Leustek T (2005) The role of 5′-adenylylsulfate reductase in controlling sulfate reduction in plants. Photosynth Res 86:1–15

Martin MN, Saladores PH, Lambert E, Hudson AO, Leustek T (2007) Localization of members of the γ-glutamyl transpeptidase family identifies sites of glutathione and Glutathione S-conjugate hydrolysis. Plant Physiol 144:1715–1732

Martin W, Rotte C, Hoffmeister M, Theissen U, Gelius-Dietrich G, Ahr S, Henze K (2003) Early cell evolution, eukaryotes, anoxia, sulfide, oxygen, fungi first (?), and a tree of genomes revisited. IUBMB Life 55:193–204

Marty L, Siala W, Schwarzlander M, Fricker MD, Wirtz M, Sweetlove LJ, Meyer Y, Meyer AJ, Reichheld JP, Hell R (2009) The NADPH-dependent thioredoxin system constitutes a functional backup for cytosolic glutathione reductase in *Arabidopsis*. Proc Natl Acad Sci USA 106:9109–9114

Maruyama-Nakashita A, Inoue E, Watanabe-Takahashi A, Yamaya T, Takahashi H (2003) Transcriptome profiling of sulfur-responsive genes in *Arabidopsis* reveals global effects of sulfur nutrition on multiple metabolic pathways. Plant Physiol 132:597–605

Maruyama-Nakashita A, Nakamura Y, Watanabe-Takahashi A, Yamaya T, Takahashi H (2004) Induction of SULTR1;1 sulfate transporter in Arabidopsis roots involves protein phosphorylation/dephosphorylation circuit for transcriptional regulation. Plant Cell Physiol 45:340–345

Maruyama-Nakashita A, Nakamura Y, Tohge T, Saito K, Takahashi H (2006) Arabidopsis SLIM1 is a central transcriptional regulator of plant sulfur response and metabolism. Plant Cell 18:3235–3251

Maruyama-Nakashita A, Nakamura Y, Watanabe-Takahashi A, Inoue E, Yamaya T, Takahashi H (2005) Identification of a novel cis-acting element conferring sulfur deficiency response in *Arabidopsis* roots. Plant J 42:305–314

Mas-Droux C, Biou V, Dumas R (2006a) Allosteric threonine synthase. Reorganization of the pyridoxal phosphate site upon asymmetric activation through S-adenosylmethionine binding to a novel site. J Biol Chem 281:5188–5196

Mas-Droux C, Curien G, Robert-Genthon M, Laurencin M, Ferrer JL, Dumas R (2006b) A novel organization of ACT domains in allosteric enzymes revealed by the crystal structure of *Arabidopsis* aspartate kinase. Plant Cell 18:1681–1692

May M, Vernoux T, Leaver C, Van Montagu M, Inze D (1998) Glutathione homeostasis in plants: implications for environmental sensing and plant development. J Exp Bot 49:649–667

Meister A (1995) Glutathione biosynthesis and its inhibition. Methods Enzymol 252:26–30

Melis A, Chen H-C (2005) Chloroplast sulfate transport in green algae – genes, proteins and effects. Photosynth Res 86:299–307

Meyer AJ (2008) The integration of glutathione homeostasis and redox signaling. J Plant Physiol 165:1390–1403

Meyer AJ, Fricker MD (2002) Control of demand-driven biosynthesis of glutathione in green *Arabidopsis* suspension culture cells. Plant Physiol 130:1927–1937

Meyer AJ, Hell R (2005) Glutathione homeostasis and redox-regulation by sulfhydryl groups. Photosynth Res 86:435–457

Meyer AJ, Rausch T (2008) Biosynthesis, compartmentation and cellular functions of glutathione in plant cells. In: Hell R, Dahl C, Knaff DB, Leustek T (eds) Sulfur metabolism in phototrophic organisms. Springer, Dordrecht, The Netherlands, pp 161–184

Meyer AJ, Brach T, Marty L, Kreye S, Rouhier N, Jacquot J-P, Hell R (2007) Redox-sensitive GFP in *Arabidopsis thaliana* is a quantitative biosensor for the redox potential of the cellular glutathione redox buffer. Plant J 52:973–986

Meyer Y, Siala W, Bashandy T, Riondet C, Vignols F, Reichheld JP (2008) Glutaredoxins and thioredoxins in plants. Biochim Biophys Acta 1783:589–600

Meyers DM, Ahmad S (1991) Link between L-3-cyanoalanine synthase activity and differential cyanide sensitivity of insects. Biochim Biophys Acta 1075:195–197

Miller AJ, Shen Q, Xu G (2009) Freeways in the plant: transporters for N, P and S and their regulation. Curr Opin Plant Biol 12:284–290

Mugford SG, Yoshimoto N, Reichelt M, Wirtz M, Hill L, Mugford ST, Nakazato Y, Noji M, Takahashi H, Kramell R, Gigolashvili T, Flugge UI, Wasternack C, Gershenzon J, Hell R, Saito K, Kopriva S (2009) Disruption of adenosine-5′-phosphosulfate kinase in *Arabidopsis* reduces levels of sulfated secondary metabolites. Plant Cell 21:910–927

Müntz K, Christov V, Saalbach G, Saalbach I, Waddell D, Pickardt T, Schieder O, Wustenhagen T (1998) Genetic engineering for high methionine grain legumes. Nahrung 42:125–127

Nakayama M, Akashi T, Hase T (2000) Plant sulfite reductase: molecular structure, catalytic function and interaction with ferredoxin. J Inorg Biochem 82:27–32

Ndamukong I, Abdallat A, Thurow C, Fode B, Zander M, Weigel R, Gatz C (2007) SA-inducible *Arabidopsis* glutaredoxin interacts with TGA factors and suppresses JA-responsive PDF1.2 transcription. Plant J 50:128–139

Nikiforova V, Freitag J, Kempa S, Adamik M, Hesse H, Hoefgen R (2003) Transcriptome analysis of sulfur depletion in *Arabidopsis thaliana*: interlacing of biosynthetic pathways provides response specificity. Plant J 33:633–650

Nikiforova VJ, Gakiere B, Kempa S, Adamik M, Willmitzer L, Hesse H, Hoefgen R (2004) Towards dissecting nutrient metabolism in plants: a systems biology case study on sulphur metabolism. J Exp Bot 55:1861–1870

Nikiforova VJ, Kopka J, Tolstikov V, Fiehn O, Hopkins L, Hawkesford MJ, Hesse H, Hoefgen R (2005) Systems rebalancing of metabolism in response to sulfur deprivation, as revealed by metabolome analysis of *Arabidopsis* plants. Plant Physiol 138:304–318

Noji M, Inoue K, Kimura N, Gouda A, Saito K (1998) Isoform-dependent differences in feedback regulation and subcellular localization of serine acetyltransferase involved in cysteine biosynthesis from *Arabidopsis thaliana*. J Biol Chem 273:32739–32745

Ohkama-Ohtsu N, Kasajima I, Fujiwara T, Naito S (2004) Isolation and characterization of an *Arabidopsis* mutant that overaccumulates *O*-acetyl-L-Ser. Plant Physiol 136:3209–3222

Ohkama-Ohtsu N, Zhao P, Xiang C, Oliver D (2007a) Glutathione conjugates in the vacuole are degraded by γ-glutamyl transpeptidase GGT3 in Arabidopsis. Plant J 49:878–888

Ohkama-Ohtsu N, Oikawa A, Zhao P, Xiang C, Saito K, Oliver DJ (2008) A γ-glutamyl transpeptidase-independent pathway of glutathione catabolism to glutamate via 5-oxoproline in *Arabidopsis*. Plant Physiol 148:1603–1613

Ohkama-Ohtsu N, Radwan S, Peterson A, Zhao P, Badr A, Xiang C, Oliver D (2007b) Characterization of the extracellular γ-glutamyl transpeptidases, GGT1 and GGT2, in *Arabidopsis*. Plant J 49:865–877

Palmieri L, Arrigoni R, Blanco E, Carrari F, Zanor MI, Studart-Guimaraes C, Fernie AR, Palmieri F (2006) Molecular identification of an Arabidopsis S-adenosylmethionine transporter. Analysis of organ distribution, bacterial expression, reconstitution into liposomes, and functional characterization. Plant Physiol 142:855–865

Pant BD, Musialak-Lange M, Nuc P, May P, Buhtz A, Kehr J, Walther D, Scheible WR (2009) Identification of nutrientresponsive Arabidopsis and rapeseed microRNAs by comprehensive real-time polymerase chain reaction profiling and small RNA sequencing. Plant Physiol 150:1541–1555

Papenbrock J, Riemenschneider A, Kamp A, Schulz-Vogt HN, Schmidt A (2007) Characterization of cysteine-degrading and H_2S-releasing enzymes of higher plants – from the field to the test tube and back. Plant Biol 9:582–588

Pasternak M, Lim B, Wirtz M, Hell R, Cobbett CS, Meyer AJ (2008) Restricting glutathione biosynthesis to the cytosol is sufficient for normal plant development. Plant J 53:999–1012

Patron NJ, Durnford DG, Kopriva S (2008) Sulfate assimilation in eukaryotes: fusions, relocations and lateral transfers. BMC Evol Biol 8:39

Pe'er I, Felder C, Man O, Silman I, Sussman J, Beckmann J (2004) Proteomic signatures: amino acid and oligopeptide compositions differentiate among phyla. Proteins 54:20–40

Pilon-Smits EAH, Hwang S, Mel Lytle C, Zhu Y, Tai JC, Bravo RC, Chen Y, Leustek T, Terry N (1999) Overexpression of ATP sulfurylase in Indian mustard leads to increased selenate uptake, reduction, and tolerance. Plant Physiol 119:123–132

Pimenta JM, Kaneta T, Larondelle Y, Dohmae N, Kamiya Y (1998) *S*-Adenosyl-L-methionine: L-methionine *S*-methyltransferase from germinating barley. Purification and lcalization. Plant Physiol 118:431–438

Ranocha P, Bourgis F, Ziemak MJ, Rhodes D, Gage DA, Hanson AD (2000) Characterization and functional expression of cDNAs encoding methionine-sensitive and -insensitive homocysteine S-methyltransferases from *Arabidopsis*. J Biol Chem 275:15962–15968

Rausch T, Wachter A (2005) Sulfur metabolism: a versatile platform for launching defence operations. Trends Plant Sci 10:503–509

Ravanel S, Block MA, Rippert P, Jabrin S, Curien G, Rebeille F, Douce R (2004) Methionine metabolism in plants: chloroplasts are autonomous for *de novo* methionine synthesis and can import *S*-adenosylmethionine from the cytosol. J Biol Chem 279:22548–22557

Rea P (2007) Plant ATP-binding cassette transporters. Annu Rev Plant Biol 58:347–375

Rebeille F, Jabrin S, Bligny R, Loizeau K, Gambonnet B, Van Wilder V, Douce R, Ravanel S (2006) Methionine catabolism in *Arabidopsis* cells is initiated by a γ-cleavage process and leads to S-methylcysteine and isoleucine syntheses. PNAS 103:15687–15692

Rennenberg H (1976) Glutathione in conditioned media of tobacco suspension cultures. Phytochem 15:1433–1434

Rennenberg H, Schmitz K, Bergmann L (1979) Long distance transport of sulfur in *Nicotiana tabacum*. Planta 176:68–74

Rinalducci S, Murgiano L, Zolla L (2008) Redox proteomics: basic principles and future perspectives for the detection of protein oxidation in plants. J Exp Bot 59:3781–3801

Robinson D (1994) The responses of plants to non-uniform supplies of nutrients. New Phytol 127:635–674

Rolland N, Droux M, Douce R (1992) Subcellular distribution of O-acetylserine(thiol)lyase in cauliflower (*Brassica oleracea* L.) inflorescence. Plant Physiol 98:927–935

Rotte C, Leustek T (2000) Differential subcellular localization and expression of ATP sulfurylase and 5′-adenylylsulfate reductase during ontogenesis of Arabidopsis leaves indicates that cytosolic and plastid forms of ATP sulfurylase may have specialized functions. Plant Physiol 124:715–724

Rouached H, Berthomieu P, El Kassis E, Cathala N, Catherinot V, Labesse G, Davidian JC, Fourcroy P (2005) Structural and functional analysis of the C-terminal STAS (sulfate transporter and anti-sigma antagonist) domain of the *Arabidopsis thaliana* sulfate transporter SULTR1.2. J Biol Chem 280:15976–15983

Rouached H, Wirtz M, Alary R, Hell R, Arpat AB, Davidian J-C, Fourcroy P, Berthomieu P (2008) Differential regulation of the expression of two high-affinity sulfate Transporters, SULTR1.1 and SULTR1.2, in *Arabidopsis*. Plant Physiol 147:897–911

Rouhier N, Lemaire SD, Jacquot J-P (2008) The role of glutathione in photosynthetic organisms: emerging functions for glutaredoxins and glutathionylation. Annu Rev Plant Biol 59:143

Ruffet ML, Droux M, Douce R (1994) Purification and kinetic properties of serine acetyltransferase free of *O*-acetylserine(thiol)lyase from spinach chloroplasts. Plant Physiol 104:597–604

Ruffet ML, Lebrun M, Droux M, Douce R (1995) Subcellular distribution of serine acetyltransferase from *Pisum sativum* and characterization of an *Arabidopsis thaliana* putative cytosolic isoform. Eur J Biochem 227:500–509

Saito K (2004) Sulfur assimilatory metabolism. The long and smelling road. Plant Physiol 136:2443–2450

Saito K, Kanda R, Kurosawa M, Murakoshi I (1994a) Overexpression of a plant cysteine synthase gene and biosynthesis of a plant specific metabolite, P-(pyrazol-l-yl)-L-alanine, in *Escherichia coli*. Can J Chem 72:188–192

Saito K, Kurosawa M, Tatsuguchi K, Takagi Y, Murakoshi I (1994b) Modulation of cysteine biosynthesis in chloroplasts of transgenic tobacco overexpressing cysteine synthase [O-acetylserine(thiol)-lyase]. Plant Physiol 106:887–895

Schachtman DP, Shin R (2007) Nutrient sensing and signaling: NPKS. Annu Rev Plant Biol 58:47–69

Schiffmann S, Schwenn JD (1994) APS-sulfotransferase activity is identical to higher plant APS-kinase (EC 2.7.1.25). FEBS Lett 355:229–232

Schmidt A (1973) Sulfate reduction in a cell-free system of *Chlorella*. The ferredoxin dependent reduction of a protein-bound intermediate by a thiosulfonate reductase. Arch Mikrobiol 93:29–52

Schmidt A (1975) A sulfotransferase from spinach leaves using adenosine-5′-phosphosulfate. Planta 124:267–275

Schmidt A (1976) The adenosine-5′-phosphosulfate sulfotransferase from spinach (*Spinacea oleracea* L.). Stabilization, partial purification, and properties. Planta 130:257–263

Schmidt A, Jäger K (1992) Open questions about sulfur metabolism in plants. Annu Rev Plant Physiol Plant Mol Biol 43:325–349

Schmidt A, Abrams WR, Schiff JA (1974) Reduction of adenosine 5′-phosphosulfate to cysteine in extracts from *Chlorella* and mutants blocked for sulfate reduction. Eur J Biochem 47:423–434

Schwenn JD (1989) Sulphate assimilation in higher plants: a thioredoxin-dependent PAPS-reductase from spinach leaves. Z. Naturforsch 44c, 504–508

Schwenn JD, Kemena A (1984) Expression of the plant sulphite reductase in cell suspension cultures from *Catharanthus roseus* L. Planta 160:151–158

Sekine K, Hase T, Sato N (2002) Reversible DNA compaction by sulfite reductase regulates transcriptional activity of chloroplast nucleoids. J Biol Chem 277:24399–24404

Setya A, Murillo M, Leustek T (1996) Sulfate reduction in higher plants: molecular evidence for a novel 5′-adenylylsulfate reductase. Proc Natl Acad Sci USA 93:13383–13388

Shen B, Li C, Tarczynski MC (2002) High free-methionine and decreased lignin content result from a mutation in the Arabidopsis S-adenosyl-L-methionine synthetase 3 gene. Plant J 29:371–380

Shibagaki N, Grossman AR (2004) Probing the function of STAS domains of the *Arabidopsis* sulfate transporters. J Biol Chem 279:30791–30799

Shibagaki N, Rose A, McDermott JP, Fujiwara T, Hayashi H, Yoneyama T, Davies JP (2002) Selenate-resistant mutants of *Arabidopsis thaliana* identify Sultr1;2, a sulfate transporter required for efficient transport of sulfate into roots. Plant J 29:475–486

Sirko A, Blaszczyk A, Liszewska F (2004) Overproduction of SAT and/or OASTL in transgenic plants: a survey of effects. J Exp Bot 55:1881–1888

Smith FW, Diatlof E (2005) Sulfate transport processes in plants. In: Saito K, De Kok LJ, Stulen I, Hawkesford MJ, Schnug E, Sirko A, Rennenberg H (eds) Sulfur transport and assimilation in plants in the post genomic era. Backhuys Publishers, Leiden, pp 3–11

Stiller I, Dancs G, Hesse H, Hoefgen R, Banfalvi Z (2007) Improving the nutritive value of tubers: elevation of cysteine and glutathione contents in the potato cultivar White Lady by marker-free transformation. J Biotechnol 128:335–343

Suter M, von Ballmoos P, Kopriva S, den Camp RO, Schaller J, Kuhlemeier C, Schurmann P, Brunold C (2000) Adenosine 5′-phosphosulfate sulfotransferase and adenosine 5′-phosphosulfate reductase are identical enzymes. J Biol Chem 275:930–936

Swamy U, Wang M, Tripathy J, Kim S, Hirasawa M, Knaff D, Allen J (2005) Structure of spinach nitrite reductase: implications for multi-electron reactions by the iron-sulfur:siroheme cofactor. Biochem 44:16054–16063

Tabe L, Hagan N, Higgins T (2002) Plasticity of seed protein composition in response to nitrogen and sulfur availability. Curr Opin Plant Biol 5:212–217

Tada Y, Spoel SH, Pajerowska-Mukhtar K, Mou Z, Song J, Wang C, Zuo J, Dong X (2008) Plant immunity requires conformational charges of NPR1 via S-nitrosylation and thioredoxins. Science 321:952–956

Takahashi H, Asanuma W, Saito K (1999) Cloning of an Arabidopsis cDNA encoding a chloroplast localizing sulphate transporter isoform. J Exp Bot 50:1713–1714

Takahashi H, Watanabe-Takahashi A, Smith FW, Blake-Kalff M, Hawkesford MJ, Saito K (2000) The roles of three functional sulphate transporters involved in uptake and translocation of sulphate in *Arabidopsis thaliana*. Plant J 23:171–182

Tejada-Jimenez M, Llamas A, Sanz-Luque E, Galvan A, Fernandez E (2007) A high-affinity molybdate transporter in eukaryotes. Proc Nat Acad Sci USA 104:20126–20130

Tennstedt P, Peisker D, Bottcher C, Trampczynska A, Clemens S (2009) Phytochelatin synthesis is essential for the detoxification of excess zinc and contributes significantly to the accumulation of zinc. Plant Physiol 149:938–948

Tomatsu H, Takano J, Takahashi H, Watanabe-Takahashi A, Shibagaki N, Fujiwara T (2007) An *Arabidopsis thaliana* high-affinity molybdate transporter required for efficient uptake of molybdate from soil. Proc Nat Acad Sci USA 104:18807–18812

Tsakraklides G, Martin M, Chalam R, Tarczynski MC, Schmidt A, Leustek T (2002) Sulfate reduction is increased in transgenic *Arabidopsis thaliana* expressing 5′-adenylylsulfate reductase from *Pseudomonas aeruginosa*. Plant J 32:879–889

Vauclare P, Kopriva S, Fell D, Suter M, Sticher L, von Ballmoos P, Krahenbühl U, den Camp RO, Brunold C (2002) Flux control of sulphate assimilation in *Arabidopsis thaliana*: adenosine 5′-phosphosulphate reductase is more susceptible than ATP sulphurylase to negative control by thiols. Plant J 31:729–740

Walker C, Boothe EJ (2003) Sulphur nutrition and oilseed quality. In: Abrol YP, Ahmad A (eds) Sulphur in plants. Kluwer Academic Publishers, Dordrecht, pp 323–340

Wang X, Stumpf DK, Larkins BA (2001) Aspartate kinase 2. A candidate gene of a quantitative trait locus influencing free amino acid content in maize endosperm. Plant Physiol 125:1778–1787

Watanabe M, Kusano M, Oikawa A, Fukushima A, Noji M, Saito K (2008a) Physiological roles of the β-substituted alanine synthase gene family in *Arabidopsis*. Plant Physiol 146:310–320

Watanabe M, Mochida K, Kato T, Tabata S, Yoshimoto N, Noji M, Saito K (2008b) Comparative genomics and reverse genetics analysis reveal indispensable functions of the serine acetyltransferase gene family in *Arabidopsis*. Plant Cell 20:2484–2496

Wawrzynska A, Lewandowska M, Hawkesford MJ, Sirko A (2005) Using a suppression subtractive library-based approach to identify tobacco genes regulated in response to short-term sulphur deficit. J Exp Bot 56:1575–1590

Wirtz M, Droux M (2005) Synthesis of the sulfur amino acids: cysteine and methionine. Photosynth Res 86:345–362

Wirtz M, Hell R (2006) Functional analysis of the cysteine synthase protein complex from plants: structural, biochemical and regulatory properties. J Plant Physiol 163:273–286

Wirtz M, Hell R (2007) Dominant-negative modification reveals the regulatory function of the multimeric cysteine synthase protein complex in transgenic tobacco. Plant Cell 19:625–639

Wirtz M, Berkowitz O, Droux M, Hell R (2001) The cysteine synthase complex from plants. Mitochondrial serine acetyltransferase from *Arabidopsis thaliana* carries a bifunctional domain for catalysis and protein-protein interaction. Eur J Biochem 268:686–693

Yang Y, Yuan JS, Ross J, Noel JP, Pichersky E, Chen F (2006) An *Arabidopsis thaliana* methyltransferase capable of methylating farnesoic acid. Arch Biochem Biophys 448:123–132

Yonekura-Sakakibara K, Onda Y, Ashikari T, Tanaka Y, Kusumi T, Hase T (2000) Analysis of reductant supply systems for ferredoxin-dependent sulfite reductase in photosynthetic and nonphotosynthetic organs of maize. Plant Physiol 122:887–894

Yoshimoto N, Takahashi H, Smith FW, Yamaya T, Saito K (2002) Two distinct high-affinity sulfate transporters with different inducibilities mediate uptake of sulfate in *Arabidopsis* roots. Plant J 29:465–473

Yoshimoto N, Inoue E, Saito K, Yamaya T, Takahashi H (2003) Phloem-localizing sulfate transporter, Sultr1;3, mediates re-distribution of sulfur from source to sink organs in *Arabidopsis*. Plant Physiol 131:1511–1517

Zeh M, Casazza AP, Kreft O, Roessner U, Bieberich K, Willmitzer L, Hoefgen R, Hesse H (2001) Antisense inhibition of threonine synthase leads to high methionine content in transgenic potato plants. Plant Physiol 127:792–802

Zhang M-Y, Bourbouloux A, Cagnac O, Srikanth CV, Rentsch D, Bachhawat AK, Delrot S (2004) A novel family of transporters mediating the transport of glutathione derivatives in plants. Plant Physiol 134:482–491

Zn – A Versatile Player in Plant Cell Biology

Stephan Clemens

Abstract During evolution, zinc was recruited for a wide range of biochemical functions in cells. About 8% of the proteins in a plant are Zn-binding. Among them are members of all six enzyme classes and myriad regulatory proteins. Zn is required in all cellular compartments. High affinity of Zn(II) to various functional groups requires a multitude of transport and chelation processes to ensure trafficking of Zn to target sites both at the cellular and the organismal level. Detailed mechanistic understanding of Zn mobilization in the soil, uptake into a plant cell, and cytosolic buffering is still limited. More is known about Zn tolerance, storage, and long-distance translocation. Molecular dissection of Zn distribution and accumulation in plants will be important also to enable breeding of higher Zn content of crops. Zn deficiency in humans is widespread and is estimated to affect more than 25% of the world's population.

1 Zn Chemistry and Biological Functions

Besides iron (Fe), zinc (Zn) is the most widely used transition metal in living systems. Recent estimates derived from proteome annotations guided by Pfam domains and structural knowledge state that in eukaryotes on average close to 9% of all proteins are Zn-binding (Andreini et al. 2006). For *Arabidopsis thaliana*, this study arrived at 8% of the proteome (about 2,400 proteins). An independent search via Pfam domains, annotation, and hand-compilation predicted 2,367 Zn proteins (Broadley et al. 2007). In accordance with these numbers, a cruder query based on annotation only retrieves from the TAIR database >1,800 *A. thaliana* proteins whose gene description contains the term "zinc" (www.arabidopsis.org). Thus,

S. Clemens
Universität Bayreuth, Lehrstuhl Pflanzenphysiologie, Universitätsstrasse 30, 95440 Bayreuth, Germany

R. Hell and R.-R. Mendel (eds.), *Cell Biology of Metals and Nutrients*, Plant Cell Monographs 17,
DOI 10.1007/978-3-642-10613-2_12, © Springer-Verlag Berlin Heidelberg 2010

prevalence of Zn in biology is now well documented owing to genome annotation and availability of large numbers of protein structures. Abundant usage of Zn is even more remarkable because Zn was in the early phases of evolution barely available for cells. In a reducing atmosphere, Zn precipitated as a sulphide and probably very few proteins contained Zn (Frausto da Silva and Williams 2001). With the increasing Zn levels after the advent of oxygenic photosynthesis usage of Zn rapidly expanded.

The wealth of Zn proteins in nature today indicates favorable chemistry of Zn for life processes. Indeed, there are several features unique in their combination to Zn among elements (Frausto da Silva and Williams 2001). Zn has a filled d shell with 10 d electrons. It is therefore occurring in only one oxidation state, i.e. it is not redox active in biological environments. Other relevant consequences are: (1) the coordination geometry of Zn(II) is more flexible compared to most other transition metal cations. When binding ligands, the Zn ion is not polarized. Its ligand field stabilization energy is zero and ligands can therefore be exchanged rapidly. Also, Zn(II) is the only transition metal ion readily entering a tetrahedral coordination. (2) The Zn(II) ion as a small ion with highly concentrated charge is a strong Lewis acid. It exhibits high binding affinity for soft bases, such as sulphide ligands, as well as for hard bases, such as amino, carboxylate, and hydroxyl ligands. According to the Irving-Williams series, only Cu(II) ions interact more strongly with ligands among the divalent transition metal micronutrients. (3) Zn(II) is kinetically comparatively labile and undergoes ligand exchange reactions more rapidly relative to other divalent transition metal cations such as Co(II), Ni(II), or Cu(II) (Berg and Shi 1996). Functions of Zn in nature can be essentially divided into two types that reflect the chemistry of Zn. As a strong Lewis acid with flexible coordination geometry and fast ligand exchange it is used as co-factor in many different types of enzymes. The combination of strong binding to various biologically important ligands with lack of redox activity explains why Zn(II) has been widely recruited for structural functions within proteins and in protein–protein interactions. Both essential roles are associated with distinct tetrahedral binding sites. Catalytic Zn is coordinated to four ligands, three of which are organic and the fourth one being a water molecule, while structural Zn is coordinated to four organic ligands. Mostly cysteine and histidine residues are serving as Zn ligands in proteins. Aspartate and glutamate are occasionally found (Andreini et al. 2006).

Zn requirement of an enzyme was for the first time shown in 1940 for carbonic anhydrase (Berg and Shi 1996). Since then the number of known Zn enzymes has expanded tremendously. All six Enzyme Commission (EC) classes contain Zn-dependent proteins. Most prominent is Zn requirement of hydrolases. More than half of all Zn enzymes apparently belong to this class. Also, 18% of all hydrolases are estimated to represent Zn proteins (Andreini et al. 2008). Other EC classes with a considerable number of Zn enzymes are oxidoreduxtases (EC1) and transferases (EC2). Fast ligand exchange and flexible coordination geometry favor catalytic activity, Lewis acidity of Zn polarizes bonds and increases electrophilicity of the substrate or ligand. Zn(II) appears to be particularly important as an attacking group in enzymes with small substrates. The highly localized charge allows formation of active sites that are comparatively nonselective and not dependent on a large number of interacting sites between enzyme and substrate (Frausto da Silva and Williams 2001).

A structural role of Zn was for the first time recognized in the transcription factor IIIA (TFIIIA). Analysis of this regulatory protein revealed a small Zn-containing domain termed "zinc finger" (Berg and Shi 1996). Today we know that hundreds of proteins containing zinc fingers of various types are encoded in each eukaryotic genome. In fact, the higher percentage of Zn proteins in eukaryotic proteomes relative to those of Bacteria or Archaea (5%–6%) is almost exclusively attributable to enrichment in proteins with Zn fingers, whereas the percentage of Zn enzymes is fairly constant throughout nature. Thus, the evolution of complex regulation in multicellular organisms has gone hand in hand with the recruitment of Zn for structural functions. Protein domains, too small to fold by themselves can easily be stabilized by the formation of a tetrahedral site, which will favor binding of Zn (II) (Berg and Shi 1996). This and the invariable valence explain why Zn is extensively utilized for structural roles. The majority of Zn-stabilized proteins are involved in gene regulation and interact with nucleic acids. A redox-active metal such as Fe or Cu would pose the enormous risk of triggering Fenton reactions leading to generation of hydroxyl radicals in vicinity of the DNA.

2 Cellular Compartmentalization of Zn

Zn is required in all cellular compartments. Compared to other essential transition metal ions Zn(II) is far more prevalent in the cytosol and nucleus (Frausto da Silva and Williams 2001). This is of course largely due to the high number of transcription factors with Zn fingers and similar domains. Another reason is the involvement of Zn enzymes in nucleic acid synthesis and maintenance, for example through RNA polymerases, DNA polymerases, helicases, histone-lysine *N*-methyltransferases, histone acetyltransferases, and histone deacetylases. This compilation underscores the importance of Zn as a redox-inactive cofactor of proteins near the DNA. Zn also has central roles in the cytoplasm. Translation initiation factors such as eIF5 are Zn-binding as well as several tRNA ligases. Highly abundant is the RING finger domain as part of ubiquitin E3 ligases (cullin-RING ligases), which play essential roles in plant signaling. Numerous repressors of hormone responses, for instance, are marked for degradation via polyubiquitination (Hotton and Callis 2008). The RING finger plays a role in the transfer of ubiquitin by E2 ubiquitin-conjugating enzymes to their targets (Matthews and Sunde 2002). RING finger domains are generally implicated in mediating protein–protein interactions (Gamsjaeger et al. 2007). Plant proteins carrying this domain include cellulose synthases (Somerville 2006).

Zn is in plant cells not only indirectly involved in proteolysis as part of the ubiquitin-proteasome pathway. Zn-dependent metallopeptidases of the FtsH family are localized in plastids and mitochondria (Adam et al. 2006). FtsH2

(AtVAR2) and FtsH5 (AtVAR1) are crucial for the development of normal chloroplasts. ILR metallopeptidases cleave auxin conjugates (Campanella et al. 2003). It should be emphasized that just as in the case of proteases, there are plant enzymes with catalytic Zn or structural Zn, or both (Marschner 1995). Prominent examples for proteins with Zn taking part in catalysis include carbonic anhydrase and the above-mentioned RNA polymerases. In Cu,Zn superoxide dismutases Zn plays a structural role and the Cu metallocenter is involved in catalysis. Alcohol dehydrogenases belong to those enzymes that carry both a structural and a catalytic Zn.

While in mammals the lysosome is a subcellular compartment exhibiting high Zn levels because of its digestive functions, many plant Zn-dependent hydrolytic enzymes are localized in the vacuole and the apoplast. Examples are the vacuolar Zn enzyme α-mannosidase, plant Zn-dependent carboxypeptidases, purple acid phosphatases containing a binuclear Fe(III)-Zn(II) or Fe(III)–Mn(II) metallocenter, and a protein family of at least five plant matrix metalloproteinases with a potential involvement in modulation of the plant extracellular matrix (Krämer and Clemens 2005). The vacuole represents in addition a major site of Zn storage and is important for tolerating an excess of Zn (see below). In *S. cerevisiae* it was demonstrated that the vacuolar Zn concentration can rise up to nearly 100 mM (Simm et al. 2007). A prerequisite for the maturation of vacuolar and apoplastic Zn proteins is delivery of Zn to the secretory pathway. Transporters of the Cation Diffusion Facilitator family (CDF) (see below) mediating Zn uptake into the ER and the Golgi have been mostly characterized in yeast and in mammalian cells (Li and Kaplan 2001; Kirschke and Huang 2003; Ellis et al. 2004). Similar roles are envisioned in plant cells, and for Mn supply to the Golgi CDF (AtMTP11) involvement was already proven (Peiter et al. 2007; Delhaize et al. 2007) In *Schizosaccharomyces pombe* the ER has been implicated also in Zn storage. Loss of an ER-localized transporter resulted in Zn hypersensitivity (Clemens et al. 2002a). An analogous contribution of the secretory pathway to metal sequestration in plant cells is not well understood, but is supported for instance by the finding, that Golgi-localized AtMTP11 does contribute to Mn tolerance (Delhaize et al. 2007; Peiter et al. 2007).

An appreciable fraction of intracellular Zn could be bound to ribosomes. Zn is a structural component of ribosomes with 6 Zn atoms being bound per ribosome. In *S. cerevisiae* this amounts to about 5% of the cellular Zn (Eide 2006). Intriguingly, a homeostatic function was found for ribosomes in *Bacillus subtilis*. A rearrangement of ribosome architecture occurs that releases Zn under Zn deficiency conditions through replacement of Zn-binding ribosomal proteins with paralogous proteins lacking a Zn-binding motif (Natori et al. 2007). No such processes have been reported yet for eukaryotic cells. To complete the list of known Zn-binding sites in cells, low-affinity interactions with lipids and DNA have to be mentioned as well as low-molecular weight chelators such as glutathione, certain amino acids (e.g. histidine) and organic acids (e.g. citrate) (see below). Zn binding to lipids has been suggested to be important for protecting membrane integrity in plant cells (Cakmak 2000).

3 Physiological Range of Zn Concentrations in Plants

Given the large number of Zn-requiring sites it is not surprising that the Zn quota of cells – i.e. the total zinc content of a cell required for its optimum growth (Outten and O'Halloran 2001) – lies in the range of 0.1–0.5 mM. This estimate applies to *E. coli* just as well as to higher eukaryotes (Eide 2006). For plants, this quota would translate into content in the order of roughly 50 μg/g dry biomass. Adequate Zn concentrations in leaves are 15–20 μg/g dry biomass according to Marschner (1995). A recent analysis of over 1,000 published studies revealed variation between plant families in shoot Zn content of about a factor of 2.5 (Broadley et al. 2007). Variation also exists at the species and cultivar level, which is important in the context of efforts to improve Zn content of edible parts in crop plants (see below).

Characteristic (visible) symptoms of Zn deficiency include chlorosis on young leaves or initial early senescence of old leaves, shortening of internodes, and reduced leaf size. Under severe Zn deficiency, plants appear stunted and exhibit reduced elongation and tip growth. Older leaves show wilting and curling with extensive chlorosis (Marschner 1995). Growth phenotypes of Zn-deficient plants have been proposed to be a consequence of disturbances in auxin metabolism. Also, dividing and elongating plant cells appear to have a higher requirement for Zn (Marschner 1995). Molecular understanding of these observations is missing. At the protein level a sensitive indication of Zn deficiency is a sharp decline in carbonic anhydrase activity. Other Zn-dependent enzymes are less affected (Marschner 1995).

Plant species and even varieties differ not only in Zn content but also in their Zn efficiency, i.e. the ability to maintain growth and yield under Zn-limiting conditions (Hacisalihoglu and Kochian 2003). Comparative studies with contrasting bean or wheat varieties suggest that Zn efficiency is primarily due to the ability of the leaves to maintain expression and activity of Zn-requiring enzymes at low total leaf Zn concentrations rather than to differences in uptake systems (Hacisalihoglu and Kochian 2003; Hacisalihoglu et al. 2003).

Zinc toxicity symptoms are observed at Zn contents above 100–300 μg/g. Again, threshold values can be quite variable between and within species. A readily detectable symptom under supraoptimal Zn supply is inhibition of root elongation. Often, young leaves display chlorosis and/or necrosis. Other scorable symptoms are a reduction in photosynthetic electron transport and photosynthetic efficiency (Marschner 1995). Primary targets of Zn toxicity are not really understood. Chlorosis is hypothesized to be a consequence of Zn-induced Fe deficiency and possibly other mineral deficiencies caused by competition between metal ions.

4 Zn Acquisition

Zn has to be acquired by plant roots from the soil solution as Zn^{2+} or Zn-chelate complexes. Most Zn in the soil is insoluble. Availability of Zn, i.e. concentration in the soil solution, is influenced by various physical, chemical and biological factors (for details see Broadley et al. 2007; Cakmak 2008). The single most relevant parameter is generally assumed to be soil pH. Increasing soil pH stimulates adsorption of Zn to soil constituents (e.g., metal oxides, clay minerals) and reduces the desorption of Zn (Cakmak 2008). Also, at high pH Zn tends to precipitate as phosphate, carbonate or oxide. Zn availability therefore vastly decreases in calcareous soils. Organic matter content of soil is another parameter. It is important to note, however, that even at a particular site a plant is growing, micronutrient availability is not at all stable. Rather, among higher eukaryotes plants are arguably subject to the widest fluctuations in metal supply. Soils undergo extreme seasonal and local variations in metal concentrations. Depending on microbial activity, water status and deposition of compounds on the soil, for example by rainwater or fertilization, soil pH and redox state can rapidly change. Also, a fluctuating water status directly influences metal ion availability because dissolved Zn ions and chelates reach the root surface by diffusion. It is therefore a necessity that metal acquisition processes are regulated by the plant to maintain an adequate influx of metal ions (Krämer and Clemens 2005) (see below).

Zn^{2+} concentrations in the soil solution vary between less than 1 nM and >1 μM (Broadley et al. 2007; Cakmak 2008). Referenced to the cellular Zn quota the ability of plant cells to accumulate Zn several hundred fold to several thousand fold over the external Zn concentration is immediately clear. Affinity of root Zn^{2+} uptake systems has in most kinetic studies been found to lie in the micromolar range. Apparently there is rather little inter- and intraspecies variation (Broadley et al. 2007). Active mobilization of Zn through plant root activities appears very likely, but concrete evidence is scarce. Plant Fe nutrition is clearly supported by active root-mediated processes. Rhizosphere pH is lowered through proton extrusion, insoluble Fe(III) is reduced to Fe(II) (=strategy I), or Fe(III)-chelating molecules (phytosiderophores) are secreted and taken up again as Fe(III) complexes (=strategy II) (Briat et al. 2007). Both lowering the rhizosphere pH and secreting ligands could also enhance Zn solubility. Actual contribution of such processes to Zn acquisition, however, has not been demonstrated unequivocally. Phytosiderophores may play a role in Zn nutrition of graminaceous species. They are released by Zn-deficient plants and Zn uptake was found to be significantly reduced in the maize *ys1* mutant defective in phytosiderophore uptake (von Wiren et al. 1996).

Nutrient availability is in addition strongly influenced by symbiotic interaction of plants with microorganisms. A beneficial role of mycorrhizal symbioses, which 70%–90% of all land plants establish (Parniske 2008), is well-documented for the macronutrient phosphorus but is suggested for micronutrient acquisition as well. Zn is a major candidate element. As fluxes are difficult to measure and molecular dissection of the symbiosis is still challenging, the contribution of symbiosis to Zn acquisition is to date mostly speculative.

Molecular understanding of plant Zn uptake is still very limited also with respect to passage from the apoplast into the symplast of root cells. The precise root cell type(s) involved in primary Zn uptake are unknown. Based on the current state of knowledge it is likely that multiple transporters contribute to Zn uptake by plant roots. Members of the ubiquitous ZIP family (**Z**RT1/**I**RT1-like **P**roteins) (TC 2.A.5) have mostly been implicated in metal uptake across the plasma membrane of prokaryotic as well as eukaryotic cells (Eide 2006). IRT1 represents the main pathway for Fe^{2+} uptake into *A. thaliana* and mediates Zn^{2+} and Cd^{2+} uptake under Fe-deficient conditions (Connolly et al. 2002; Vert et al. 2002). Much less is known about the physiological role of the other 16 ZIPs in *A. thaliana* (Hanikenne et al. 2005). Early characterization in *S. cerevisiae zrt1zrt2* mutant cells demonstrated Zn^{2+} uptake activity through mutant complementation for several ZIPs. Also, ZIP genes were shown to be up-regulated under conditions of Zn deficiency in roots, which is why they have been implicated in Zn acquisition (Grotz et al. 1998; Talke et al. 2006). Similar results have been obtained for ZIPs from rice and other plant species (reviewed in Colangelo and Guerinot 2006; Krämer et al. 2007). However, clear demonstration of functional roles of individual ZIPs is barely available. Apparently, neither insertion mutant nor overexpressor lines for individual ZIP genes show scorable phenotypes related to Zn nutrition (Lin et al. 2009). At least in grasses, Zn may also be taken up as a Zn-phytosiderophore complex (see above). The phytosiderophore uptake system YS1 in maize (Curie et al. 2001) operates as a metal-phytosiderophore proton cotransporter and can mediate cellular uptake of complexes of the phytosiderophore 2′-deoxymugineic acid with Fe(III), Ni(II), Zn(II), Cu(II) (Schaaf et al. 2004).

5 Cellular Zn Homeostasis

Biological functions of Zn ions are in part explained by their high affinity to various functional groups. Obviously the interaction with proteins and other molecules has to be tightly controlled. The necessary processes of regulated uptake, chelation, trafficking, storage, and efflux are referred to as Zn homeostasis. They have to orchestrate the specific delivery of metal ions to their respective apometalloproteins (Krämer and Clemens 2005; Grotz and Guerinot 2006). Furthermore, metal ions have to be moved to target cells, tissues and organs distant from the site of uptake without being sequestered by other available sites along the way (Clemens et al. 2002b). A crucial question of Zn homeostasis is: how big a fraction of total cellular Zn is available as free, hydrated metal ions, or labile Zn, available for binding by newly synthesized zinc metalloproteins. From binding affinities of bacterial metal sensors it was inferred that the concentration of free Zn ions in the cytosol is in the femtomolar range (Outten and O’Halloran 2001), meaning essentially the absence of free Zn ions. This estimate, however, is not generally accepted for eukaryotic cells (Eide 2006). Regardless, even a labile Zn pool in the nanomaolar range as proposed is in sharp contrast to the cellular Zn quota of 0.1–0.5 mM (see above). Thus, there is clearly Zn buffering through Zn-binding proteins and high-affinity low molecular weight chelators which prevent uncontrolled interaction. The low concentration of free Zn immediately poses the question as to how Zn ions reach their numerous target sites. For Cu it is known that specialized metallochaperone pathways ensure delivery of Cu ions to Cu proteins and into cellular compartments

(O'Halloran and Culotta 2000; Huffman and O'Halloran 2001). These mechanisms are highly conserved and found also in plant cells (Puig et al. 2007). The existence of analogous pathways for Zn, however, is considered unlikely (Grotz and Guerinot 2006; Eide 2006). Cellular Zn sites are about 20-fold more abundant than Cu sites. There are no indications yet that a corresponding multitude of Zn trafficking pathways is expressed. Thus, a Zn metallochaperone network probably does not exist. From that the existence of a small labile Zn pool can be deduced. Given the high degree of conservation in Zn usage across kingdoms this assumption is valid for practically all cell types.

Inside the cytosol, incoming Zn^{2+} ions are likely to undergo controlled binding, either to metal buffering proteins or to low-molecular-weight chelator molecules. The latter may include glutathione (GSH), which has a high affinity for Zn at cytoplasmic pH values, GSH-derived molecules (phytochelatins) and the nonproteinogenic amino acid nicotianamine, synthesized through condensation of three molecules of *S*-adenosylmethionine by nicotianamine synthases (NAS) (Ling et al. 1999). When considering first a root cell, there are three principal fates of a Zn(II) ion. It can be integrated into a Zn-requiring site, can be sequestrated (in the vacuole or possibly in other compartments) and later remobilized, or it can symplastically move to the stele and be loaded into the xylem (Clemens et al. 2002b). Especially under conditions of high Zn supply CDF transporters (in plants often called MTPs) protect the cytosol from Zn overload. In *A. thaliana*, tonoplast-localized transporters MTP1 and MTP3 mediate in different root cell types vacuolar sequestration of Zn and contribute thereby to Zn tolerance (Kobae et al. 2004; Desbrosses-Fonrouge et al. 2005; Arrivault et al. 2006). A similar yet distinct function has been assigned to a transporter of the major facilitator family. ZIF1 is also tonoplast-localized and contributes to Zn tolerance. It is hypothesized to transport Zn-chelate complexes into the vacuole (Haydon and Cobbett 2007). ZIF1 and MTP1 show additive effects, i.e. are parts of independent pathways. Sequestered Zn can be remobilized from the vacuole. Candidate proteins are transporters of the Nramp family. AtNramp3 is hypothesized to function in the mobilization of the micronutrients Fe, Mn and Zn and has been shown to be (together with AtNramp4) essential for seed germination under Fe-limited conditions (Lanquar et al. 2005).

Essential both intracellularly and for symplastic passage (e.g. to the xylem parenchyma cells) are chelators of Zn to suppress aberrant binding. Prime candidate for this function is nicotianamine (NA). NA binds several transition metals with high affinity (Benes et al. 1983; Callahan et al. 2006). In vivo Zn binding by NA is to date mostly supported by indirect evidence. Expression of various nicotianamine synthase isoforms partially complemented Zn hypersensitivity of mutant yeast strains of *S. cerevisiae* and *S. pombe* (Weber et al. 2004; Becher et al. 2004). NA is apparently involved in Fe, Cu, Zn and Ni homeostasis in plants (Stephan et al. 1996; von Wiren et al. 1999; Hell and Stephan 2003). NA deficiency causes intercostal chlorosis especially in young leaves, indicating defective Fe transport to developing tissues. Zn and Cu contents are reduced. Conversely, NAS overexpression resulted in 2.5-fold higher Zn levels and 1.9-fold higher Fe levels in young leaves (Takahashi et al. 2003). Recently it was demonstrated that NA is synthesized not only in plants but also in certain filamentous fungi such as *Neurospora crassa* (Trampczynska et al. 2006).

Part of the NA physiology in plants appears to be the transport of NA–metal complexes mediated by YSL transporters. These represent a subfamily of oligopeptide transporters related to YS1 from maize (see above). In *A. thaliana* there are 8 YSLs and evidence is growing that at least some of these are indeed NA–metal complex transporters, involved, for instance, in metal delivery from the vasculature (Le Jean et al. 2005; Schaaf et al. 2005; Le Jean et al. 2005; Waters et al. 2006).

Recent observations identified phytochelatins (PCs) as potential Zn chelators in plants and *S. pombe* (Tennstedt et al. 2009). PCs, peptides of the general structure (γ-Glu–Cys)$_n$–Gly ($n = 2$–11), are nonribosomally synthesized from GSH in a transpeptidase reaction by phytochelatin synthases (PCS) (Cobbett and Goldsbrough 2002). PC accumulation helps various organisms to survive in the presence of toxic cadmium or arsenic concentrations. This has been well-documented through the isolation and generation of plant, yeast and nematode mutants deficient in PC synthesis (Clemens et al. 1999; Ha et al. 1999; Vatamaniuk et al. 2001). Zn exposure was now found to elicit considerable PC accumulation in roots and shoots. PC deficiency resulted in reduced Zn accumulation in Zn-exposed plants and in a loss of Zn tolerance (Tennstedt et al. 2009).

Organic acids have also repeatedly been discussed as binding partners for transition metals (Rauser 1999). However, Zn–citrate or Zn–malate complexes display low stability constants and are therefore likely to act as binding partners only in compartments such as vacuoles and the xylem which contain fewer competing molecules and show a low pH, which reduces the stability of Zn complexes with thiol or amino groups (Callahan et al. 2006).

6 Long-distance Transport and Accumulation of Zn

Translocation of Zn from the root to the shoot via the xylem requires efflux from the symplast into the apoplast. Long distance transport from leaves to storage or reproductive organs occurs via the phloem and therefore possibly symplastically. Efflux activities involved here could be transport out of cellular compartments into the cytosol (see below). In some cases such as the loading of seeds with micronutrients, efflux from the phloem into the apoplast is needed.

Best-understood Zn efflux activity in plant cells is dependent on $P1_B$-type ATPases. Plasma membrane-localized HMA2 and HMA4 have a crucial role in the translocation of Zn(II) from the root to the shoot of *A. thaliana* (Hussain et al. 2004). Compared to the wild-type, the *A. thaliana hma2-2hma4-1* double mutant accumulates only about half the Zn concentrations in the shoot and approximately twofold higher Zn concentrations in the root. Shoots therefore display severe symptoms of Zn deficiency, which can be rescued by watering with extra Zn. Transcriptional activity was localized to the vascular tissue of roots consistent with a function in xylem loading. The nonessential toxic metal Cd is also being translocated to the shoot in an HMA2/HMA4-dependent manner (Wong and Cobbett 2009), highlighting the limited specificity of most plant metal transporters. Another example for this is the contribution of IRT1 to Zn uptake (see above).

The speciation of Zn(II) during transport in the xylem is unknown. An equilibrium is likely to exist between various Zn species, for example free aqueous Zn^{2+} ions, Zn-chelates and Zn bound to cell walls lining the xylem (Krämer and Clemens 2005). The transfer of Zn from the xylem into the phloem system can be assumed to incorporate a key transport step resembling cellular uptake. Inside the phloem, Zn undergoes mass flow-driven movement from the source (mainly photosynthetic leaves) toward sink tissues. In the phloem, Zn is likely to move as a Zn–NA complex, or possibly bound to proteins.

Zn translocated to the shoot has to be taken up again into the symplasm. Principally the same mechanisms as described for root cells are assumed to be operational, i.e. uptake across the plasma membrane, cytosolic chelation, transport into the secretory pathway, plastids and mitochondria, and storage in the vacuole. Little is known, however, about the contribution and exact physiological role of individual transporter proteins. Also, the relative weight of different chelators and the interference from other micronutrients can today not be described quantitatively.

Of particular importance for human health and nutrition is the loading of Zn and other transition metals into the seed. The micronutrient intake of a large proportion of the World's population is insufficient. Zn deficiency in humans is widespread and is estimated to affect more than 25% of the world's population (Maret and Sandstead 2006). According to a recent WHO report (http://www.who.int/whr/2002/), Zn deficiency ranks fifth among the most important health risk factors in developing countries and 11th worldwide. Studies have shown that Zn supplementation can significantly decrease child mortality in developing countries (Black 2003; Welch and Graham 2004). Since plants are a major entry point for essential micronutrients into the food chain, breeding for increased Zn (and Fe) content (=biofortification) has become an important goal (Grotz and Guerinot 2006; Uauy et al. 2006). The molecular basis for the observed diversity in crop plants, however, is unknown. Understanding of what controls metal accumulation, localization and binding forms will be essential for devising strategies to improve human micronutrient nutrition. The seed is symplastically isolated from the mother plant. Thus, Zn has to leave the symplast through an efflux system, localized in transfer cells and other maternal cells and be taken up into filial cells such as the aleurone layer in cereals (Palmgren et al. 2008). Candidate proteins for these processes are again $P1_B$-ATPases and ZIPs, respectively. Storage in vacuoles likely involves MTPs. An unknown fraction of micronutrients loaded into the seeds is mobilized from senescing leaves. In *A. thaliana*, a reduction of Zn content in leaves during senescence by about 50% was measured. For comparison, macronutrients N, P and K were reduced by 80%–90% (Himelblau and Amasino 2001). Mobilization requires the transport of metal ions into companion cells of the vasculature, either symplastically or apoplastically. Subsequently, metals are symplastically released from the companion cells into the phloem. Zn mobilization appears to depend upon YSL transporters and therefore likely involves Zn–NA complexes. The double mutant *ysl1ysl3* displays symptoms of Fe deficiency and reduced levels of Fe, Zn and Cu in seeds, at least partially due to reduced mobilization from senescing leaves (Waters et al. 2006).

7 Zn Toxicity and Tolerance

Supraoptimal Zn supply can lead to Zn toxicity. One major mechanism of toxic action is unregulated high affinity binding of Zn to sulphur-, nitrogen- and oxygen-containing functional groups in biological molecules, resulting in the inactivation of metabolites or proteins. Metal-induced disruption of electron transport chains and of enzymatic reactions can lead to secondary oxidative stress. In addition, oxidative stress can be caused indirectly by a depletion of reduced glutathione, the major redox buffer in plant cells, through formation of metal–glutathione complexes (Krämer and Clemens 2005; Clemens 2006). A second mode of toxicity is competition of metal ions, leading to displacement of essential cofactor metal cations, for example Mg^{2+} (e.g. in chlorophyll), Mn^{2+} and Fe^{2+}. As mentioned above, an excess of Zn^{2+} can cause Fe or Mn deficiency (Marschner 1995).

Plants possess basal metal tolerance that can be regarded as a consequence of various homeostatic mechanisms since components of metal distribution and buffering are key tolerance factors. Obvious examples for the connection between metal trafficking mechanisms and tolerance are $P1_B$-type ATPases. These metal pumps, e.g. *A. thaliana* HMA2 and HMA4, are as discussed essential for the translocation of Zn from the root to the shoot. At the same time they confer Zn tolerance, as the efflux activity in root cells not surrounding the vasculature reduces cytosolic Zn overload (Verret et al. 2004; Mills et al. 2005). These multiple roles of $P1_B$-type ATPases have also been found in humans (Bertinato and L'Abbe 2004). Similarly, various MTP transporters in *A. thaliana* such as MTP1 and MTP3 are most likely important for metal storage and/or trafficking as well as tolerance. The same applies to the major facilitator ZIF1, which transports Zn-chelate complexes into the vacuole. PCs as potential Zn chelators also contribute to tolerance. A homeostatic function is postulated yet remains to be demonstrated (Tennstedt et al. 2009). Thus, tolerance is at the cellular level a question of capacities to buffer, sequestrate (mostly in the vacuole) and efflux Zn ions. In addition there can be preferential storage in certain cell types such as trichomes (Brune et al. 1995; Sarret et al. 2006). A contribution of the cell wall to binding excess Zn has been proposed but is not proven yet.

8 Cell Biology of Zn Hyperaccumulation

As discussed above, the Zn content of cells and organisms is very stable throughout nature. A notable exception is Zn hyperaccumulating plants. Their leaf Zn content can be >10,000 μg/g d.w., i.e. more than 100-fold that of "normal" plants

(Baker 1989). Obviously, such hyperaccumulators are Zn hypertolerant as well. Zn hyperaccumulation – which is a constitutive trait displayed even on sites with average Zn levels in the soil – has over the past years been intensively studied in two model systems closely related to *A. thaliana*, *A. halleri* and *Thlaspi caerulescens*, and has been fueled by the concept of biofortification (see above). Comparative transcriptome studies revealed constitutive high expression of metal homeostasis genes in *A. halleri* and *T. caerulescens* relative to nonhyperaccumulators such as *A. thaliana* (Weber et al. 2004; Becher et al. 2004; van de Mortel et al. 2006). This finding suggested that an altered regulation of metal homeostasis is underlying the evolution of hyperaccumulation. The genes more strongly expressed encode, for instance, metal transporters (HMAs, MTPs, ZIPs, Nramps) and enzymes involved in chelator synthesis (NAS). Indeed, for one of the candidate genes (HMA4) a major role of its elevated expression level in Zn hyperaccumulation was proven recently (Hanikenne et al. 2008). Efficient translocation of Zn to the shoot is dependent on strong Zn efflux activity around the xylem. Other postulated differences between hyperaccumulators and nonhyeraccumulators are: possibly more efficient uptake into the root, reduced sequestration in root cells, stronger buffering capacity to detoxify excess Zn, higher storage capacities in leaf cell vacuoles (Krämer et al. 2007; Verbruggen et al. 2009). Binding partner in vacuoles could be malate (Sarret et al. 2002). Transgenic approaches involving RNAi-mediated intervention in Zn hyperaccumulators will be required to dissect the actual relevance of these processes.

9 Regulation of Zn Homeostasis

Plants are exposed to extreme fluctuations in Zn availability. In agriculture, Zn deficiency is far more relevant than Zn toxicity. Large areas of agricultural soils are micronutrient-deficient because of low concentrations or low availability of micronutrients. Zn deficiency is at least among cereals the most serious mineral deficiency. It is common, for example, in soils in the Middle East, India and in parts of Australia, America and Central Asia. Many of the regions with Zn-deficient soils are also the regions where Zn deficiency in the human population is widespread (Cakmak 2008).

Coordination of transport activities and chelator synthesis for the various metals is essential. Changes in micronutrient availability have to be integrated with growth and developmental processes. Transcriptional responses to both micronutrient deficiency and excess are well-documented. Under Zn-deplete conditions ZIP genes are induced (see above). Since these early reports, microarray studies have identified numerous additional genes that specifically respond to micronutrient deficiencies at the transcript level (Wintz et al. 2003; Talke et al. 2006; van de Mortel et al. 2006). Many of them encode metal homeostasis factors previously mentioned including ZIP metal transporters, MTP gene family, $P1_B$-type ATPase transporters (HMA gene family), nicotianamine synthases (NAS), and Yellow Stripe1-like proteins (YSLs).

Metal sensing and signal transduction pathways mediating such transcriptional responses have to be postulated. For Zn, these pathways are completely unknown in plants. There is, for instance, no indication that a Zn-sensing transcription factor similar to *S. cerevisiae* ZAP1 exists. Similarly, no *cis* elements have been identified yet in genes up- or down-regulated upon changes in Zn supply.

Another possible level of regulation is posttranslational modification of metal homeostasis protein stability and/or subcellular localization. Again, evidence in plants is missing when considering Zn. ZIP transporters in *S. cerevisiae* and mammalian cells have been demonstrated to undergo Zn-stimulated endocytosis upon resupply of Zn to Zn-deficient cells (Eide 2006). Analogous processes appear very likely for plant cells given the documented posttranslational modification of IRT1 (Connolly et al. 2002).

10 Perspectives

Interest in fundamental biological questions of metal homeostasis has been rapidly increasing over the past 10–15 years. During this period many potential factors of plant Zn metabolism have been identified. The majority of these are transporter proteins. Large families are implicated in metal homeostasis. Physiological functions of individual components, however, are understood at the molecular detail only in a few cases. For example, we do not know precisely which protein(s) mediate(s) the uptake of Zn(II) from the rhizosphere into the root symplast. Our insight into cell specificity of Zn accumulation or other aspects of Zn homeostasis such as re-mobilization is extremely limited.

Also, an unknown but certainly significant number of molecular players are yet to be discovered. We can only make predictions about Zn-dependent proteins and their metal requirements. Structural information is missing as well as knowledge about the process of Zn insertion into proteins. Regulation of Zn homeostasis is practically not understood at all. These huge gaps are at least partly due to the fact that Zn homeostasis is genetically underexplored. Very few mutants showing defects in Zn tolerance or distribution have been isolated. Natural diversity is – with the exception of seed mineral content (Vreugdenhil et al. 2004) – scarcely documented. Poor genetic dissection can probably be explained by a lack of good, i.e. sensitive and easily scorable, markers for metal status.

A major question is speciation of metal ions in the cytoplasm and in other compartments of plant cells as well as in the extracellular space. Moreover, we have only a very limited understanding of whether and how metal specificity is achieved by plants. A number of metal transporters appear to transport multiple metals. Metal chelators such as nicotianamine are not specific, but form complexes of different stability with a range of metal ions. Many examples of apparent competition between metal ions are known. We need to know binding affinities of all players in Zn homeostasis and to perform in vivo imaging of Zn-chelate complexes. Only then will we be able to undertake any quantitative description and possibly manipulation of Zn fluxes.

References

Adam Z, Rudella A, Van Wijk KJ (2006) Recent advances in the study of Clp, FtsH and other proteases located in chloroplasts. Curr Opin Plant Biol 9:234–240

Andreini C, Banci L, Bertini I, Rosato A (2006) Zinc through the three domains of life. J Proteome Res 5:3173–3178

Andreini C, Bertini I, Cavallaro G, Holliday GL, Thornton JM (2008) Metal ions in biological catalysis: from enzyme databases to general principles. J Biol Inorg Chem 13:1205–1218

Arrivault S, Senger T, Krämer U (2006) The *Arabidopsis* metal tolerance protein AtMTP3 maintains metal homeostasis by mediating Zn exclusion from the shoot under Fe deficiency and Zn oversupply. Plant J 46:861–879

Baker AJM (1989) Terrestrial higher plants which hyperaccumulate metallic elements – a review of their distribution, ecology and phytochemistry. Biorecovery 1:81–126

Becher M, Talke IN, Krall L, Krämer U (2004) Cross-species microarray transcript profiling reveals high constitutive expression of metal homeostasis genes in shoots of the zinc hyperaccumulator *Arabidopsis halleri*. Plant J 37:251–268

Benes I, Schreiber K, Ripperger H, Kircheiss A (1983) Metal complex formation by nicotianamine, a possible phytosiderophore. Experientia 39:261–262

Berg J, Shi Y (1996) The galvanization of biology: a growing appreciation for the roles of zinc. Science 271:1081–1085

Bertinato J, L'Abbe MR (2004) Maintaining copper homeostasis: regulation of copper-trafficking proteins in response to copper deficiency or overload. J Nutr Biochem 15:316–322

Black RE (2003) Zinc deficiency, infectious disease and mortality in the developing world. J Nutr 133:1485S–1489S

Briat J-F, Curie C, Gaymard F (2007) Iron utilization and metabolism in plants: physiology and metabolism. Curr Opin Plant Biol 10:276–282

Broadley MR, White PJ, Hammond JP, Zelko I, Lux A (2007) Zinc in plants. New Phytol 173:677–702

Brune A, Urbach W, Dietz KJ (1995) Differential toxicity of heavy-metals is partly related to a loss of preferential extraplasmic compartmentation – a comparison of Cd-stress, Mo-stress, Ni-stress and Zn-stress. New Phytol 129:403–409

Cakmak I (2000) Possible roles of zinc in protecting plant cells from damage by reactive oxygen species. New Phytol 146:185–205

Cakmak I (2008) Enrichment of cereal grains with zinc: agronomic or genetic biofortification? Plant Soil 302:1–17

Callahan DL, Baker AJ, Kolev SD, Wedd AG (2006) Metal ion ligands in hyperaccumulating plants. J Biol Inorg Chem 11:2–12

Campanella JJ, Larko D, Smalley J (2003) A molecular phylogenomic analysis of the ILR1-like family of IAA amidohydrolase genes. Comp Funct Genomics 4:584–600

Clemens S (2006) Toxic metal accumulation, responses to exposure and mechanisms of tolerance in plants. Biochimie 88:1707–1719

Clemens S, Bloss T, Vess C, Neumann D, Nies D, Zur Nieden U (2002a) A transporter in the endoplasmic reticulum of *Schizosaccharomyces pombe* cells mediates zinc storage and differentially affects transition metal tolerance. J Biol Chem 277:18215–18221

Clemens S, Kim EJ, Neumann D, Schroeder JI (1999) Tolerance to toxic metals by a gene family of phytochelatin synthases from plants and yeast. EMBO J 18:3325–3333

Clemens S, Palmgren MG, Krämer U (2002b) A long way ahead: understanding and engineering plant metal accumulation. Trends Plant Sci 7:309–315
Cobbett C, Goldsbrough P (2002) Phytochelatins and metallothioneins: roles in heavy metal detoxification and homeostasis. Annu Rev Plant Physiol Plant Mol Biol 53:159–182
Colangelo EP, Guerinot ML (2006) Put the metal to the petal: metal uptake and transport throughout plants. Curr Opin Plant Biol 9:322–330
Connolly EL, Fett JP, Guerinot ML (2002) Expression of the IRT1 metal transporter is controlled by metals at the levels of transcript and protein accumulation. Plant Cell 14:1347–1357
Curie C, Panaviene Z, Loulergue C, Dellaporta S, Briat J, Walker E (2001) Maize yellow stripe1 encodes a membrane protein directly involved in Fe(III) uptake. Nature 409:346–349
Delhaize E, Gruber BD, Pittman JK, White RG, Leung H, Miao YS, Jiang LW, Ryan PR, Richardson AE (2007) A role for the AtMTP11 gene of *Arabidopsis* in manganese transport and tolerance. Plant Journal 51:198–210
Desbrosses-Fonrouge AG, Voigt K, Schröder A, Arrivault S, Thomine S, Krämer U (2005) *Arabidopsis thaliana* MTP1 is a Zn transporter in the vacuolar membrane which mediates Zn detoxification and drives leaf Zn accumulation. FEBS Lett 579:4165–4174
Eide DJ (2006) Zinc transporters and the cellular trafficking of zinc. Biochim Biophys Acta 1763:711–722
Ellis CD, Wang F, MacDiarmid CW, Clark S, Lyons T, Eide DJ (2004) Zinc and the Msc2 zinc transporter protein are required for endoplasmic reticulum function. J Cell Biol 166:325–335
Frausto da Silva JJR, Williams RJP (2001) The biological chemistry of the elements. The inorganic chemistry of life, 2nd edn. Oxford University Press, New York
Gamsjaeger R, Liew CK, Loughlin FE, Crossley M, Mackay JP (2007) Sticky fingers: zinc-fingers as protein-recognition motifs. Trends Biochem Sci 32:63–70
Grotz N, Fox T, Connolly E, Park W, Guerinot ML, Eide D (1998) Identification of a family of zinc transporter genes from *Arabidopsis* that respond to zinc deficiency. Proc Natl Acad Sci USA 95:7220–7224
Grotz N, Guerinot ML (2006) Molecular aspects of Cu, Fe and Zn homeostasis in plants. Biochim Biophys Acta 1763:595–608
Ha SB, Smith AP, Howden R, Dietrich WM, Bugg S, O'Connell MJ, Goldsbrough PB, Cobbett CS (1999) Phytochelatin synthase genes from *Arabidopsis* and the yeast *Schizosaccharomyces pombe*. Plant Cell 11:1153–1163
Hacisalihoglu G, Hart JJ, Wang YH, Cakmak I, Kochian LV (2003) Zinc efficiency is correlated with enhanced expression and activity of zinc-requiring enzymes in wheat. Plant Physiol 131:595–602
Hacisalihoglu G, Kochian LV (2003) How do some plants tolerate low levels of soil zinc? Mechanisms of zinc efficiency in crop plants. New Phytol 159:341–350
Hanikenne M, Krämer U, Demoulin V, Baurain D (2005) A comparative inventory of metal transporters in the green alga *Chlamydomonas reinhardtii* and the red alga *Cyanidioschizon merolae*. Plant Physiol 137:428–446
Hanikenne M, Talke IN, Haydon MJ, Lanz C, Nolte A, Motte P, Kroymann J, Weigel D, Krämer U (2008) Evolution of metal hyperaccumulation required cis-regulatory changes and triplication of HMA4. Nature 453:391–395
Haydon MJ, Cobbett CS (2007) A novel major facilitator superfamily protein at the tonoplast influences zinc tolerance and accumulation in *Arabidopsis*. Plant Physiol 143:1705–1719
Hell R, Stephan UW (2003) Iron uptake, trafficking and homeostasis in plants. Planta 216:541–551
Himelblau E, Amasino RM (2001) Nutrients mobilized from leaves of *Arabidopsis thaliana* during leaf senescence. J Plant Physiol 158:1317–1323
Hotton SK, Callis J (2008) Regulation of cullin ring ligases. Annu Rev Plant Biol 59:467–489
Huffman D, O'Halloran T (2001) Function, structure, and mechanism of intracellular copper trafficking proteins. Annu Rev Biochem 70:677–701

Hussain D, Haydon MJ, Wang Y, Wong E, Sherson SM, Young J, Camakaris J, Harper JF, Cobbett CS (2004) P-type ATPase heavy metal transporters with roles in essential zinc homeostasis in *Arabidopsis*. Plant Cell 16:1327–1339

Kirschke CP, Huang LP (2003) Znt7, a novel mammalian zinc transporter, accumulates zinc in the Golgi apparatus. J Biol Chem 278:4096–4102

Kobae Y, Uemura T, Sato MH, Ohnishi M, Mimura T, Nakagawa T, Maeshima M (2004) Zinc transporter of *Arabidopsis thaliana* AtMTP1 is localized to vacuolar membranes and implicated in zinc homeostasis. Plant Cell Physiol 45:1749–1758

Krämer U, Clemens S (2005) Functions and homeostasis of zinc, copper and nickel in plants. Topics Curr Genet 14:215–271

Krämer U, Talke IN, Hanikenne M (2007) Transition metal transport. FEBS Lett 581:2263–2272

Lanquar V, Lelievre F, Bolte S, Hames C, Alcon C, Neumann D, Vansuyt G, Curie C, Schröder A, Krämer U, Barbier-Brygoo H, Thomine S (2005) Mobilization of vacuolar iron by AtNRAMP3 and AtNRAMP4 is essential for seed germination on low iron. EMBO J 24:4041–4051

Le Jean M, Schikora A, Mari S, Briat JF, Curie C (2005) A loss-of-function mutation in AtYSL1 reveals its role in iron and nicotianamine seed loading. Plant J 44:769–782

Li L, Kaplan J (2001) The yeast gene MSC2, a member of the cation diffusion facilitator family, affects the cellular distribution of zinc. J Biol Chem 276:5036–5043

Lin Y-F, Liang H-M, Yang S-Y, Boch A, Clemens S, Chen C-C, Wu J-F, J-Li H, Yeh K-C (2009) Arabidopsis IRT3 is a zinc-regulated and plasma membrane localized zinc/iron transporter. New Phytol 182:392–404

Ling HQ, Koch G, Bäumlein H, Ganal MW (1999) Map-based cloning of chloronerva, a gene involved in iron uptake of higher plants encoding nicotianamine synthase. Proc Natl Acad Sci USA 96:7098–7103

Maret W, Sandstead HH (2006) Zinc requirements and the risks and benefits of zinc supplementation. J Trace Elem Med Biol 20:3–18

Marschner H (1995) Mineral nutrition of higher plants (2nd edn). Academic press Amsterdam

Matthews JM, Sunde M (2002) Zinc fingers – folds for many occasions. IUBMB Life 54:351–355

Mills RF, Francini A, Ferreira da Rocha PS, Baccarini PJ, Aylett M, Krijger GC, Williams LE (2005) The plant P1B-type ATPase AtHMA4 transports Zn and Cd and plays a role in detoxification of transition metals supplied at elevated levels. FEBS Lett 579:783–791

Natori Y, Nanamiya H, Akanuma G, Kosono S, Kudo T, Ochi K, Kawamura F (2007) A fail-safe system for the ribosome under zinc-limiting conditions in *Bacillus subtilis*. Mol Microbiol 63:294–307

O'Halloran T, Culotta V (2000) Metallochaperones, an intracellular shuttle service for metal ions. J Biol Chem 275:25057–25060

Outten CE, O'Halloran TV (2001) Femtomolar sensitivity of metalloregulatory proteins controlling zinc homeostasis. Science 292:2488–2492

Palmgren MG, Clemens S, Williams LE, Krämer U, Borg S, Schjorring JK, Sanders D (2008) Zinc biofortification of cereals: problems and solutions. Trends Plant Sci 13:464–473

Parniske M (2008) Arbuscular mycorrhiza: the mother of plant root endosymbioses. Nat Rev Microbiol 6:763–775

Peiter E, Montanini B, Gobert A, Pedas P, Husted S, Maathuis FJM, Blaudez D, Chalot M, Sanders D (2007) A secretory pathway-localized cation diffusion facilitator confers plant manganese tolerance. Proc Natl Acad Sci USA 104:8532–8537

Puig S, Andres-Colas N, Garcia-Molina A, Penarrubia L (2007) Copper and iron homeostasis in *Arabidopsis*: responses to metal deficiencies, interactions and biotechnological applications. Plant Cell Environ 30:271–290

Rauser WE (1999) Structure and function of metal chelators produced by plants. Cell Biochem Biophys 31:19–48

Sarret G, Harada E, Choi YE, Isaure MP, Geoffroy N, Fakra S, Marcus MA, Birschwilks M, Clemens S, Manceau A (2006) Trichomes of tobacco excrete zinc as zinc-substituted calcium carbonate and other zinc-containing compounds. Plant Physiol 141:1021–1034

Sarret G, Saumitou-Laprade P, Bert V, Proux O, Hazemann JL, Traverse AS, Marcus MA, Manceau A (2002) Forms of zinc accumulated in the hyperaccumulator *Arabidopsis halleri*. Plant Physiol 130:1815–1826

Schaaf G, Ludewig U, Erenoglu BE, Mori S, Kitahara T, von Wiren N (2004) ZmYS1 functions as a proton-coupled symporter for phytosiderophore- and nicotianamine-chelated metals. J Biol Chem 279:9091–9096

Schaaf G, Schikora A, Haberle J, Vert G, Ludewig U, Briat JF, Curie C, von Wiren N (2005) A putative function for the Arabidopsis Fe-phytosiderophore transporter homolog AtYSL2 in Fe and Zn homeostasis. Plant Cell Physiol 46:762–774

Simm C, Lahner B, Salt D, LeFurgey A, Ingram P, Yandell B, Eide DJ (2007) *Saccharomyces cerevisiae* vacuole in zinc storage and intracellular zinc distribution. Eukaryot Cell 6:1166–1177

Somerville C (2006) Cellulose synthesis in higher plants. Annu Rev Cell Dev Biol 22:53–78

Stephan UW, Schmidke I, Stephan VW, Scholz G (1996) The nicotianamine molecule is made-to-measure for complexation of metal micronutrients in plants. Biometals 9:84–90

Takahashi M, Terada Y, Nakai I, Nakanishi H, Yoshimura E, Mori S, Nishizawa NK (2003) Role of nicotianamine in the intracellular delivery of metals and plant reproductive development. Plant Cell 15:1263–1280

Talke IN, Hanikenne M, Krämer U (2006) Zinc-dependent global transcriptional control, transcriptional deregulation, and higher gene copy number for genes in metal homeostasis of the hyperaccumulator *Arabidopsis halleri*. Plant Physiol 142:148–167

Tennstedt P, Peisker D, Böttcher C, Trampczynska A, Clemens S (2009) Phytochelatin synthesis is essential for the detoxification of excess zinc and contributes significantly to the accumulation of zinc. Plant Physiol 149:938–948

Trampczynska A, Böttcher C, Clemens S (2006) The transition metal chelator nicotianamine is synthesized by filamentous fungi. FEBS Lett 580:3173–3178

Uauy C, Distelfeld A, Fahima T, Blechl A, Dubcovsky J (2006) A NAC gene regulating senescence improves grain protein, zinc, and iron content in wheat. Science 314:1298–1301

van de Mortel JE, Almar Villanueva L, Schat H, Kwekkeboom J, Coughlan S, Moerland PD, Loren V, van Themaat E, Koornneef M, Aarts MGM (2006) Large expression differences in genes for iron and zinc homeostasis, stress response, and lignin biosynthesis distinguish roots of *Arabidopsis thaliana* and the related metal hyperaccumulator *Thlaspi caerulescens*. Plant Physiol 142:1127–1147

Vatamaniuk O, Bucher E, Ward J, Rea P (2001) A new pathway for heavy metal detoxification in animals. Phytochelatin synthase is required for cadmium tolerance in *Caenorhabditis elegans*. J Biol Chem 276:20817–20820

Verbruggen N, Hermans C, Schat H (2009) Molecular mechanisms of metal hyperaccumulation in plants. New Phytol 181:759–776

Verret F, Gravot A, Auroy P, Leonhardt N, David P, Nussaume L, Vavasseur A, Richaud P (2004) Overexpression of AtHMA4 enhances root-to-shoot translocation of zinc and cadmium and plant metal tolerance. FEBS Lett 576:306–312

Vert G, Grotz N, Dedaldechamp F, Gaymard F, Guerinot M, Briat J, Curie C (2002) IRT1, an *Arabidopsis* transporter essential for iron uptake from the soil and for plant growth. Plant Cell 14:1233–1243

von Wiren N, Klair S, Bansal S, Briat JF, Khodr H, Shioiri T, Leigh RA, Hider RC (1999) Nicotianamine chelates both FeIII and FeII. Implications for metal transport in plants. Plant Physiol 119:1107–1114

von Wiren N, Marschner H, Römheld V (1996) Roots of iron-efficient maize also absorb phytosiderophore-chelated zinc. Plant Physiol 111:1119–1125

Vreugdenhil D, Aarts MGM, Koornneef M, Nelissen H, Ernst WHO (2004) Natural variation and QTL analysis for cationic mineral content in seeds of *Arabidopsis thaliana*. Plant Cell Environ 27:828–839

Waters BM, Chu HH, Didonato RJ, Roberts LA, Eisley RB, Lahner B, Salt DE, Walker EL (2006) Mutations in *Arabidopsis* yellow stripe-like1 and yellow stripe-like3 reveal their roles in metal ion homeostasis and loading of metal ions in seeds. Plant Physiol 141:1446–1458

Weber M, Harada E, Vess C, von Roepenack-Lahaye E, Clemens S (2004) Comparative micro-array analysis of *Arabidopsis thaliana* and *Arabidopsis halleri* roots identifies nicotianamine synthase, a ZIP transporter and other genes as potential metal hyperaccumulation factors. Plant J 37:269–281

Welch RM, Graham RD (2004) Breeding for micronutrients in staple food crops from a human nutrition perspective. J Exp Bot 55:353–364

Wintz H, Fox T, Wu YY, Feng V, Chen W, Chang HS, Zhu T, Vulpe C (2003) Expression profiles of *Arabidopsis thaliana* in mineral deficiencies reveal novel transporters involved in metal homeostasis. J Biol Chem 278:47644–47653

Wong CKE, Cobbett CS (2009) HMA P-type ATPases are the major mechanism for root-to-shoot Cd translocation in *Arabidopsis thaliana*. New Phytol 181:71–78

Index

MIX
Papier aus verantwortungsvollen Quellen
Paper from responsible sources
FSC® C105338

If you have any concerns about our products,
you can contact us on
ProductSafety@springernature.com

In case Publisher is established outside the EU,
the EU authorized representative is:
Springer Nature Customer Service Center GmbH
Europaplatz 3, 69115 Heidelberg, Germany

Printed by Libri Plureos GmbH
in Hamburg, Germany